CBRN

SURVIVING CHEMICAL, BIOLOGICAL, RADIOLOGICAL AND NUCLEAR EVENTS

PIERO SAN GIORGIO
& CRIS MILLENNIUM

CBRN

SURVIVING **CHEMICAL**, **BIOLOGICAL**, **RADIOLOGICAL** & **NUCLEAR** EVENTS

ARKTOS
LONDON 2020

ORIGINAL TITLE *NRBC: survivre aux événements nucléaires, radiologiques, biologiques et chimiques* (2016)

ISBN 978-1-912975-66-2 (Paperback)
978-1-912975-67-9 (Hardback)
978-1-912975-68-6 (Ebook)

TRANSLATION Roger Adwan

EDITING John Bruce Leonard
Roger Adwan

COVER & LAYOUT Tor Westman

Arktos.com fb.com/Arktos @arktosmedia arktosmedia

CONTENTS

BIOLOGICAL RISKS

CHEMICAL RISKS

CBRN INCIDENTS: REACTIONS AND PROTECTION

KITS AND EQUIPMENT

CRIS MILLENNIUM

Cris Millennium has been working in the 'Nuclear-Radiological-Biological-Chemical' field since 2003. For several years, he served in the national CBRN unit of the National Gendarmerie, where he particularly contributed to the development of the scientific police deployed in contaminated environments before being selected to join the GIGN.[1] For a six-year period, he acted as the head of the CBRN cell of this prestigious counterterrorism unit, offering advice to the authorities, defining detection and protection equipment and providing personnel training.

Since 2013, he has been active as a CBRN advisor for a certain foreign government. In addition to his established experience in genuine toxic environments (including VX, sarin, radiation, etc.) and his multiple specialised CBRN degrees, Cris Millennium holds a master's degree in biology, a postgraduate degree in forensic science and a master's degree in 'CBRN-E risk and threat management'.

PIERO SAN GIORGIO

For more than twenty years, Piero San Giorgio was responsible for the emerging markets of Eastern Europe, the Middle East and Africa in the domain of high-tech industry. Since 2005, he has been preparing for the collapse of the economy and studying different means of survival. His first book, entitled *Survivre à l'effondrement économique*,[2] was originally published in 2011 and is now an international bestseller. This success has turned Piero San Giorgio into a prominent 'spokesperson' for the survivalist movement in the French-speaking world.

---

1 TN: GIGN, or *Groupe d'intervention de la Gendarmerie nationale*, is the elite police tactical unit of the French National Gendarmerie. Its missions and roles are many and comprise counter-terrorism, hostage rescue, the surveillance of national threats, the protection of government officials, and eradicating organised crime.

2 TN: *Survive the Economic Collapse*, Radix, 2013.

Translator's Foreword

Considering the fact that the French language uses a different acronym for the book's title compared to the English rendering, it may, at first sight, seem illogical, or at least inappropriate, to keep to the original chapter and section order, thus respecting the French acronym 'NRBC'. This decision was, however, imposed upon me by the book's content and the numerous references to later sections and chapters made by the authors themselves. It would therefore have been completely nonsensical to reorganise the whole book only to suit the English acronym 'CBRN' while simultaneously disregarding the reference-related confusion that would result from such restructuring. In the absence of a better solution, the most reasonable option was to retain the original text in its entirety and structure and simply adapt the book's title to reflect the English norm.

Roger Adwan

Preface by Dmitry Orlov[1]

We live in a dangerous world, a world rendered even more dangerous by the incessant technological developments which, for the most part, bring their own share of unexpected consequences before which technology itself is often left powerless.

The danger is all the greater when one is unable to identify it: our senses are ill-suited to detecting chemical contaminants and completely helpless when it comes to perceiving radioactivity or radiological contamination. As for the threat posed by microscopic pathogens, the ability of our immune system to resist them has, paradoxically, been compromised by improved hygiene and the use of antibacterial soap and antibiotics.

These dangers are exacerbated by other risks faced by our world, both present and future ones. The permanent safety of many of our industrial technologies — those that manage radioactivity, virulent organisms and toxic substances — is based on our assumption of perpetual *social stability*. In the case of long-lived radionuclides such as uranium or plutonium, the period of social stability necessary to ensure their safe storage and their isolation from the environment is

1 AN: Dmitry Orlov is a Russian engineer and writer and the author of books explaining how the collapse of our societies will impact our lives. His most famous books include *Reinventing Collapse: The Soviet Example and American Prospects*, New Society Publishers, 2008; and *The Five Stages of Collapse*, Le Retour aux Sources editions, 2016. https://fr.wikipedia.org/wiki/Dmitry_Orlov.

expected to last for thousands of years. History, however, teaches us that human societies never last that long.

Whenever a civilisation crumbles, what naturally follows is a dark age during which the population collapses, knowledge and literacy become a rarity, urban centres are abandoned and the few survivors must discover means of subsistence at much more primitive levels, all by themselves and without resorting to advanced technologies.

When will this happen?

Well, between you and me, we already live in an age when nation-states vanish ever more rapidly, millions of refugees roam the planet, and the financial systems that enable the very existence of globalised industrialised civilisation are in such a pitiful state that central banks are forced to use such bizarre devices as negative interest rates coupled with the unlimited issuance of fiat money.

But nothing lasts forever, and one should thus not ignore the possibility that we — or our children — may experience periods of great uncertainty, confusion and chaos. Whatever happens, we all want to live happily and in good health and be filled with a sense of accomplishment, and we strive to provide the same for our children to enjoy. None of this, however, is possible without peace of mind, and knowing that the prospects for this well-being are ever uncertain, we are burdened with anxiety. Some of this anxiety is due to helplessness resulting from our conditioned obedience and an ignorance which we ourselves have actually desired: we have been taught to trust experts when it comes to our well-being and not to question them too much. But where will all these experts be once cities fall under the yoke of a people in revolt and become too dangerous to approach? To control our anxiety, we must learn to identify the risks and be prepared to face them.

This book describes the risks which we are least able to apprehend using our common sense, powers of perception and instincts. They belong to the domain of experts and, without sophisticated know-how and specialised equipment, one finds oneself completely defenceless when facing them. Even if we truly wanted to be able to deal with them, we could not, most of the time, even detect their presence. Nevertheless, by reading the information contained in this book and considering the option of a modest investment in protective equipment and detection (an investment that can sometimes be beyond the means of a single family, but remains conceivable at the level of a small community), one can overcome one's anxiety and regain the ability to lead a fulfilling life.

The threats are many, but the greatest of them all is simply embodied by panic. When people fall ill and none can determine the cause, society may well experience sudden collapse. And yet, panic could still be avoided should some well-informed people be able to explain to others what is happening and what they should and should not do.

You, too, can become one of these people. So, do not panic, and read this book!

Dmitry Orlov, Beaufort,
South Carolina, USA

Dedicated to our families and all those
who, through their daily commitment,
protect us against all nuclear, radiological,
biological and chemical tragedies.

Introduction

Life on Earth is increasingly threatened with destruction by disasters such as global warming, nuclear war, a new genetically engineered virus and other dangers.

— British Astrophysicist STEPHEN HAWKING, 2007

Formerly called NBC (Nuclear, Biological and Chemical) by the military, this acronym included elements that allowed combatants to fight against this threat. Ranging from knowledge of the risk to the use of protection and detection means, NBC defence focused on Nuclear, Biological or Chemical weapons of mass destruction.

Today, considering the manner in which the threat has evolved, an additional letter has been added — the letter 'R', in reference to the radiological aspect.[1] The objective is no longer merely to allow for events of the 'nuclear explosions' type (Hiroshima, Nagasaki), but also to take into account new, smaller-scale risks. The dissemination of radioactive products (contaminations) and the use of sources emitting invisible rays that may prove fatal fall into this category. The current

1 AN: A radiological phenomenon is a physical manifestation involving radioactivity. However, unlike a nuclear phenomenon, it does not lead to chain reactions, as is the case with an atomic explosion or one in a nuclear power plant.

acronym 'NRBC' ('CBRN'[2] in English) is now internationally recognised and used by many governments as well as NATO.[3]

Although this awareness is recent, it must be emphasised that the use of biological or chemical agents dates back to the dawn of history. At times, their use has influenced the very outcome of a battle or even that of an entire war, and has, in most cases, left its mark upon the populations in question.

Several centuries before Christ, the Assyrians used rye ergot[4] to poison water wells in the desert. Solon of Athens, meanwhile, had a large amount of hellebore[5] roots dumped into the river supplying water to the besieged city of Kirra.[6] Many other examples of resorting to animal or human corpses abound throughout history. The most infamous case is certainly the one where the Tatars catapulted plague-ridden corpses during the siege of the Genoese trading post of Caffa, in 1346. The disease decimated assailants and defenders alike, and the blockade was lifted: the survivors returned home, allowing the 'evil' to spread quickly. This event acted as the source of the Black Death which ravaged Europe until 1352. About a third of the population, i.e. twenty-five million people, perished — an unprecedented disaster!

Smallpox, meanwhile, was used in South America by the Spanish conquistadors to decrease the resistance of the indigenous Incas. Later on, the English and, subsequently, the French gave the Indian tribes of North America blankets infected with this very same virus. It goes without saying that this constituted a genuine hecatomb for those

2 AN: The official term is 'CBRN-E', with the 'E' signifying enhanced 'explosive' threat. The topic of explosives deserves its own, separate book and goes beyond the scope which the authors wish to cover in this book.

3 AN: The North Atlantic Treaty Organisation.

4 AN: A parasitic fungus known as *Claviceps Purpurea*, which causes hallucinations, necrosis, the gangrene of human extremities and may lead to death.

5 AN: A poisonous plant with very powerful purgative properties.

6 AN: This city, located in Crimea, is nowadays called Feodosiya.

peoples, none of whom had ever been confronted with this biological agent before!

As regards the use of chemicals, convincing results were also obtained, even if the consequences were limited in comparison with the pandemics mentioned above. During sieges or armed conflicts, most of the great civilisations (Rome, Persia, China, etc.) occasionally used toxic fumes containing sulphur, arsenic, antimony and sometimes poisonous plants. These were also used as a simple poison smeared upon the tip of an arrow or spear. The Celts, for example, used to coat their weapons with aconite (*Aconitum Napellus*) sap when hunting wolves. It is therefore not surprising that one of the common names which this plant goes by is 'wolfsbane'.

With the technological evolution of society, considerable progress has been made in the field of science. This has allowed humans to synthesise new and more harmful chemicals and modify living organisms to create more dangerous bacteria and viruses. During World War I, German chemist Fritz Haber suggested using chlorine against the Allies in Ypres (Belgium). The result exceeded military expectations: 15,000 soldiers were put out of action and about 5,000 succumbed to the effects of this gas… What then followed was a chemical weapon race: chlorine, phosgene and hydrogen cyanide[7] were produced in industrial quantities. New compounds, each more devastating than the last, were synthesised: vesicants[8] such as yperite (better known as mustard gas) and neurotoxins[9] such as sarin or VX were thus created.

Today, many countries still keep this type of chemical warfare agents, although they are banned by international conventions.[10] The

7 AN: Zyklon-B is an example of this.

8 AN: Products that cause chemical burns, vesicles and other skin and respiratory tract lesions.

9 AN: Chemical agents impacting the nervous system: a single drop can kill an individual!

10 AN: The Convention on the Prohibition of the Development, Production, Stockpiling and Use of Bacteriological (Biological) and Toxin Weapons and on Their Destruction, which entered into force on 26 March 1975; and the

implementation of these agreements, however, has allowed for the launching of a major decommissioning programme that will lead to the complete disappearance of these toxic agents. As of 31 December 2013, 80% of the 72,531 tonnes of declared global stocks have been destroyed. This process, however, is not always free of 'snags', and some nations have already made headlines because of their own arsenal (North Korea, Iran, Libya, Syria, etc.). As for the United States and Russia, in addition to their remaining chemical weapons, they possess particularly virulent viruses such as smallpox,[11] which is ranked among the most dangerous ever.

However, it is interesting to note that these militarised agents are not the only threat. Indeed, it is enough to look around to realise that the world is full of deadly compounds or substances. Industry and research laboratories are, for instance, among the largest users of toxic products.

To be perfectly honest, the danger may very well be lurking around the corner of your home. Think of the 3,500 people who died[12] overnight as a result of a methyl isocyanate leak (a chemical gas used for pesticide synthesis) at the Union Carbide plant site in Bhopal, India, in 1984! And what about the survivors of Chernobyl or Fukushima? Were they prepared for such an eventuality? Will the populations that have lost everything additionally have to contend with delayed health effects? And should a new virus with a mortality rate akin to that of Ebola and a transmission as smooth as that of flu spread across the world, would you still take the Tube or use any other public means of transport?

Convention on the Prohibition of the Development, Production, Stockpiling and Use of Chemical Weapons and on their Destruction, which came into effect on 29 April 1997.

11 AN: The smallpox virus was said to have been eradicated in 1980 following vaccination campaigns conducted on a global scale. If the USA and Russia had not kept it in their laboratories, it would have disappeared off the face of the Earth.

12 AN: With nearly 25,000 deaths twenty years later, following complications or other illnesses caused by people's exposure to the gas.

Unfortunately, the threat is not limited to these few celebrated examples. People who lived through the Cold War remember the danger of nuclear war hovering over our world's nations. In spite of the fact that this page has now been turned, the world's nuclear arsenal remains impressive, with more than 16,000 warheads.[13] In the current context of renewed international tensions, the possibility of new wars cannot be ruled out. Such a hypothesis could indeed revive the use of these weapons. What would happen if ill-intentioned people gained possession of certain radiological materials or biological/chemical agents? And in the event that our society were to fall into chaos following an economic collapse, what would happen to such sensitive equipment and facilities as nuclear power plants, heavy industries, laboratories, and so on?

In the face of these few examples, several questions inevitably arise:

- How would you respond if such an event were to occur in the vicinity of your own home?
- Would you be able to understand what is happening?
- Would you be up to the task of protecting yourself and those closest to you against such a threat?

During a natural disaster, people quickly grasp the essence of the danger in question, although they do not always respond appropriately. Take floods and fires — everyone knows what water and fire are. Could you say the same of gamma radiation, suffocating gases or an emerging virus?

In the case of a CBRN event, it is important to understand what is happening in order to make choices that will save your life.

---

13 AN: The estimates are: the USA: 7,300; Russia: 8,000; the United Kingdom: 225; France: 300; China: 250; India: 110; Pakistan: 120; North Korea: 10; and Israel: 75 to 400, depending on the sources.

Without claiming to be exhaustive, this book will nonetheless address various threats in the nuclear, radiological, biological and chemical fields. The section entitled 'Scientific Foundations', which is an extension of a very complex subject, allows one to acquire a decent understanding of the phenomena involved. The stories that follow will be commented on and linked to events proving that reality can sometimes exceed fiction. The stated goal is to explain to the readers the main mechanisms involved, as well as the behaviour to adopt (or avoid). Last but not least, several chapters will be dedicated to the means enabling us to fight against these threats, including the material aspect, i.e. which protection or detection equipment to use.

Intended for a novice audience or those seeking to deepen their knowledge in the field, this book should therefore provide readers with a new outlook and increased understanding of CBRN risks and threats. Depending on their environment or their convictions, everyone can prepare for the possibility of such an event and, who knows, perhaps even learn to SURVIVE!

NUCLEAR AND RADIOLOGICAL RISKS

I never cease to marvel at the fact that, every single day, my body is traversed by more than 6,000 billion neutrinos resulting from nuclear reactions that occur uninterruptedly inside the sun.

— American astrophysicist Lawrence M. Krauss

You cannot consider yourself a real country unless you have your own beer brand and airline. It does help if you have a football team, or nuclear weapons, but you need, at least, a brand of beer.

— American Musician Frank Zappa (1940–1993)

1. Radioactivity

A scientific discovery is worthless if it
cannot be explained to a waitress.

— British physicist Ernest Rutherford (1871–1937)

Radioactivity is a natural phenomenon that occurs in the deepest part of matter, at the atomic level.[1] Unstable nuclei will strive to recover stable states by splitting and/or emitting highly energetic particles and radiations. As we shall see at a later point, these invisible radiations can have beneficial uses from the human perspective (medicine, research, etc.) or, on the contrary, be responsible for terrible catastrophes…

Radioactivity was first brought to human attention about 120 years ago. In March 1896, Henry Becquerel, a physics professor at the Museum of Natural History in Paris, discovered that uranium emitted radiation, which was unobservable to the naked eye but capable of making imprints upon photographic plates. And it was Pierre and Marie Curie who, a few years later, defined the phenomenon and named it 'radioactivity', giving, furthermore, two particularly radioactive elements their names: polonium (18 July 1898) and radium (26 December 1898).

1 AN: See the following chapter entitled 'Scientific Foundations'.

With regard to our society's time scale, this new science is therefore very recent. Indeed, since these radiations are invisible to us, they had simply remained undetected.

Natural Radioactivity

It thus seems obvious that, in order to exist, radioactivity did not wait for man to discover it and has been part of the universe since the beginning of time. Nuclear reactions, for example, occur within each and every star. The resulting cosmic and solar rays flood space, in which all stellar objects are immersed. The energy they bear could cause considerable damage to any living being. Protected by our planet's magnetic field and its atmosphere, however, man can enjoy his 'peaceful' life on Earth, remaining unaware of this danger. In addition, it also seems to be the case that radioactivity has certainly played an important role in the evolution of species. Indeed, since ionising radiation can cause mutations in living cells, it may well be the main agent in the process of evolution!

Thanks to the technological level achieved by our modern society, it is now possible to enter this phenomenon's very sphere of influence. The higher the altitude, for instance, the higher the amount of radiation. It is thus a well-known fact that a trip by plane will expose you to additional radiation. A negligible amount indeed, yet still a measurable one!

For the astronauts that are sent to circle our planet or land on the Moon, the dose of received radiation is remarkably higher. Even if it becomes relatively significant, it does not cause immediate death, nor, it would seem, in the years that follow. The case of the two most famous American astronauts is a good example of this: Neil Armstrong, the first man to walk on the moon, died on 25 August 2012, at the age of eighty-two. To this very day, his now eighty-five-year-old comrade, Buzz Aldrin, has continued to increase his calls for the conquest of space. On the other hand, things could be different if one considers an expedition to Mars. With the technology currently at our disposal, such a journey would take six to eight months and the ship's occupants would completely exit the protection zone generated by our own planet, thus receiving high doses of radiation that would, this time around, probably have a more or less long-term negative impact on their health.

Let us, however, return to Earth. Due to certain components that it contains, our planet itself is naturally radioactive. When it was created about 4.5 billion years ago, it thus emitted far more ionising radiation than it does today. Since ancient times, however, many radioactive elements have gradually disintegrated and become stable. One thing that the general public is not very aware of is that the energy released by this physical phenomenon helps to keep our planet's core in a molten state, which is of paramount importance. Indeed, it is this liquid magma that gives birth to the Earth's magnetic field and enables continental drift… Without radioactivity, our continents would certainly have evolved differently and life would probably have followed a different path…

In contrast with what it was like at the time of its creation, today's Earth has, on the whole, rather low levels of radiation on its surface. These vary in accordance with the geographical zones and depend, among other things, on the soil's specific composition (with granitic regions, for instance, emitting more radiation than alluvial plains). Although low, the radiations present remain, however, sufficient for all of our planet's living beings to be immersed in radioactivity.[2]

---

2 AN: And yet, one must not forget the destructive nature it can represent for every living being. For example, an inversion of the Earth's magnetic poles would leave our world exposed to cosmic and solar radiation for a period of several decades (in the best-case scenario). Under these conditions, life on our planet would certainly suffer great damage. Fortunately, although such catastrophes have taken place in the past, they only occur on a geological timescale. One can thus only hope that, if any of these phenomena were to occur, the authorities would take the necessary preventive and informative measures.

Man's Use of Radioactivity

There are, however, more than just drawbacks to ionising radiation. When tamed by man, it acquires many a scope of application including:

Energy

The production of electricity in nuclear power plants is probably both the most famous and most dreaded use of radioactivity. It involves fission reactions (during which large uranium or plutonium atoms are broken down, creating other smaller elements) that release a colossal amount of energy. Comparatively, the fission of all the atoms contained in a single kilogramme of uranium-235 is equivalent to the combustion of 2,500,000 kilogrammes of coal! The heat generated by these nuclear phenomena is transformed into electricity in EDF[3] power stations, which use Pressurised Water Reactor (PWR) technology. Today, fifty-eight of them produce more than 75% of the electric power in France. In comparison to this, the proportion of electricity generated by the United States' 99 reactors corresponds to 20% of the country's electricity consumption.

Industry

- Food sterilisation is a good, little-known and yet widespread example of such application. In general, gamma radiation[4] is used to destroy microorganisms, insects and parasites present on/in vegetables, fruit, cereals, fish, meat… Generally considered safe for consumers, this process enables a much better conservation of products.

3 TN: *Électricité de France*, a nuclear electric power generation company.

4 AN: See the next chapter entitled 'Scientific Foundations'.

- Industrial radiography involves the use of X-rays or gamma rays, which can serve to detect defects in various metal parts or the sensitive welds of airplanes, pipelines, buildings, etc.

- Leak detection and level gauges. Radioactive radiation is used in processes that range from detecting cracks in a dam or filling a can of beer to controlling the level of a chemical tank.

- Satellite power supply. Systems that mostly make use of plutonium-238 and, in a few rarer cases, cobalt-60 or strontium-90[5] can provide electricity for many years to come, without the necessity of any sort of maintenance.

Medicine

Our understanding of radioactivity has revolutionised medicine. Is there anyone who has never been taken to a hospital's emergency department to have an X-ray done? Many of us have a family member or friend who has been examined using radioactive markers (nuclear medicine) or has been treated for cancer using radiotherapy or brachytherapy. The numbers speak for themselves. About sixty million radiology examinations are performed every year in France. In addition to this, more than 500,000 nuclear medicine tests are performed annually and approximately 100,000 patients undergo radiation therapy on a yearly basis.

Science

Radioactivity is used in many sciences. Its properties make it indispensable in fields that stretch from the study of the universe (astrophysics) to that of DNA (molecular biology) or our understanding of

5 TN: All of which are radioactive isotopes. For further clarity, isotopes are defined as variants of a particular chemical element that differ in the number of neutrons they include and their atomic mass, but not in their chemical properties or their respective number of protons.

our own history (archaeology). For the latter, carbon-14 dating can assign dates to organic objects that are less than 50,000 years old, which, as I am certain you will agree, can be very useful when establishing the chronology of events.

The examples mentioned above are bound to make people aware of the fact that the radiological spectrum is present all around us. In actual fact, the greatest danger comes from man himself. Recent history is fraught with accidents that have caused considerable damage and others whose admittedly less media-reported consequences have been equally devastating. Furthermore, when one considers the current international context, in which inter-community tensions are exacerbated and the prevailing economic system is simply on the verge of collapse, threats of terrorism or malicious intent are added to the risk of accidents, not to mention the possibility of a third world war.

2. Scientific Foundations

The fortuitous concourse of atoms
is the origin of all that is.

— Greek philosopher DEMOCRITUS, 460–370 BC

All bodies are transparent to this agent.
For brevity's sake, I shall use the expression
"rays" and, to distinguish them from others
of this name, I shall call them "X-rays."

— German physicist WILHELM RÖNTGEN, 1845–1923

The purpose of this chapter is to provide a foundation of minimal scientific knowledge to enable readers to design and interpret the phenomena and risks involved. In order for the content to remain within the reach of the general public, some summaries had to be made and only the key elements essential to people's understanding are thus presented.

The Structure of Matter

Matter is an assembly of tiny foundational bricks, completely invisible to the naked eye, called atoms. Each of these atoms comprises a nucleus around which electrons (extremely tiny negatively charged particles) gravitate at vertiginous speeds. The first representations of the atom used the planetary model (the so-called Rutherford model), which was based on an analogy with the sun and the planets — a central orb representing the nucleus, surrounded by smaller spheres representing the electrons. Later on, the famous physicist Niels Bohr completed the concept. One, however, had to wait for the birth of quantum mechanics for reality to appear somewhat different. Nevertheless, the old representation has the merit of facilitating understanding.

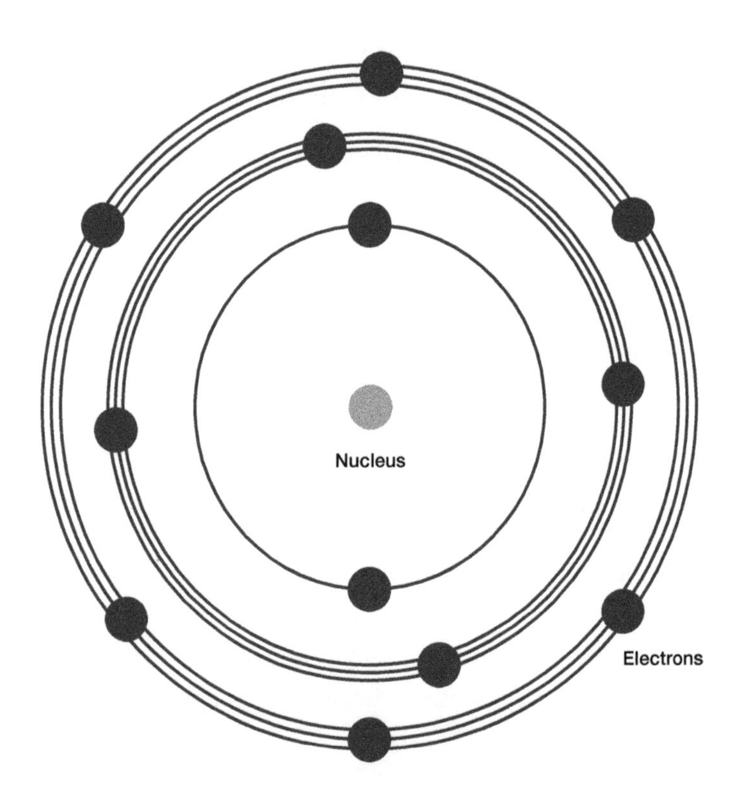

Figure 1. Bohr's Model

Let us now analyse the nucleus. In actual fact, the latter is not a single and inseparable orb, but an assembly of protons (positive particles, which accurately counterbalance the electrons' negative charge) and neutrons[1] (with no electric charge, but endowed with a mass identical to that of a proton). These two types of tiny 'particles' can come together to form a ball (the nucleus) into which most of an atom's mass is concentrated. The number of protons within the nucleus is of paramount importance. Indeed, it is this very number that shall determine the nature of the chemical element[2] thus composed.

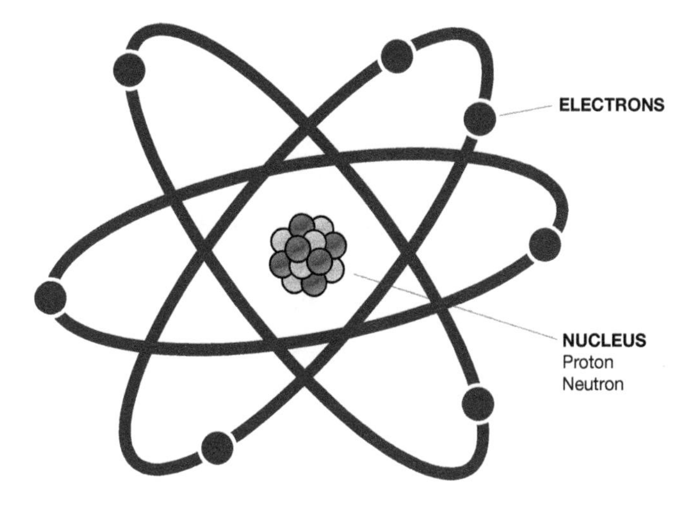

Figure 2. The Atom (Nucleus = Protons + Neutrons)

MATTER IS A LOT OF VOID!

It turns out that atoms are largely[2] made of vacuum. This information is essential because this free space plays an important role in

1 AN: With the exception of hydrogen, which only comprises one proton in its nucleus (and no neutrons).

2 AN: The expression 'chemical element' refers to a set of atoms with the same number of protons in their nucleus and with identical chemical properties. They can be represented by a single symbol. To date, there are 118 elements, of which 94 are found in a natural state. Hydrogen (H), for example, has only one proton, whereas iron (Fe) has 26 and lead (Pb) 82.

the propagation of radiation. As we will see later, this will determine the way one should protect oneself.

So as to better understand this notion of 'void', it is possible to relate the distances existing in this world of the infinitely small to scales that are more understandable from our perspective: if we were to enlarge the nucleus of a hydrogen atom (a proton) to the size of a pinhead, its electron would be represented by a speck of dust moving at about a 100-meter distance from it (i.e. the length of a football stadium). Another comparison: if the same hydrogen nucleus, located in Paris, formed a ball as tall as a man of 1.70 meters, its electron would be a ball of less than a millimetre whose trajectory would travel south of Sicily, towards the island of Malta…

These two examples illustrate the fact that emptiness is an essential part of matter. In fact, it even represents its largest share! Depending on the elements, of course, one encounters variations. Aluminium and lead atoms thus have neither identically sized nuclei nor the same number of electrons. At a macroscopic level, this difference seems quite obvious when one takes hold of a lead bar and an aluminium bar of the same size, each in a different hand. Try it, it's an astounding experience!

Things are actually not quite that simple, however. Indeed, most elements can comprise a different number of neutrons while keeping their proton count unchanged. In this case, the name remains the same[3] and they retain identical chemical properties. They are then known as isotopes, some of which will turn out to be radioactive. In order to define them, a certain number of protons and neutrons is associated with them, which determines their respective mass number.

3 AN: The sole exception is hydrogen, which is given a different name according to the number of neutrons that its isotopes possess.

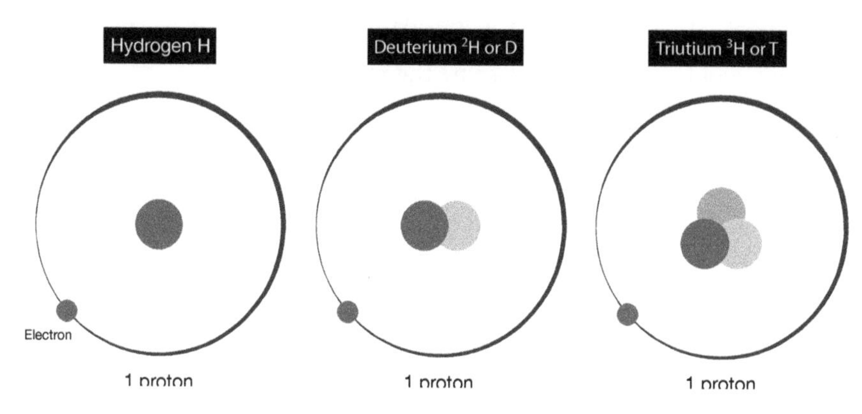

Figure 3. Hydrogenic isotopes

So as to better understand this isotope notion, let us give two more examples:

1. **Carbon** (an element comprising 6 protons)

   In nature, the main isotope (98.89%) is carbon 12 (^{12}C). This means that its nucleus has 6 protons (which is obligatory if it is to bear the name 'carbon') and 6 neutrons in it, resulting in a total of 12 (the mass number). ^{12}C is stable. A small proportion of this element (1.11%) is found in the form of carbon-13 (6 protons + 7 neutrons) and is also stable. Last but not least, a tiny part (0.0000000001%) turns out to be radioactive—this is carbon-14 (which, this time, comprises 8 neutrons).

   The carbon element thus includes 3 isotopes.

2. **Uranium** (an element comprising 92 protons)

   In the case of uranium, there are 26 isotopes, all of which are radioactive. The main isotope (99.27%) found in nature is uranium-238 (238 = the mass number). Its nucleus contains 92 protons (which allows it to be called 'uranium') and 146 neutrons. The second in terms of quantity (0.72%) is uranium-235. The latter is well-known because it is used for the manufacture of certain atomic bombs (i.e. nuclear fission bombs). The other 24 isotopes exist only in the form of traces, since they share the remaining 0.01%.

CARBON-14 DATING

Carbon-14 (^{14}C) dating enables us to determine more or less accurately the age of an organism within a period of time varying between 5,000 and 50,000 years.

The principle is a relatively simple one: carbon is mainly constituted by the isotope ^{12}C, which is stable. However, a small part exists in the form of the isotope ^{14}C, which is, by contrast, radioactive (meaning that it disintegrates little by little).

Depending on its half-life,[4] half of the ^{14}C carbon present in the air is supposed to disappear every 5,730 years. However, this is not the case, because it is recreated, permanently and in low proportion, through cosmic radiation in the upper atmosphere. A balance is thus formed, and the ratio between these two isotopes therefore remains constant over the period of time that is of interest to researchers.

Since this coefficient is fixed, every living entity is in harmony with the external environment and has, therefore, the same ^{14}C carbon content. Following its death, however, this isotope is no longer renewed. Under such circumstances, its proportion decreases because it disintegrates little by little. Measuring **the ^{14}C carbon/^{12}C carbon ratio** thus makes it possible to determine when the life of the organism came to an end. The less ^{14}C remains in the fossil for us to date, the older the death.

This technique also allows us, for instance, to find out precisely when an archaeological object of organic origin[5] was manufactured.

4 AN: See the chapter entitled 'Scales and Units'.

5 AN: Anything that stems from a living being, such as a skeleton, wood, etc.

Radioactivity

Of all the elements present in nature, there are many that possess unstable isotopes, meaning that they will disintegrate at their own pace (known as radioactive decay) so as to gain stability. In order for them to do this, several means are available to them: their nucleus may break, eject particles or perhaps generate electromagnetic radiation. These isotopes with unstable nuclei can be of natural or artificial origin. They are said to be radioactive and will emit health-threatening ionising radiation (capable of tearing electrons out of atoms). The main reason for this instability is usually either the presence of too many protons and/or neutrons constituting the nucleus or an imbalance in their ratio.

There are four main types of radiation:

- **'Alpha' Radiation**

  Compared to other ionising rays, these are highly charged (two positive charges) and relatively heavy particles (bringing together two protons and two neutrons). They are ejected at high-energy levels. This 'alpha' ejection phenomenon can only occur in the case of heavy nuclei. The goal is obviously to make the latter more stable. The emission of this alpha corpuscle commonly uses up most of the energy released, while the rest can be discharged as gamma radiation (see below).

- **'Beta' Radiation**

  These are small particles the size of an electron. They are called 'electrons' when carrying a negative charge and 'positrons' when carrying a positive charge.

- **Neutron Radiation**

  As the name suggests, this radiation involves the emission of neutrons, i.e. of electrically neutral particles with a mass identical to that of protons. This phenomenon is quite rare in nature, but can be encountered following certain human activities such as the production of electricity in nuclear power plants.

- **'Gamma' and 'X' Radiation**
  What we are dealing with here is the emission of electromagnetic radiation. Having the same nature as light, these rays are endowed with a much higher energy. They are thus referred to as photons. Their main characteristics are that they are massless and have no electrical charge. This type of radiation has a very wide spectrum. This phenomenon makes it possible to de-energise an unstable atom that has a surplus of energy and may also accompany the disintegrations mentioned above. 'Gamma rays' and 'X-rays' are therefore of the same type, but their names differ because of their origin: whereas gamma rays emanate from the nucleus, X-rays stem from electrons.

THE NOTION OF RADIOACTIVE FILIATION

When a radioactive isotope gives birth to another nucleus which is also radioactive, the phenomenon is called 'radioactive filiation'. Some unstable elements will thus disintegrate into a more or less complex cascade (a disintegration chain) before becoming a chemical element whose with a stable nucleus. For example, natural uranium (U_{238}) will eventually result in lead-206 (Pb_{206}), which is stable and therefore non-radioactive. The path will however be a very long one and will lead through many unstable isotopes, the most famous one being undoubtedly radon, a radioactive gas that is frequently found in granitic zones.

Scales and Units

Radioactivity is nowadays a well-known physical phenomenon. Different scales and units have been created to characterise it:

- **Activity**

The activity of a radioactive sample is the number of disintegrations that unstable nuclei undergo within it every second. The international activity unit is the becquerel (Bq). Since it is a very small unit, multiples will be used most of the time.

- 1 kilobecquerel = 1,000 Bq
- 1 megabecquerel = 1 million Bq
- 1 gigabecquerel = 1 billion Bq
- 1 terabecquerel = 1,000 billion Bq

The former measurement unit was the curie (Ci), which corresponded to the activity of one gram of radium, a naturally occurring element in the soil (alongside uranium). The curie is a much larger unit than the becquerel—in fact, 1 Ci = 37,000,000,000 Bq (37 billion Bq).

- **Radioactive Decay and Half-Life**

The activity of a radioactive sample decreases with time. Indeed, since unstable nuclei undergo gradual transformation, there is less and less ionising radiation to emit. This phenomenon is known as radioactive decay. The half-life is the time it takes for a sample to lose half of its activity. Each radioisotope thus has a half-life of its own. It remains constant and does not vary in accordance with external conditions such as temperature, pressure, etc.

RADIOISOTOPES	RADIOACTIVE HALF-LIFE	ORIGIN	EXAMPLES OF USE/ PRESENCE
Oxygen-15	2 minutes	Artificial	Medical Imagery
Iodine-131	8 days	Artificial	One of the many kinds of nuclear reactor waste
Cobalt-60	5.27 years	Artificial	Gammagraphy, radiotherapy

RADIOISOTOPES	RADIOACTIVE HALF-LIFE	ORIGIN	EXAMPLES OF USE/ PRESENCE
Radium	1,600 years	Natural	Formerly used in radiotherapy and luminescent paint
Plutonium-239	24,100 years	Artificial	Fuel for nuclear power stations, atomic bombs, etc.
Uranium-238	4.5 billion years	Natural	Fuel for nuclear power plants

The chart above highlights the fact that important variations exist between the half-lives of different elements. Thus, although it only takes eight days for iodine-131 to lose half of its reactivity, uranium-238 requires 4.5 billion years (which is roughly the age of the Earth itself) to attain the same result. This means that at the time of our planet's birth, there was twice as much of this radioisotope present.

- **Absorbed Dose (D)**

  This is the amount of radiation absorbed by an organism or object. The international unit is the gray (1 Gy = 1 joule/kg of irradiated material), which replaced the rad in 1986 (1 Gy = 100 Rads). Simplifying things slightly, this value gives you an idea of the amount of radiation received.

- **Equivalent Dose (H)**

  This quantity derives from the absorbed dose. It is expressed in **sieverts** (1 Sv = 1 joule/kg of irradiated material). It takes into account the nature of the ionising radiation impacting the body or object in question. Indeed, depending on its nature, the impact on one's health will be amplified to one extent or another. A weighting factor (Wr), which is equivalent to a 'damage modifier', is thus applied to the different radiations.

RADIATION TYPE	ALPHA (α)	BETA (β)	NEUTRONS (n)	GAMMA (γ)/X
Weighting Factor	20	1	5 to 20 (depending on the energy)	1

Be careful, though: these weighting factors only come into play when ionising radiation is effective. For example, alphas that are likely to cause twenty times more damage (Wr = 20) than gammas or betas will only do so if the target is 'within range' and unprotected (with a distance of less than a few centimetres[6] and the absence of a screen blocking the radiation, such as the skin, which suffices to stop alpha rays).

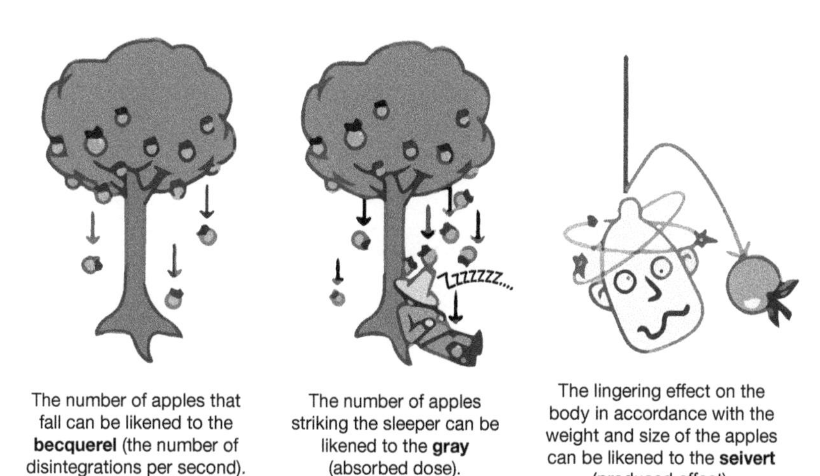

The number of apples that fall can be likened to the **becquerel** (the number of disintegrations per second).

The number of apples striking the sleeper can be likened to the **gray** (absorbed dose).

The lingering effect on the body in accordance with the weight and size of the apples can be likened to the **seivert** (produced effect).

Figure 4. Radioactivity Measurement Units (Source: CEA)

- ## Dose Rate

For reasons of practicality and evaluation, it is convenient to use the concept of a dose rate, i.e. the absorbed dose per unit of time expressed in gray per hour (Gy/h) or the equivalent dose in sievert per hour (Sv/h). In order to properly comprehend these quantities,

6 AN: See the paragraph entitled 'Scopes and Means of Exposure'.

we can compare them with what happens with a tap. Imagine it allowing a trickle of water to flow, with a flow rate of three litres per hour. If we place a bowl under it, we will recover a litre and a half after half an hour (= the collected dose). If the container remains there for a period of two hours, six litres will be obtained. The phenomenon is identical for dose rates resulting from ionising radiation. Therefore, if a person spends two hours in a place where the surrounding radioactivity is three millisieverts per hour (flow rate = 3 mSv/h), they will receive a dose of 6 mSv. Knowing the dose rate, which can be indicated, in real time, by a radiation detection device, it thus becomes possible to assess the amount of radiation absorbed by an individual depending on the duration of their presence.

Multiples and Submultiples

PREFIX	SYMBOL	MULTIPLYING FACTOR
exa	E	10^{18} = 1 000 000 000 000 000 000
peta	P	10^{15} = 1 000 000 000 000 000
tera	T	10^{12} = 1 000 000 000 000
giga	G	10^{9} = 1 000 000 000
mega	M	10^{6} = 1 000 000
kilo	k	10^{3} = 1 000
hecto	h	10^{2} = 100
deca	da	10^{1} = 10
deci	d	10^{-1} = 0.1
centi	c	10^{-2} = 0.01
milli	m	10^{-3} = 0.001
micro	μ	10^{-6} = 0.000 001
nano	n	10^{-9} = 0.000 000 001
pico	p	10^{-12} = 0.000 000 000 001
femto	f	10^{-15} = 0.000 000 000 000 001
atto	a	10^{-18} = 0.000 000 000 000 000 001

Units and Equivalences

QUANTITY	IS* UNITS	FORMER UNITS	RELATIONSHIP BETWEEN THE UNITS
Activity	Becquerel (Bq)	Curie (Ci)	1 Ci = 3.7 x 10^{10} Bq = 37 GBq 1 Ci = 37 billion Bq
Absorbed Dose	Gray (Gy)	Rad	1 Gy = 100 Rad 1 Rad = 1/100 Gy = 10 mGy
Equivalent Dose	Sievert (Sv)	Rem	1 Sv = 100 Rem 1 Rem = 1/100 Sv = 10 mSv

* International System

3. Exposure to Radioactivity

The Azores anticyclone should block the potential arrival of the radioactive plume (of Chernobyl).

— Brigitte Simonetta, weather presenter,
Journal télévisé de 20 heures,[1] Antenne 2, 29 April 1985

Scopes and Means of Exposure

Since radioactive radiation consists either of various particles (alpha, beta, or neutrons) or of energy waves, their properties are different. Depending on their nature, therefore, they will have specific spans.

RADIATION TYPE	AIR DISTANCE COVERED
Alpha	A few centimetres
Beta	A few metres
Neutron	Several hundred metres
Gamma/X	Several hundred metres[2]

1 TN: The 8.00 p.m. Television News.

2 AN: Electromagnetic radiation (γ and X) is not stopped but only attenuated by the air. Its scope in this environment is therefore virtually infinite. However, since the received dose does vary in an inversely proportional manner to the square of the distance, it is generally acknowledged that even in the case of

Of course, this chart only offers a rough approximation of the distances.[3] It does, however, allow one to bear in mind the scope of the different radiations and, therefore, the extent of the danger zone. It is applicable when the radioactive source is intermittent and small-scale. The calculation becomes more complex when it comes to other configurations such as an extensive contamination of powder substances or liquids or a cloud of radioactive vapours or gases…

Note: Depending on its form, the radioactive source is said to be:

- a sealed source — when the structure and the packaging prevent, in normal use, any and all dispersion of radioactive material.

Figure 4. Sealed Pu^{238} source contained in a cardiac pacemaker. (Source: Andra)

- a non-sealed source — when it can easily be disseminated throughout the surrounding environment. A good example is that of easily-opened containers comprising radioactive powder, liquid or gases…

---

relatively sizeable activity sources, the dose becomes negligible after a few hundred meters.

3 AN: Many factors can cause the range to vary, one of the most important being the actual energy of the radiation.

Figure 4.1. Non-sealed source containing radioactive powder formerly used as a beauty product. (Source: Wikipedia Commons)

It is therefore perfectly understandable that, depending on the actual nature of the source, the different ionising rays will not reach the body in the same way. There are thus two main types of radiation exposure:

External exposure

As suggested by the name, the radioactive source remains outside the body in this case. It can assume various forms such as dust (a non-sealed source) or inseparable fragments (a sealed source). As a result of this, depending on the situation, it is possible to distinguish two modes of external exposure, namely:

- **Remote irradiation:**[4] although the radioactive source (usually a sealed one) lies at a certain distance from the individual, it constantly emits its radiation flow. A person can thus be irradiated, i.e. suffer damage as a result of the action of this ionising radiation

4 AN: Here, the term 'irradiation' is used to distinguish contact-free external exposure from that which involves contact (contamination). Technically speaking, however, talking about an irradiated organism means that its cells have been damaged by radiation, regardless of their source (external or internal).

(mainly the 'gamma' kind[5] — see the range chart) without the need to actually touch the radioactive element.

- **External contamination:** In this mode of exposure, an individual comes into contact with radioactive particles (dust, liquids, etc.). The latter can be found on the skin, in the hair and so on. Any person that has thus been contaminated could pass this contamination on to his family and disperse it across any places he passes through.

Note: An irradiated organism that does not show traces of contamination will not 'irradiate' others, meaning that it will not radiate ionising radiation. We can compare this phenomenon to a burn: a flame causes damage, but once it is removed, its action is stopped. In addition, if someone touches this type of injury, they will not get burned. The same applies to irradiation.

Internal exposure (or internal contamination)

This mode of exposure occurs when radioactive elements enter the body. The most common routes of entry are through breathing, ingestion, injury or even the skin and/or eyes. In this case, not only are the radio-contaminants directly in contact with living cells, but they are likely to remain in the body longer than in the case of external contamination. The obvious consequence is that they will cause continuous damage throughout the whole time they remain inside the body.

Exposure Sources

It is important to remember that our environment is bathed in radioactivity. This varies according to many criteria including, in particular,

5 AN: Within their own range, beta rays can also cause superficial burns. As for neutrons, they are not very present in nature. However, should any sort of emission arise from human action, they can cause irradiation even across longer distances.

soil composition and altitude. In France, it is thus estimated that 60% of the radioactivity received is of natural origin. As for the rest, it mainly stems from medical care or examinations (scanners, X-rays, etc.).

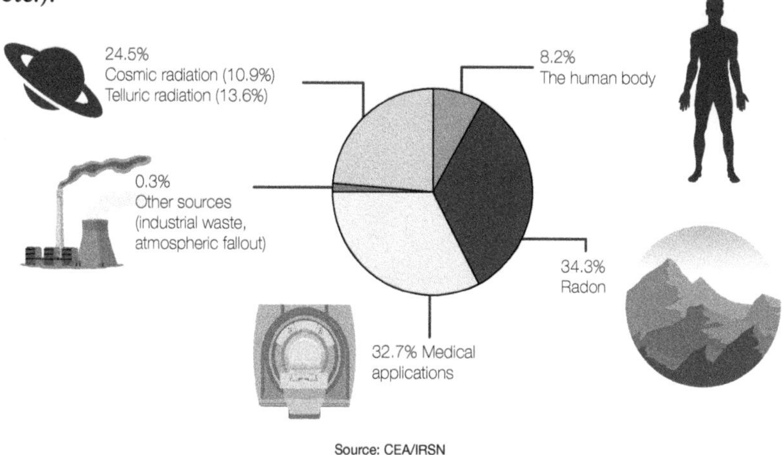

Figure 7. The Sources of Radioactivity in France

Based on a US study, the following chart of public exposure to ionising radiation shows that people generally receive a total annual dose of about 620 mRem (6.2 mSv). Natural sources of radiation account for about 50 percent of this total and man-made sources for the remaining 50 percent.

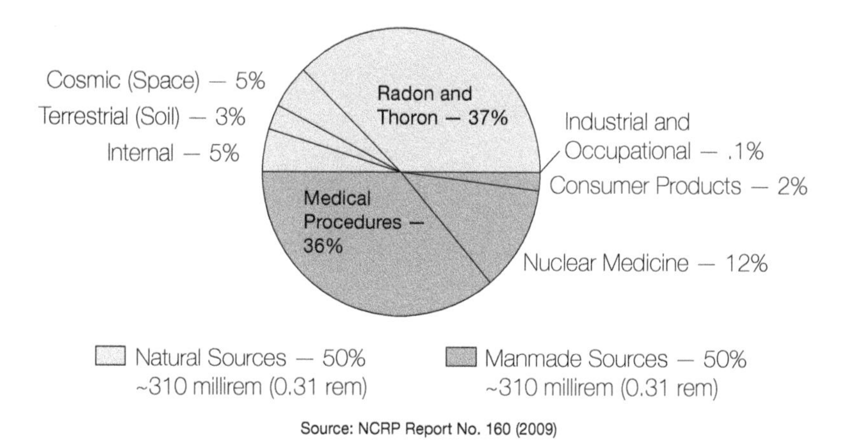

Figure 7.1. Sources of Radiation Exposure in the United States

Note: One can notice that the radiation dose received from medical procedures (including nuclear medicine) is a little bit higher in the US than in France.

The following figure gives examples of natural or artificial radioactivity doses received on various occasions, such as a Paris/New York round trip or a year spent in La Paz.

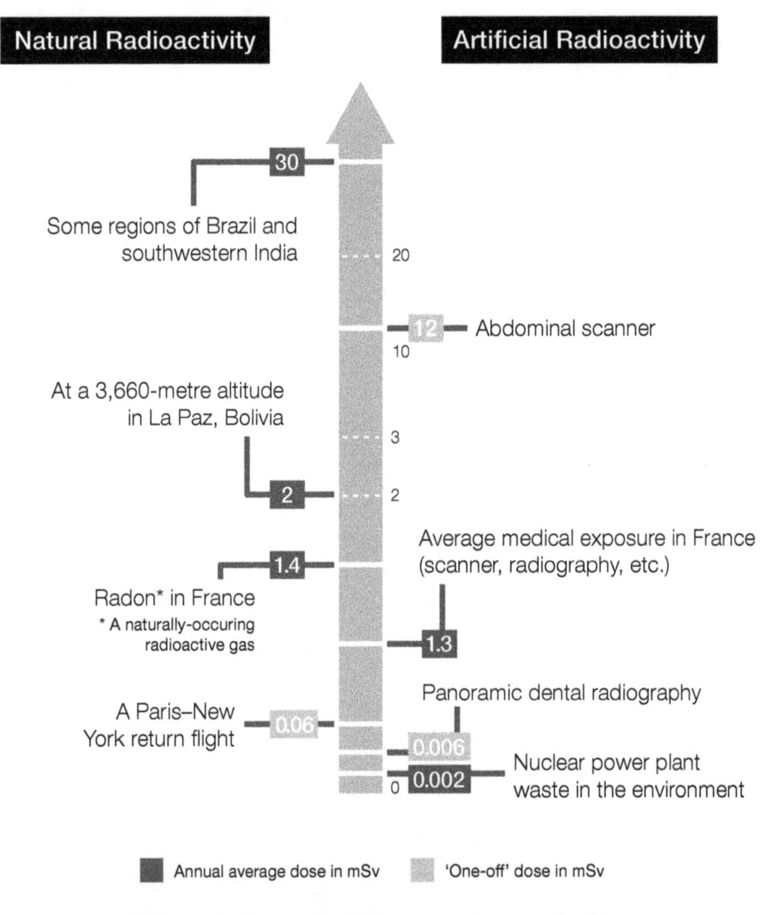

Figure 8. Examples of Exposure Sources for Man
(Source : *Institut de Radioprotection et de sûreté nucléaire*[6])

6 TN: The Institute of Radioprotection and Nuclear Safety.

Health Impact

As regards the impact of ionising radiation on our health, it results from a complex process: the transfer of energy onto living matter will generate free radicals inside our cells or act directly upon our proteins, DNA, etc., damaging them.

Although the body has its own means of repair, these can sometimes be flawed, or they can be overwhelmed by excessive radiation.

Two major categories of health effects thus surface:

- **Long-term effects** (known as aleatory or stochastic effects)
  When the doses of received radiation remain low, there is a slight possibility of incurring long-term effects (years later) in the form of leukaemia and other kinds of cancer. It is, however, impossible to predict the consequences of low-dose exposure for a given individual.

- **Immediate effects** (so-called non-stochastic or deterministic ones)
  When the received doses of ionising radiation are significant, effects will inevitably occur within a relatively short period of time following exposure. Reddening, burns, and vomiting are the main visible manifestations. The immune system and internal organs are also impacted. In this case, there is a correlation between the symptoms' occurrence rate and severity on the one hand and the irradiation level on the other.

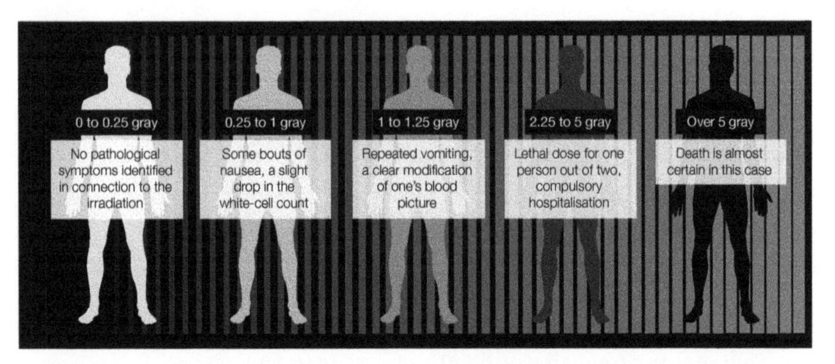

Figure 9. '50 Shades of gray' — The effects of Homogenous Irradiation (Source: CEA/Yuvance)

NUCLEAR INCIDENT OR ACCIDENT: THE INES SCALE

In order to measure the severity of a civil nuclear incident or accident, the international nuclear event ranking scale, known as the INES (International Nuclear Event Scale), is conventionally used. It comprises eight gravity levels ranging from 0 to 7. Created by analogy with the Richter scale (used for earthquakes), it is a useful information tool for both the media and the public. The events reported are evaluated according to three criteria:

- The degradation of defence-in-depth (impact on security barriers) — no human consequences are necessarily generated.
- On-site impact (involving potential consequences for the facility and its workers), devoid of any and all external effect.
- Off-site impact on people, property and the environment.

The more serious and widespread the consequences of an event become, the higher its rating (see the chart below).

- Level 1 to 3 events are classified as 'incidents'.
- Those belonging to higher levels (4–7) are, by contrast, referred to as 'accidents'.
- The seventh and last level corresponds to major accidents such as Chernobyl in 1986 and Fukushima in 2011.

INES

TYPE	OFF-SITE INCIDENCE	ON-SITE INCIDENCE	DEGRADATION OF DEFENCE-IN-DEPTH
7. Major Accident	Extensive release — far-reaching impact on human health and the environment.		
6. Serious Accident	Significant release of radioactive material likely to require the implementation of planned countermeasures.		
5. Accident (entailing off-site risk)	Limited release of radioactive material likely to require the implementation of some planned countermeasures.	Severe damage to the reactor core or radiological barriers.	
4. Accident (entailing no off-site risks)	Minor waste: public exposure within the prescribed limits.	Significant reactor or radiological barrier damage, or lethal exposure of a worker.	Defence loss and contamination.
3. Serious Incident	Very limited release, with public exposure representing a mere fraction of the prescribed limits.	Severe contamination or pronounced effects on the health of a worker.	Narrowly evaded accident. Loss of defensive lines.
2. Incident	No consequences.	Significant contamination or overexposure of a worker.	Incident with a significant failure of safety devices.
1. Anomaly		No consequences.	Anomaly stemming from an authorised operation mode.
0. Deviation			Unimportant anomaly from a safety point of view.

Conclusion

Radioactivity is a natural phenomenon in which our entire environment is submersed. Indeed, the latter is permanently exposed to more or less significant levels of irradiation (mainly due to cosmic and telluric rays, as well as radon). We ourselves are, in fact, slightly radioactive, since our bodies contain radioactive elements such as potassium-40. Nevertheless, for more than a century now, humans have begun to use and artificially create these ionising radiations in various fields of application, across a broad spectrum, ranging from atomic bombs and nuclear electricity-generating power stations to simple dental X-rays.

Nowadays, many causes could lead to situations presenting radiological risks that would endanger you, your family, your neighbourhood, your city or even the whole planet. The most unlikely but certainly most terrifying one would be a nuclear war. However, as history has shown, some natural catastrophes (Fukushima) or major accidents (Chernobyl) can also have significant consequences for both our environment and people's health. On a smaller scale, we must not, moreover, exclude actions of the terrorist kind or those resulting from malice or even plain and simple negligence.

If the authorities were to one day announce a leak or radioactive contamination in your proximity, would you feel ready to face this event? Were you yourself to discover one day that a nuclear or radiological event had taken place, would you have the required resources to deal with it?

Once you have covered the technical part of this book on nuclear and radiological risks, you will be able to understand the dangers you might face and immediately ask yourself the following basic questions:

- What is the nature of the source? Where is it and what shape does it take?
- What types of radiation are involved?

- What mode of exposure are you facing?

As we shall see in the section entitled 'CBRN Incidents: Reactions and Protection', answering these questions will allow one to take appropriate measures. In the meantime, we suggest that you read through the next chapter and acquaint yourself with the different possible scenarios, which will first be described and then analysed to help you understand the risks involved.

4. Scenarios

Only mortal danger is colourless.

— Russian writer VLADIMIR NABOKOV (1925–1977)

Remote Irradiation

Fiction

Sitting with his face glued to the window of the TER,[1] Frédéric watched as the landscape passed by in silence. His features betrayed his pensiveness and tiredness. Things were not all perfect, but he had, at least, been feeling relieved since he'd signed this six-month CDD[2] for a new job. Admittedly, although the work pace was indeed brisk, his major interest lay in learning a job that was quite sought after and which he could specialise in — that of a welder.

The announcement of the next stop drew him out of his lethargic state. He slowly awoke to reality, rose to his feet and headed for the

1 TN: *Transport express régional*, typically shortened to TER, is the brand name used by the SNCF, the French national railway company, to refer to the rail service run by the regional councils of France.

2 TN: *Contrat de travail à durée déterminée*, i.e. a fixed-term employment contract.

exit door. Once outside, he took a deep breath. As had been his habit for two months, he went to the place where Stéphane, his comrade and team leader, was waiting for him.

— 'Hi Stéphane,' he said, as he got into the company's 4x4.

— 'Hello, Fred. How are you today? Doing well?'

— 'Yeah, yeah. I just hope you're not going to ask me for the impossible, as usual!'

— 'Nothing insurmountable, don't worry! A team checked the gas pipes last night. There's apparently a flaw in one of the welds. You'll have to take a look at that when you get there… and repair it, of course!'

— 'One single flaw in the whole thing? Then my friends and I have done a good job. Pretty good for a bunch of beginners.'

— 'Yeah, but we're still behind schedule. We'll have to get a move on it.'

— 'Obviously. I'd have been surprised if we didn't.'

Having arrived on the site, Frédéric kitted himself out, took his gear and went down to the basement. Following the complicated path laid out by the lamps that had been installed for this very purpose, he advanced for several minutes through the pipe-framed labyrinth.

At the end of his itinerary, something caught his attention: a small metal piece shining in the darkness.

Curious, Frédéric went closer to it and grabbed it. He had never seen anything like it. What could it be?

Never mind, he'd worry about that later. He put the object in his pocket and then continued for three or four meters until he reached the spot where the defective weld was located. Conscientiously, he unpacked his equipment and, having examined things quickly, began his repair work. The pipe was difficult to access, but after half an hour of struggle, Fred sat up, smiling.

He was proud of his work. He was going to become a skilled welder!

Satisfied, he found Stéphane and declared that everything was now in order. Having simultaneously been given his new instructions, he returned to the building site and finished his day without any problems.

Back in his own apartment, Frédéric hugged his girlfriend Jamie, cuddled their eighteenth-month-old son a bit and went straight into the shower.

Unpleasant headaches overwhelmed him at this point and a throbbing pain spread across his lower back, causing him discomfort. A rash had, furthermore, appeared at the top of his right buttock. He asked Jamie to examine the lesion, and they both concluded that it was an insect bite. Hardly a surprise, considering the basements in which he worked!

Leaving his belongings in the dirty linen bin, Fred then joined his family at the dinner table. After putting their little boy to bed and spending a carefree evening, the couple went to bed early.

An hour later, however, once they had just fallen asleep, they were awakened by someone stubbornly ringing their doorbell.

Frédéric got up and went to open the door.

To his surprise, he found Stéphane standing there, accompanied by a man he did not know.

'Good evening,' said the foreman. 'I'm sorry to bother you at this time, but I really need to know whether you didn't happen to pick anything up on the job site today?'

— 'What a strange question! What on earth do you mean? I can assure you that I have not stolen anything!'

— 'No, that's not what I mean. In fact, the night shift team have just realised that the radioactive source of our gammagraph is missing. We spent the entire evening looking for it, but it's no longer there. We reckon someone picked it up and took it by accident.'

Immediately, Frédéric remembered the small metallic piece he had put in his back pocket that very morning.

— 'Wait a second,' he said before heading to the bathroom.

He returned with the said sample in hand.

—'Throw it on the floor, into the corner there! Immediately!' the stranger suddenly told him in a loud voice.'

Casting a questioning look at the two men, Fred obeyed.

Attracted by the noise, meanwhile, Jamie had come into the room.

—'Everybody out!' the man continued.

—'My son is in the other room,' said the young woman.

—'Take him and get out of the apartment without wasting a moment's time, please.'

Jamie complied and took her child into her arms before following the group to the outside.

—'Can we at least be told what's going on?' she asked.

—'Your husband has brought a radioactive source into your home.'

—'And what if he has?'

—'It is likely that he has been irradiated. It may be the same for yourself and your child…'

—'Irradiated? But we didn't feel a thing!'

—'That's normal. These radiations are imperceptible to the human senses. And that's precisely the problem… In any case, I shall warn the authorities, as well as the specialists at Percy Hospital. Stéphane will have you looked at first thing tonight.'

Frédéric cast worried glances at his wife.

—'What's in store for us?' he asked.

—'I think the doctors are best able to inform you. They will perform some tests and should be able to quickly estimate the doses you have taken in.'

—'And what about our flat?'

—'I'll stay here and take care of everything. A specialised team will be on the premises in ten minutes' time. The radioactive source will be placed in a container and transferred to a safe location.'

—'Can we come back again?'

—'Yes, of course you can. Once the source has been removed, there should be no danger.'

—'In this case, just follow me!' intervened Stéphane. 'The sooner we are at the hospital, the faster the doctors can take care of you.'

The next day, after an entire night of medical examinations, Frédéric felt really exhausted.

He glanced at his wife and little boy, who were sleeping in the bed next to his.

He smiled, but concerns were gnawing at him. The fact that he had had blood samples taken and had been placed into machines whose existence he hadn't even suspected until then only added to his anxiety.

In addition, he did not feel well at all. His nausea, urge to vomit and headaches never left him...

Suddenly, he picked up the sound of a discussion drawing closer as the slapping of footsteps resounded through the corridor.

Frédéric hoped it was the professor in charge of the case. Suddenly, his heart skipped a beat. He had just heard his own name!

He reached out and gently woke Jamie. She emerged from her sleep just as a group of doctors entered the room.

—'Hello,' began the man presumably in charge. 'My name is Professor Damien. I hope last night was not too difficult for you. If you need anything, do not hesitate to ask the nursing staff or myself.'

—'What we would particularly be keen to hear about, actually, is our state of health,' whispered Frédéric in a low voice.

—'Yes, of course. That's actually why I'm here...'

A heavy silence filled the room.

For a few long seconds, the doctor let his eyes wander, as he looked successively at each of the three members of the family.

—'I have just analysed all the data, and I'm afraid that the results aren't good for everyone present.'

It was as though a leaden shroud had fallen over the gathering. As if paralysed by the announcement to come, Frederic and Jamie hardly dared to breathe. On the one hand, they did want to know the truth without delay, but on the other, they dreaded it.

Finally, the professor turned to the woman and her child.

—'In your case, madam, and your son's, the received doses of radiation are significant. The good news is that they are not likely to cause any serious damage. You may feel a bit tired in the days to come, but nothing more. If you wish, you can go home today, but you will have to come to see us once a week for medical monitoring.'

Jamie let out a sigh of relief and hugged her son tightly. After a few seconds, however, she turned to Frédéric, as anxiety overwhelmed her.

'As for you, sir,' said the professor, 'things are a little more complicated. High doses of gamma radiation went through you for a period of several hours. Although this may not be very visible at the moment, severe damage is likely to occur in the coming weeks.'

—'What do you mean by severe damage?' asked Frédéric in a weak voice.

—'At first, it is likely that the rash on your buttocks will evolve to the point of significant necrosis. Your cells have been too irradiated to survive.'

—'Necrosis?'

—'I fear that a hole is gradually appearing in your bodily tissues, closest to where the radioactive source was. It may extend until a part of your behind has been eliminated.'

—'Anything else I need to know?' asked Frédéric with a lump in his throat.

—'Your pain may well intensify. Many of your internal organs have suffered significant damage. Your immune system is also very weak. It is truly imperative for us to keep you here.'

—'I do not understand. Simple headaches and reddening of the buttocks, and you're telling me that I'm in a state that justifies hospitalisation!'

— 'I'm sorry, but you must understand that your case is very serious, much more serious than it seems to be. The consequences of the strong irradiation that you have suffered may not yet be clearly perceptible, but this will change over time… The damage is already in place in your body and, little by little, the consequences will surface.'

FOUR DAYS LATER

Frédéric listened to the doctors' report without commenting on anything.

Already at first glance, his white blood cell count had dropped drastically. Provided he had understood things correctly, this meant that his body was losing its ability to defend itself against microbes. However, he was not too concerned. After all, he was in a hospital. If he were to catch an infection, he would have more medicine available than needed.

In fact, what disturbed him most was the huge blister on his buttock. Indeed, it kept growing, day after day. His entire right buttock was also swollen as a result of inflammation. What is more, the pain he felt refused to cease, despite the painkillers he was taking.

THREE WEEKS LATER

Frédéric was weary, and his condition deteriorating. He was now experiencing tremendous fatigue.

As regards his buttocks, things were not getting any better. The skin had begun to fall off, and in some places, the flesh that had been reddened by the inflammation had taken on a black shade that did not inspire confidence at all… In all honesty, what he was experiencing was a genuine nightmare. The doctors had not been mistaken in their diagnosis…

THREE MONTHS LATER

When Jamie opened the door and kissed Frédéric's lips, her boyfriend did not have the strength to smile.

—'Hello,' he said in a weak voice.

—'How are you feeling today?'

'Worse than yesterday, unfortunately,' he murmured. 'Even though I didn't think that was even possible.'

—'And your leg?'

Without a word, Frederic slowly raised the sheet to reveal his inner limbs, until he had uncovered the bottom of his back.

Jamie stood there, petrified. After a few seconds, she could not help but look away: the doctors had continued to clean the wound and get rid of the necrotic tissue. At this point, half of the buttocks and the back of the thigh had been removed.

—'Don't worry, I'm here. I'll always be here!' she let out in a whisper.

The man felt his eyes grow misty as a few teardrops ran down his cheeks. A mere three months earlier, he had been delighted to have found a new job. He had imagined himself becoming a good welder and perhaps even specialising in some field and working on oil rigs. Unfortunately, it only took a few hours for his life to change. And why? Because he had picked up a small piece of metal, just a tiny radioactive source which he knew nothing about but which would trigger atrocious and irreversible consequences in his own body… if he survived at all!

Depressed, he let out a deep sigh. His extreme fatigue and ceaseless pain did not help matters, either. At the end of his strength, he managed to slightly turn his head towards his wife and painfully utter:

—'The doctors told me this morning that they would have to amputate my leg and half of the hip… And, in all likelihood, the bladder as well.'

A lump formed instantly in Jamie's throat. The young woman was unable to voice a single word. With a gentle gesture, she ran her hand through her husband's hair.

Frédéric closed his eyes: his wife and little boy were the only things that still gave him the strength to go on living…

The Facts

This fictional account is based on a real event that took place in Yanango, Peru, in February 1999. For obvious reasons (location, environment, etc.), the story differs somewhat from the facts. It does, however, follow the same general chronology and includes the essential elements characterising the incident.

HOW THE ACTUAL EVENTS UNFOLDED

The accident occurred on 20 February 1999 at the Yanango hydroelectric power plant, 300 kilometres from Lima. A thirty-seven-year-old welder, performing an emergency repair, found the gammagraph's[3] radioactive source on the ground and, unaware of what it was, placed it in his back pocket. With his work day over (and still carrying the tiny ionising object), he returned home and joined his family, namely his wife and three children (who were respectively eighteen months, seven years and ten years old at the time).

No sooner had the night begun than a staff member of the company which owned the radiographical device showed up at his door. The man was dropping in on all the employees, one at a time, in search of the missing radioactive source. The welder, understanding the nature of the problem, retrieved the radiation source from the bathroom's laundry bag and handed it to his colleague. The source was immediately put aside and, in accordance with an emergency protocol ensuring work safety, reintroduced into the apparatus which it should never have left in the first place.

3 AN: A gammagraph is a device that allows for the examination metallic parts, thanks to the presence of a radioactive source that can exit its housing for a given amount of time, thus enabling its user to obtain a radiography.

An alert was forwarded to the relevant national authorities, who would then report the situation to the IAEA[4] two weeks later.

The next day, the employee and his family travelled to Lima Hospital to have various tests performed. The man was subsequently admitted and kept in hospital. After three months of medical care, the continued deterioration of the patient's condition prompted the authorities to transfer him to the burn unit at the Percy military hospital in Clamart, a few kilometres from Paris. Despite various therapies using the most modern techniques, the progress was not satisfactory.

Six months after carrying the radioactive item in his pocket, the welder underwent an amputation of his right leg. Multiple complications — infections of all kinds, serious damage to several organs, large fluctuations in the number of white blood cells and platelets, and necrosis surfacing in new places — contributed to a persistent deterioration of the patient's condition.

Sinking into depression, the welder returned to Peru on 17 October (about eight months after being exposed to the source), where he has been given further medical care. And yet, his health continues to deteriorate. The presence of his family, however, gave him the morale and the strength to survive despite his condition's continued worsening.

CONSEQUENCES

A person (the welder) suffered a high degree of radiation and had one of his legs amputated six months after exposure. The many complications left him in critical condition. The welder's wife, who only underwent short exposure while sitting on her husband's trousers (in which he kept the source), only absorbed slight radiation. A lesion (inflammation, desquamation) appeared on the man's buttock after two weeks and was fully resorbed after several months of care. All three children were exposed to low doses that did not result in visible and immediate effects.

---

4 AN: IAEA = International Atomic Energy Agency, which falls under the aegis of the UN. Founded in 1957, its headquarters are located in Vienna, Austria.

Analysis

The investigated item (4 millimetres in diameter x 8 millimetres in length), which the welder had picked up, contained iridium-192, a highly radioactive radionuclide and a powerful emitter of gamma and beta rays. This irradiating element is encountered in sealed form, which means that under normal conditions, it is not supposed to cause any contamination to its environment. On the other hand, its ability to emit radiation remains intact. Its activity on an average day of use is estimated at 1.37 TBq (37 Ci).

The IAEA and the Institute for Radiation Protection and Nuclear Safety have given an estimate of the doses received by the source's carrier. The results are quite impressive:

LOCALISATION	ESTIMATED DOSES[5]
Skin (contact with the source)	> 1,000 Gy
Tissues (2–3 cm under the skin)	100 Gy
Sciatic nerve	25–30 Gy
Femur	20 Gy
Femoral artery	10–15 Gy

Note: These are localised doses and not doses that impact the entire organism (as is the case with those presented in Figure 8). In view of the clinical signs, the link between the received dose and the damage caused seems obvious: the more severe the irradiation, the more quickly the effects are manifested and the more significant they become. In this case, two elements are essential:

- **The Length of the Exposure Period**

  It is logical that the amount of time spent near the radiating source influences the received dose. Imagine, for instance, moving your hand right above the flame of a lighter, then repeating the action

5 AN: Other estimates suggest higher absorbed doses.

while holding the position for several seconds: the result will obviously not be the same. The principle is identical when exposed to a source of radiation. The consequences are thus completely different when comparing the welder, who kept the radioactive item in his pocket for several hours, and his wife, who sat on it for approximately ten minutes.

- **Distance from the Source**

Due to their very limited ability to penetrate tissues, beta rays will mainly cause damage to the skin. On the other hand, gamma rays will inflict deep damage, since they are able to penetrate the whole body without any difficulty. Distance, however, remains a determining factor: the further away the radioactive source is, the lower the received dose. To ascertain this, it is enough to compare the 1000 Gy (the most optimistic evaluation) that came into contact with the skin with, for example, the estimated 20 Gy at the femur… Even if there is a phenomenon of gamma-ray attenuation as the rays pass through bodily tissues, the difference in the figures is mainly accounted for by the distance variation from the source itself. Furthermore, following a careful examination of the manner in which the damage afflicting the victim progressed, it turns out that not all the damage occurred at the same time. The first lesions surfaced at the point of contact with the radioactive element and, later on, in more remote areas. This confirms the fact that, in the case of large doses, the more the area is irradiated, the more rapidly the effects appear. On a slightly larger scale, the influence of the distance factor continues to be felt. In the victim's apartment, the children, who were in the adjoining room,[6] received negligible doses compared to their mother, who sat on the radiation source for several minutes while it was still in her husband's pants.

---

6 AN: Another factor, that of protection resulting from the presence of screens (walls), plays a role in such cases. This aspect is developed in the chapter entitled 'CBRN Incidents: Reactions and Protection'.

To conclude, the Yanango incident was caused by the loss and subsequent collection of a tiny radioactive source of iridium-192. The irradiation experienced by the bearer was truly significant and would cause massive damage to his body. As long as he kept the radioactive element on his own person, it posed a potential danger to others as well, in the sense that the (mainly gamma) radiations stemming from this iridium source would impact people in the vicinity. As soon as it was removed, however, its action ceased and the former bearer no longer spread radiation. Exposure time and distance[7] are thus the main factors to consider for this type of incident.

THE *DIRTYBIRD* SCENARIO

The term *DirtyBird* is the code name that armies give to any scenario in which a satellite containing radiological components (intended for its own energy production) loses altitude, exits its orbit and falls back to Earth, scattering contaminating debris. A real case occurred on 24 January 1978, when the Soviet satellite *Cosmos 954* fell on Canadian soil.

In order to prepare for this kind of prospect, the Swiss Army conducted a simulation of such a scenario[8] in November 2014, a simulation that involved more than 1,000 soldiers of the NBC 10 defence battalion, in collaboration with the Spiez-based NBC 1 defence laboratory, around twenty organisations of civilian first responders from the Geneva area, and approximately 150 German soldiers of the NBC Bundeswehr School. The objective of the exercise was not only to test the entire chain of command, the introduction of the different players and the specific capabilities required to solve a complex radiological scenario, but also to assess

7 AN: When a person wants to protect themselves from a possible risk of irradiation, they can also make use of screens to achieve this. Details can be found in the chapter entitled 'CBRN Incidents: Reactions and Protection'.

8 AN: Project D-CH ABC FTX 14, *EclairaGE*—N.4/2015.

the proper functioning of procedures and equipment. The exercise showed that the relevant elements of the Swiss troops can be mobilised quickly (twelve hours for army elements, twenty-four hours for the NBC 10 battalion) and that foreign aid, whether military or civil, can also be integrated into potential transborder crisis management. It has thus been established that effective collaboration between civilian organisations (which have the necessary means to deal with potentially contaminated people) and military personnel (possessing advanced detection, protection and decontamination equipment) is one of the key elements of success when it comes to limiting both contamination and the number of possible victims.

Radioactive Contamination

Fiction

The sun was already descending below the horizon and darkness was beginning to seep into the premises when the two friends reached the building's side entrance.

Alex was the first to step forward and, bending slowly, examined the door carefully—the lock had long been broken. It had to be said that the hospital had been abandoned quickly once the authorities had stopped paying the employees.

The man exhaled silently. He missed the life he had had as a mechanic before the crisis, but what could he do about it? He did not even understand how the situation could have deteriorated so much—and so fast! All these things were beyond him… During the past two years, he had, at least, managed to get out of trouble and properly provide for his family.

Focusing on his task once again, he took another step, wincing as pieces of glass shattered with a grinding sound under the sole of his shoe.

He froze for a moment.

Under no circumstances did he want to attract attention and be caught by members of the local gang. Hearing no sound, he quickly nodded to Sebastian before pushing the door open.

Once again, the grinding sounds made him frown, but he did not stop this time. Accompanied by his comrade, he plunged into the unlit corridor, lit his headlamp and headed for the stairs.

During his last visit to the premises, he had spotted a large metallic device anchored in one of the basement rooms. Today, he hoped to be able to take a lot of the objects and resell them on the black market.

Having reached his destination, he began to smile: nothing had changed. The object he desired still dominated the centre of the abandoned premises.

Sebastian and Alex took their tools out of their backpacks and got down to work.

After an hour of relentless effort, the two accomplices stopped — they had finally dismantled the different objects that interested them.

Not wanting to remain on the premises any longer, they gathered their loot and headed for their lair. Once there, they could tranquilly disassemble whatever parts were still combined.

As soon as they arrived in the tiny studio next to Alex's house, the two men hurriedly unpacked the items they had found and continued dismantling them. After two hours of persistent work, however, they had to stop, for fatigue began to afflict them as violent nausea and terrible headaches now overwhelmed them.

On the next day, when Alex got out of bed, he did not feel any better. On the contrary, his condition seemed to have actually deteriorated. He had, in fact, vomited all night. In addition, his hands were all red, as if sunburnt.

It was thus a great relief when Corinne, his wife, returned in the middle of the afternoon accompanied by Gerard, the paediatrician whose clinic had previously been located two hundred meters from their home, before the crisis set in.

Alex smiled as the doctor approached. Since the two men had done each other many favours in recent months, trust had been established between them and strong bonds now united them.

Diagnosing a case of gastroenteritis, Gerard reassured his friend: in a few days, he would be back on his feet, as if nothing had happened. Before leaving, however, he repeated the usual instructions based on common sense: wash your hands, do not eat anything unfamiliar, keep an eye open for expired food, avoid bloated tins…

The week was a long one for Alex. Although the vomiting had stopped after the first day, he could not swallow anything at all. In addition, his stools were tinged red. Fortunately for him, his two little daughters gave him courage.

Finally, after five days of ordeal, the symptoms receded little by little. Slowly finding his appetite and recovering his strength, he decided to resume his task where he had left it.

Going to pick up his friend Sebastian, he was surprised to learn that he had also experienced nausea and vomiting. Could they have touched some rat poison while in the hospital or inhaled some stagnant chemicals in the basement?

In the end, it did not matter much any more. They would finish dismantling the different parts that day and sell them on the parallel market the next.

After an hour of effort, Alex suddenly shouted. Having pierced a metal capsule, he had just released a fine sort of powder, which spread both on the ground and on his tools. In the semi-darkness of his makeshift workshop, it emitted a deep blue glow.

— 'Hey! Come and have a look at this!'

'Never seen anything like it,' answered Sebastian.

— 'I think we could use it to save on lighting.'

— 'Yeah, some well-positioned pellets could indeed be very useful.' Alex flashed a broad smile.

— 'Do you think this powder is valuable?'

— 'No idea!'

— 'I'm sure of it! Here, take a little.'

— 'Thanks,' replied Sebastian. 'Try to sell this mysterious powder, and I'll take care of the rest. We'll see who gets paid more.'

— 'Looks like it's our lucky break, at last!'

His eyes alight, he thought of his two little angels who had comforted him when he had found himself writhing in pain the previous week. What a gift it would make for them! A phosphorescent midnight blue painting substance to wear as makeup! They would be over the moon!

Happy beyond measure, Alex left Sebastian and ran to show his family this strange treasure.

In the evening, he invited a small number of friends over to celebrate the event. Before the meal, however, he decided to rest a bit. Once more, he was overwhelmed by fatigue, as nausea returned to torment him… To top it all off, a bad case of diarrhoea and intestinal pain burdened him again.

Unable to get up, he asked Corinne to apologise to his guests for his indisposition. Indeed, his condition was getting worse from one minute to the next, and he couldn't sit at the table in that state, of course! Feeling so bad that he could not even think, he slipped into his bed and spent the whole night shivering and moaning.

The next day, still in a deplorable state, he asked his wife to sell the capsule containing the blue powder. Corinne returned in the late afternoon with a chicken, a bunch of carrots, a smaller cabbage and… some money! What an excellent surprise!

His buyer intended to use the powder to make jewellery. Of course, the young woman suspected the man of planning to make a huge profit from the luminescent material. However, she was still satisfied with her side of the bargain. She now had enough to provide her family with food for days to come!

No sooner had she returned home than she rushed to her husband's bedside. Alex's condition had once again worsened. Things were apparently more serious than they seemed!

Taking some of the powder she had kept, she used a fingertip to gently put some on the door handle.

—'Even when it's dark, you can now find the toilet without wasting a moment's time,' she whispered in his ear.

Alex nodded and then curled up again, his hands clutching his stomach.

During the forty-eight hours that followed, Corinne felt her anxiety intensify little by little. Her husband was not recovering. Nausea, vomiting, headaches and diarrhoea seemed to have latched onto him. In addition, his palm had now doubled in size.

When, on the morning of the third day, Gerard rang the doorbell, part of her anguish dissipated, as if by magic. She greeted the former paediatrician with relief and took him to see Alex.

After a few minutes, however, her stomach was in knots. Seeing the doctor's tense face, she realised that something was wrong.

—'So?'

—'I'm not sure. It's a shame I don't have the means to perform a blood test!'

—'Tell me! What's the problem?'

—'What Alex is exhibiting are symptoms of massive irradiation!'

Corinne stood there for a moment, utterly speechless.

She had seen some reports on the topic several years earlier. The events in question had taken place in the vicinity of a nuclear power station that had exploded, if her memory served her right.

—'I don't understand,' she finally said. 'How is that even possible?'

—'No idea. For the moment, it's only a theory. I'll come back tomorrow with a friend who used to work for EDF. I think he can help.'

—'You're scaring me. How long will Alex be in such a state?'

This time, it was Gerard who said nothing.

—'So?'

—'Hard to say. Before doing anything else, I will need to either confirm or invalidate my assumption.'

—'And what can *I* do to help him recover?'

—'Try to give him soup. If he holds it in, it should rehydrate him—he needs that.'

On the next day, Gerard arrived accompanied by Étienne, his comrade who had worked in a nuclear power station in the past. When he rang the doorbell, Corinne slowly came to open.

'Damn,' he said. 'Looking at you, I'd say you haven't slept all night!'

'And you'd be right,' said the young woman. 'Vomiting and diarrhoea! I'm afraid Alex has passed this mess on to me'.

—'How is he?'

—'No improvement. I'm even under the impression that he's losing his hair.'

The two men gave each other a grave, meaningful look.

—'Well, do not panic. Étienne has brought a radiation detector. We will quickly find out whether my theory is correct.'

Gerard's friend took out a small device, similar to a hairdryer, and began to move it slowly in front of him.

As it approached Corinne's hands, an increasingly sharp sizzle could be heard.

—'What does this mean?' asked the distraught woman.

—'You've recently touched something unusual—what was it?' came Étienne's short question.

—'Nothing special … Oh, but I have! The blue powder that Alex found in the old hospital…'

Corinne dropped into her chair.

—'Oh my God!' she whispered. 'No! No, not that!'

Suddenly, she rose to her feet and ran into the other room, screaming, 'Jessica!'

A few minutes later, she reappeared with a more or less ten-year-old blonde girl by her side. Étienne and Gerard froze when they noticed that the girl's eyes, cheekbones and mouth area were all stained blue.

—'O.k., the situation is worse than I thought,' Gerard eventually said. 'Where is this powder? We must put it away somewhere, no matter what!'

—'I don't have it any more. I sold it two days ago...'

—'Damn it! If this substance is highly radioactive, it could contaminate the entire area!'

—'But is there nothing we can do? Like use a bit of alcohol or bleach?'

Gerard let out a long sigh before answering.

—'I don't know what kind of substance we're dealing with exactly, Corinne, but I can assure you that no disinfectant will work against it.'

—'So, for Alex and Jessica...'

—'You mustn't waste a single second and visit a hospital complex that is still in use.'

—'You know we can't afford to cover such expenses; they won't even let us in.'

—'It would be best to have them come here,' Étienne said. 'Otherwise, we'd risk contaminating the emergency unit.'

Once again, silence filled the room.

Standing up straight, Gerard took out a pair of latex gloves and a surgeon's mask from his pocket.

—'I'm going to examine Alex's hand more carefully,' he said in a dry voice.

As he approached his friend, he felt his eyes tear up. Memories of his lectures on the topic of contamination and radiation suddenly flooded his mind.

With a heavy heart, he slowly realised the extent of the damage. In two or three weeks, Alex would certainly have to undergo amputation, at the elbow in the best of cases. However, this would not alter the damage already done to the internal organs. The dose of received radiation seemed to be really significant, and, at first sight, beyond any and all possible remission.

Gerard felt his heart go frozen cold. If he was right, his friend would die without anyone being able to do anything about it.

Gently, he sat on the edge of the bed. Slowly, he turned his head and met Alex's gaze. Despite his suffering, the latter immediately understood the fate that awaited him.

For a long minute, neither one of them moved, after which time Gerard finally got up. He could not find the words to comfort his friend. Who would provide for this family if the father died?

With his mind continuing to analyse the situation in spite of itself, the paediatrician then realised with horror the actual extent of the disaster — it was not only Alex who had been irradiated and contaminated, but also Corinne, Jessica, and certainly their other little girl. With all the powder on the girls' fingers and faces, the amount taken in by their bodies could be fatal. Not to mention Sebastian, the eternal friend, who had allegedly spread his own share of radioactive substance.

Gerard took his friend's hand into his own and clasped it. It had just dawned on him that the entire area was doomed. This house was irretrievable, as was that of the mysterious buyer and of other people still… Soon, the news of a radioactive contamination would spread like wildfire and a state of panic would gradually take hold of the population. Hundreds of people would be quarantined, others simply killed… It was fear that caused such reactions — this he knew, for he had already experienced it!

Slowly, Gerard closed his eyes, as if to drive out his memories and apocalyptic visions… But to no avail.

At this point, one thing was certain: this area, which had managed to organise itself and survive the great economic crisis, was going to plummet into utter chaos.

~

The Facts

The piece of fiction related above is based on a real-life event that occurred in September 1987 in Goiânia, Brazil. Obviously, the environment is not the same, and there are differences between the two versions. In addition, the stated purpose of the narrative is to focus on the essential points, not to offer an exhaustive synthesis of the serious Goiânia incident.

HOW THE ACTUAL EVENTS UNFOLDED

In 1985, a private radiotherapy institute moved to its new premises. Following certain court litigations, some materials were left behind in the old building.

On 13 September 1987, two individuals entered the building, seized the caesium therapy unit and transported it home so as to dismantle it. During the first evening alone, both experienced vertigo, vomiting and diarrhoea.

On 18 September 1987, after successfully releasing the caesium capsule, they sold the lot to a scrap dealer living nearby. The latter, very excited, invited his family and friends to show them his new acquisition.

On 25 September, the radioactive material and accessories around it were acquired by another merchant.

On 29 September 1987, seeing that everyone had fallen ill, the wife of the first scrap dealer recovered the phosphorescent blue powder and brought it to the local hospital. The substance was then put aside and analysed the next day. The result was unequivocal and confirmed the material's high radioactivity. It was, however, already too late to prevent contamination from spreading across the city.

Indeed, three days after being stolen from the abandoned premises, the source was pierced and began to leak. From that moment on, caesium powder was scattered everywhere. In addition to this, the phenomenon was exacerbated once the capsule was completely opened while in possession of the first scrap dealer.

CONSEQUENCES

There were four fatal casualties. The scrap dealer's six-year-old niece died of widespread infection on 23 October 1987, a month after ingesting a little caesium powder while eating her sandwich. As for the scrap merchant's wife, she succumbed on 23 October 1987, at the age of thirty-seven. Last but not least, two of his employees, aged eighteen and twenty-two respectively (and including the one who had managed to fully open the capsule), passed away in October 1987.

All showed similar lesions: the swelling of certain body parts, internal bleeding, hair loss, respiratory problems and damage to the lymphatic system... Following a study conducted by the Brazilian authorities, it turned out that forty-six people had received high doses. Although 'only' four of them eventually died, twenty further cases showed radiation-related symptoms and required treatment.

The scrap dealer himself survived, despite receiving an extreme radiation dose of 7 gray. However, he subsequently fell prey to depression and alcoholism and died of cirrhosis in 1994.

The man who had been the first to dismantle the device in the old hospital saw a burn develop on his hand, and soon the latter began to swell. A month later, he had to have his arm amputated.

As soon as the authorities became aware of the radiological accident, contamination checks were carried out: 112,000 people were examined, of whom 249 proved to be carriers of external and/or internal contamination; eighty-five houses were found to be heavily contaminated (seven, considered unrecoverable, were completely demolished, and the resulting rubble and soil were moved out of Goiânia); 200 inhabitants were evacuated; a total of 3,500 m^3 of contaminated waste (the equivalent of 275 lorries) was extracted and transported to a disposal area twenty kilometres from the city.

The psychological impact and crowd movements are also major elements to be taken into account. For instance, riots broke out at the burial of the six-year-old girl, when 2,000 people tried to prevent the interment (although the coffin was both reinforced and sealed).

In the end, it is obvious that the consequences of such an event are disastrous. In the case of Goiânia, however, the authorities' speed of action and international aid (the sending of experts and materials) made it possible to 'limit the damage'.

And yet, one must be careful and bear in mind that this class 5 accident on the INES scale was due to irradiation and contamination resulting from *only 93 grams of caesium-137*. Just try to imagine the consequences in a country plagued by chaos (a civil war, economic collapse, etc.), where no authority would take this type of accident into account. Knowing that caesium-137 has a half-life of about thirty years, meaning that its activity is halved every thirty years, it would potentially continue to irradiate a population lacking any and all awareness of this danger.

Analysis

The source that triggered the accident was initially used for radiotherapy-based cancer treatment at Goiânia hospital. Its characteristics were:

- 93 grams of caesium-137 (in the form of caesium chloride powder).
- A capsule (= sealed source) to house the caesium in, protected by a shielding layer of lead and steel.
- A total activity of 50.9 TBq (= 1375 Ci) at the time of the theft.

Caesium-137 is a potent emitter of gamma and beta radiation. In the case of Goiânia, the source's emission rate at a distance of one meter was more than 4.5 Gray/h. During the theft and the attempt to dismantle the radiotherapy instrument, the man received massive doses of gamma rays (as the beta rays were stopped by the metallic envelope).[9] These were all the stronger on account of the limited distance. For example, the dose received at 10 cm from the radioactive element is 100 times higher than the one absorbed at a 1-metre distance.[10]

In the case of the individual who disassembled the capsule while it was still inside the device, this factor was of paramount importance. Indeed, his hand drew closer to the source at the time of the disassembly. This is why the first visible symptoms (the reddening and subsequent swelling of the fingers and palm) surfaced at this point. However, even if the amount of gamma rays was highest at the extremity of his upper limbs, the rest of the body was also irradiated by high doses. Vomiting, headaches and diarrhoea were thus the direct consequences.

To this distance factor, one must, of course, add the actual duration of the exposure. The more time one spends near the source, the higher the dose received. As previously described, these two notions are crucial in cases of external irradiation.

Concerning the Goiânia incident, however, other phenomena come into play the moment the capsule was opened and the caesium powder spread. The beta radiation, which had, up to that point, remained confined in its metal housing, was then added to the gamma rays. These accentuated the damage (mainly the radiological skin burns) for all those who remained in their zone of influence (a few meters by air), all the more so because the distance is so short.

Now imagine the additional damage caused by the use of powder as make-up, as people actually contaminate themselves with highly radioactive materials!

Nevertheless, this phenomenon can become even more critical: it is enough for the particles to enter the body to cause more damage. For example, if ingested, they would pass through the mouth, the trachea,

---

9 AN: So as not to complicate the reasoning process, any potential braking radiation (producing X-rays) is not taken into account.

10 AN: See the chapter entitled 'CBRN Incidents: Reactions and Protection'.

the stomach, the intestines... In the process, a more or less significant part of the radioactive substance, depending on the actual nature of the compound, would thus be absorbed, allowing it to extend its ravages to the liver, the kidneys, the blood, the bone marrow... What is more, there would be nothing left to stop or reduce the radiation: no more protective skin, virtually no distance, a long exposure period. All the factors would thus converge to cause maximal damage.

The example of the six-year-old is an overwhelming one. It was enough for a tiny amount of powder to find its way into the sandwich she ate to render the outcome fatal. In short, as everyone can understand, internal contamination is to be avoided at all costs.

Unfortunately, however, radioactive substances are, in most cases, invisible to the naked eye. Without proper detection devices, it is impossible to know if someone has been contaminated or not. Anyone that has had any contact with materials emitting ionising radiation (in the form of unsealed sources) could thus, both easily and without realising it, pass these dangerous elements on to those around them. The opportunities are many: a handshake, a kiss or a touch involving a loved one who has already been afflicted... Worse still, indirect and involuntary contact allows contamination to spread in a much more devious fashion. All it takes is for you to sit in a bus, open a door or lean against the handrail of an escalator once a contaminated person has passed through.

Since radioactivity is impossible to eradicate using chemical means (bleach, antiseptics, acids, etc.), the decontamination of any area or person is conducted by removing any and all trace of the radioactive substance. And the Goiânia incident highlights just how difficult this is to achieve: 93 grams of caesium caused a contamination that required the evacuation of 275 lorryloads of waste and the destruction of seven houses...

It may also be the case that on some battlefields, which are becoming increasingly urban, radiological contamination is caused by the residues of high-speed anti-tank shells with a depleted uranium[11] tip (120mm, 30mm, etc.), which, sprayed by the impact, scatter contaminating elements throughout the environment. This depleted uranium is far less dangerous than plutonium, for instance, yet could still cause localised contamination and impact populations that remain unaware of the danger, as witnessed in Iraq, Serbia, Kosovo, Afghanistan and Libya.

11 AN: Such uranium comprises lower amounts of the U-234 and U-235 isotopes than those found in natural uranium.

5. Nuclear Power Plants

There is no plausible alternative to nuclear energy if we want to save civilisation.

— British scientist and author James Lovelock

Given the very complexity of the major accident central to this section, it seemed useful to replace fiction with a chapter explaining nuclear power plants in detail. In addition, the description of the tragic Fukushima event is so eloquent that it is perfectly self-sufficient.

The Nuclear Centre of Electricity Production (CNPE)

A detailed description of how a nuclear power plant functions is, of course, beyond the scope of the current work. However, the elements presented below allow for a better understanding of the phenomena involved and the possible risks incurred.

Against all expectations, the operating principle of a CNPE[1] is not fundamentally different from the first coal-fuelled power plants constructed in the early twentieth century. One heats a liquid in order to generate steam and drive a turbine on which alternators are mounted.

1 AN: Later, the simpler (but, from a purist's point of view, unsuitable) term of 'nuclear power plant' will be used.

Although this may seem reductive, this last step is similar (principle-wise, at least) to the one that occurs in your bike's dynamo, enabling your light to work.

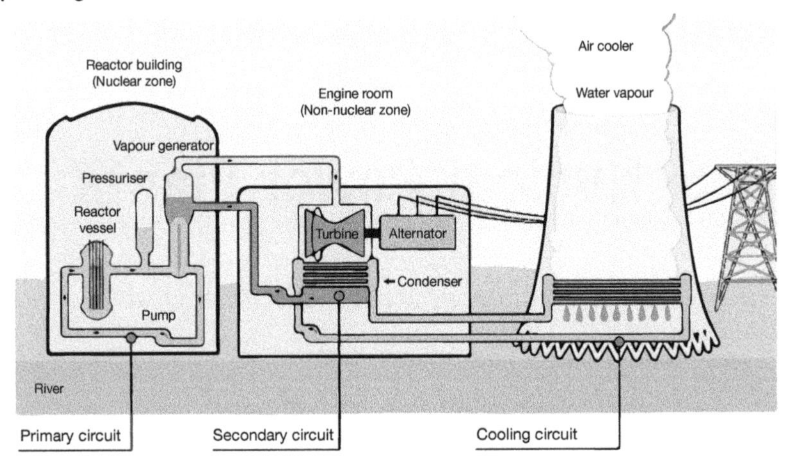

Figure 10. Scheme Displaying the Functioning of a Nuclear Centre of Electricity Production

The differences, however, are many, including the generated power, the disproportion of the systems and structures, and the type of consumed 'fuel'. Whereas it is your legs that do all that is required to turn the wheel, in this instance, a fuel with a huge energy potential is used to turn water into steam and drive the turbines. This 'fuel' does obviously have very particular properties. It must produce a nuclear fission reaction, meaning that the atoms will split and become other (mostly radioactive) elements while producing a considerable amount of energy. Keeping such a phenomenon contained under optimal security conditions is not an easy task, of course. One must not only control the nuclear reaction itself and keep the radioactivity confined, but also ensure the cooling of the whole (when we see the clouds of steam coming out of the cooling towers, we understand that it is by no means a small matter). Other constraints are imposed as well, including the management of the fuel supply and the disposal of waste…

NUCLEAR FUELS

The fission reaction that takes place in a nuclear reactor mirrors the following scheme: an atom splits, creating new elements (radioactive fission products) while releasing several neutrons, alpha, beta and gamma radiation, as well as huge amount of energy.

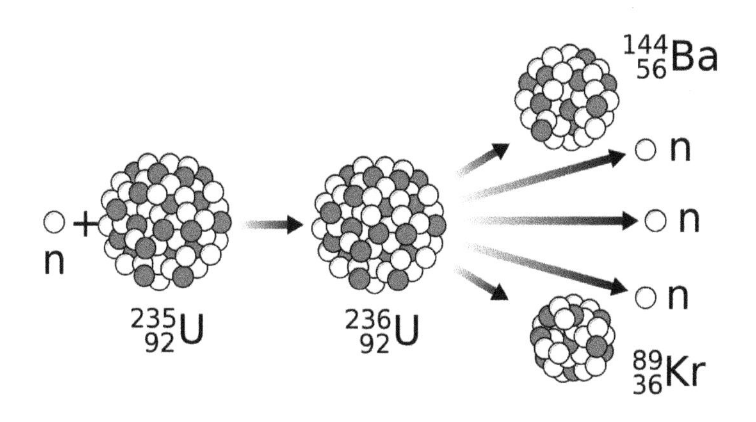

Figure 11. The Fission Process

The fuel used in nuclear electricity production centres consists of *fissile material* (which can spontaneously split and trigger the reaction described above) as well as *fertile material*, i.e. that which has the capacity to absorb one or more neutrons and become, in turn, fissionable, thus continuing the fission process — hence the term *chain reaction*.

A total of 99.3% of *natural uranium* is, for example, composed of the U-238 isotope, leaving 0.7% to the U-235 isotope. Uranium-238 is fertile, whereas uranium-235 is naturally fissile. Some nuclear sectors do resort to natural uranium, but most make use of enriched uranium, i.e. the kind in which the rate of fissile material (u-235) is increased (usually by 3 to 5%). This enrichment process is not an easy one and requires a technical know-how that only a handful of our world's countries actually have mastered.

In order to make the nuclear fission reaction efficient and easier to control, many CNPE concepts have emerged, some of which are still in use today. Depending on the path one chooses, there are thus differences in the fuel composition, the nature of the moderator (which slows neutrons down and increases the efficiency of the reaction), and the coolant used (which transports heat).

Nuclear reactors can be classified using several methods. The most common ones are either by coolant or moderator materials.

The following list gives examples of reactor types used in nuclear power plants in commercial operation:

- **Pressurised Water Reactor (PWR)**—this technology is the most common one. It is used in France (fifty-eight reactors), China, Japan, Russia and in the United States (sixty-five reactors).
- **Boiling Water Reactor (BWR)**—technology used in the United States (thirty-four reactors), Japan (e.g. in Fukushima), Sweden.
- **Pressurised Heavy Water Reactor (PHWR)**—technology used in Canada (CANDU—*Canadian deuterium uranium*—Reactor) and India.
- **Advanced Gas-cooled Reactor (AGR)**—these are gas-cooled reactors using graphite as the neutron moderator and carbon dioxide as a coolant. They are mainly found in the United Kingdom (fourteen reactors).
- **Light Water Graphite Reactor or RBMK** (from Russian 'Reaktor-Bolshoy Moshchnosti Kanalnyy')—this was the case of the Chernobyl nuclear power plant.
- **Fast Neutron reactor (FBR)**—in this type of reactor, liquid sodium is used as a coolant. Only two such reactors are currently in operation in Russia.

This list is not exhaustive and constant technological evolution paves the way for new concepts to appear. Currently, the different sectors use various fuels such as natural uranium, enriched uranium or MOX (*Mixed OXide fuel*).

WHAT IS MOX?

MOX is a recycled fuel primarily used in commercial reactors across Europe, as well as in 10% of the world's reactors. It is comprised of a mixture of recycled plutonium[2] (3 to 12%, extracted from the spent fuel of reactors) and depleted uranium.

The use of this type of fuel has been subject to repeated criticisms, due to both its hazard and its questionable economic benefit.

What is interesting to note is that following the end of the *Manhattan Project*,[3] a proliferation of ideas and concepts arose to create electricity-producing nuclear power plants. Most of these studies have led to the construction of prototypes.

Then, in the same way that natural evolution selects the fittest species for survival, only the technologies which yield the most convincing results and present minimal risks and technical constraints have survived. Nevertheless, it is now clear that not all sectors are equivalent to one another in terms of dependability. The example of Chernobyl has shown, in the past, that a major incident could impact the entire planet. Similarly, although reputed to be safer, the Fukushima power plant illustrates the sadly dramatic fact that zero risk does not actually exist.

---

2 AN: Source: Areva — http://www.areva.com/FR/activites-1951/qu-estce-que-le-combustible-mox-les-etapes-de-sa-fabrication.html.

3 AN: The Manhattan Project was a secret project launched by the United States, the United Kingdom and Canada in 1939. It resulted in the creation of the very first atomic bomb.

These accidents, in addition to those of lesser gravity, have made it possible—and still do—to change the legislations and regulations governing this type of structure. As a result, the safety and security rules are nowadays stricter and the technical constraints are becoming increasingly significant. Nevertheless, not all countries and manufacturers are advancing at the same speed. It must be acknowledged that the financial and political stakes are now enormous. It can sometimes be 'advantageous' to sacrifice risk prevention measures or promote outdated concepts in order to offer a lower price.

FUEL ASSEMBLY IN FRENCH NUCLEAR POWER PLANTS

The enriched uranium used as fuel (in the form of uranium oxide UO_2) is compacted into *pellets* with a diameter of less than 1 cm. These pellets, whose number ranges from 265 to 300, are then stacked up in *zirconium tubes* to form the *fuel rods*, whose length varies between 4 metres and 4.80 metres, depending on the power of the reactor. These rods are then grouped around guide tubes and constitute a set known as a '*fuel assembly*', into which the control rods[4] are inserted. 264 rods are necessary to obtain a complete assembly and between 157 and 205 of them feed the core. Last but not least, considering that a pellet contains 8.3 grammes of UO_2, the average mass of fuel present in a reactor can be estimated to total one hundred tonnes.

4 AN: Also called control rods, they are made of materials that absorb neutrons. They can slide and fit into the framework of the fuel rods and thus limit or even stop the chain reaction.

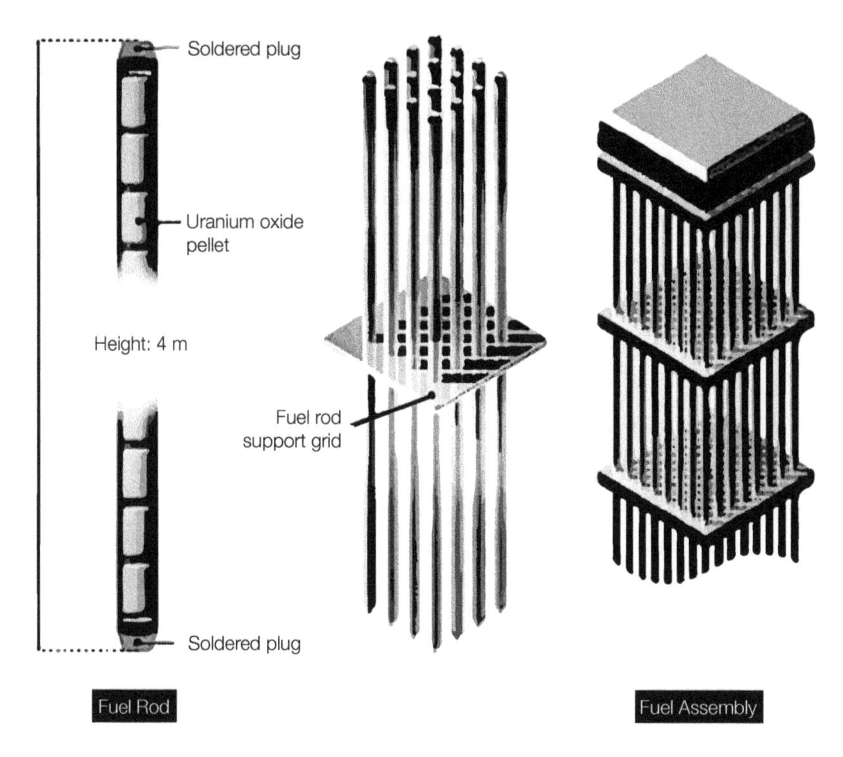

Figure 12. Fuel Assembly

When an assembly is worn out, it still contains some initial fissile material (less than 1% uranium-235), newly created plutonium and many highly radioactive compounds. The spent fuel is then put on hold for several years in the storage pool in order to give it time to cool and allow the low half-life ionising elements to decrease. It is then treated[5] for the purpose of recovering uranium-235 and plutonium (when the MOX sector is active) and subsequently classifying the residues according to their activity and radioactive decay. Only a small part of the spent fuel will end up as ultimate waste. However, although these compacted and vitrified products

5 AN: Some countries do not treat their spent fuel, but simply store it.

represent only 1% of the initial volume, they will retain extremely high levels of radioactivity for thousands, even millions of years...

Nuclear power plants generally have several walls to prevent unwanted leakage or rejection. In the case of a reactor stemming from the ordinary water sector, three barriers constitute the physical obstacles allowing the confinement of radioactive materials:

1. **Fuel rod cladding** (first barrier)
   Fuel rod cladding is a zirconium alloy that tightly envelops the uranium pellets and resulting fission products.

2. **The primary circuit envelope** (second barrier)
   Consisting of stainless steel, it contains the reactor core (fuel rods) and the water[6] that enables cooling (the primary circuit). Its thickness allows it to maintain the whole's tightness despite the high pressure involved (more than 150 bars).[7]

3. **The containment building** (third barrier)
   This is the characteristic building of nuclear power plants, similar to a dome. It contains the primary circuit and its envelope and is the ultimate defence in the event of leakage. Its structure can differ according to the model in question: a single concrete wall covered with an internal metal layer, a double concrete wall comprising an intermediate vacuum, etc.

6 AN: This is the case of ordinary water reactors, such as the ones in France and Fukushima.

7 TN: The bar is an internationally unapproved unit of pressure.

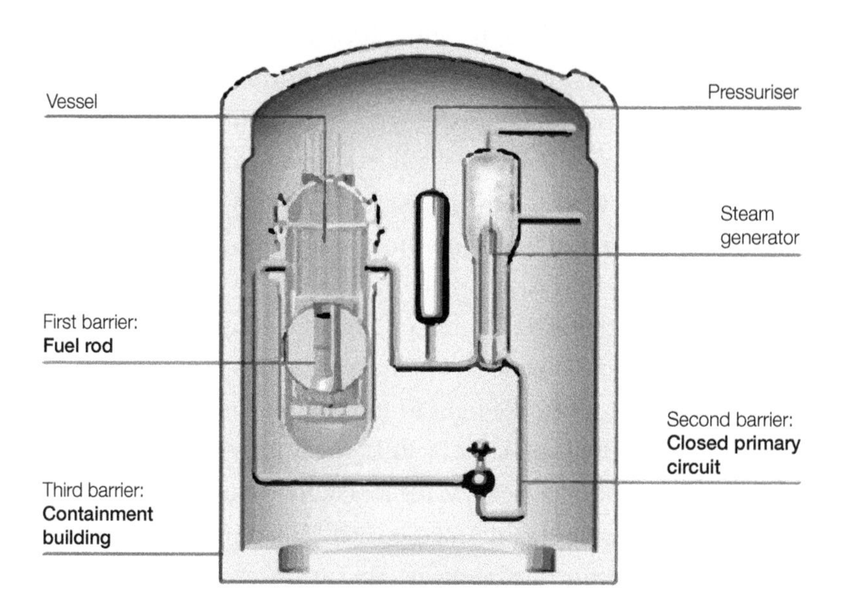

Figure 13. The 3 Security Barriers between the Fuel and the Environment (Source: EDF)

The power plants based on new designs, including the EPR proposed by the consortium led by EDF and Areva (which was initially called *European Pressurised Reactor*, and then re-baptised *Evolutionary Power Reactor*), have brought various improvements such as the redundancy of the safety systems or corium recovery tanks. This model is based on third-generation reactors whose functioning follows the principle of pressurised water supply.

Three Miles Island

On 28 March 1979, an accident occurred at the Three Miles Island plant in Pennsylvania (United States).

Following a sequence of both design-related and human error, as well as material failure, a meltdown process was initiated in the core of reactor number 2.

Everything began at 4 a.m. with a breakdown of the pumps that supplied the cooling system with water. The logical result was a rise in temperature and thus an increase in pressure within the primary circuit. After a few seconds, automatic safety actions came into play: the opening of a relief valve (in order to reduce the pressure), the stopping of the turbine, followed by that of the reactor (through the insertion of the control rods into the core).

The pressure therefore decreased, but the discharge valve that should have closed automatically remained locked in its open position. In addition to this, the light in the control room indicated the opposite to the technician. Little by little, the primary circuit was thus emptied. What thus occurred was the loss of the second safety barrier.

Following inaccurate information and disruptive phenomena resulting from (unsuspected) leakage, the operators thought that everything had returned to normal and manually stopped the safety fuel injection system. Design defects, human negligence (the closing of the cooling valves of the emergency system, when they should have been opened instead) and inappropriate responses from the operators themselves eventually led to the partial meltdown of the fuel (and thus the loss of the first safety barrier).

Highly contaminated vapours and liquids were thus released in large quantities into the containment building (the third and final protective barrier), dangerously building up pressure. Fortunately, the dome held out.

In the end, about 40% of the fuel melted and 20% found its way to the bottom of the reactor vessel. Since the containment structure had resisted, the consequences for the environment were minimal: the discharges stemmed mainly from the sustainment of the primary circuit's effluent pumping system.

In the end, the accident, which could have had dire consequences, was given a level-5 categorisation on the INES scale. Two days later, the US Nuclear Regulatory Commission (NRC) admitted that a core meltdown was indeed possible (although it had, until then, been considered impossible!).

In addition, all major nuclear power plant manufacturers proceeded to reassess their designs, as operators reviewed their procedures and public authorities re-examined their emergency and evacuation devices.

Fukushima

On 11 March, 2011, at 2:46 p.m., an 8.9-magnitude earthquake occurred at sea, to the east of the Honshu region, Japan's main island. The tremors caused the automatic shutdown of the nuclear reactors of the Fukushima-Daiichi plant through the insertion of the control rods into the cores.

The earthquake triggered a tidal wave of exceptional power that devastated the Japanese coast. The human casualty toll was a terrible one and the damage considerable: almost 20,000 people were killed or went missing, with more than 300,000 buildings left damaged or destroyed. As for the homeless, they were in their tens of thousands…

The wave responsible for this disaster struck the Fukushima-Daiichi power plant less than an hour after the earthquake. At this point of the coast, it reached a height of twelve to fifteen meters (depending on which source one believes). The 5.7 metre anti-tidal wave walls were thus ineffective and the water invaded the site, causing significant damage, especially in the cooling systems. Throughout the day, TEPCO[8] staff, rescue services and the Japanese military worked hard to prevent any and all deterioration of the situation.

The next morning, TEPCO begins to inject water into reactor number 1. Later, it is forced to enable the release of water steam whose pressure build-up threatens to trigger the explosion of the containment building. In the early afternoon, radioactivity levels up to eight

8 AN: TEPCO, or the *Tokyo Electric Power Company*, is the power company in charge of the plant.

times higher than normal are detected near the plant, and up to 1,000 times higher in the control room itself. TEPCO reports that the cooling water level in reactor 1 has dropped dramatically and that a core meltdown may be in progress. It also confirms the presence of a radioactive leak towards the outside. Around the end of the afternoon, an explosion devastates the top of the building containing the reactor. The images taken on the spot cause the whole world a great deal of concern, as radioactivity levels continue to increase. Twelve hours after the start of the events, the accident is given a level-4 classification according to the INES scale.

Despite all the efforts made by both TEPCO and governmental departments, the pressure inside the reactor continues to grow and threatens to seriously damage the latest safety systems still in operation. Fearing a potential loss of these elements, new steam discharges are authorised. In the evening, the manager has seawater injected into the reactor in an attempt to compensate for the lack. Subsequently, he also adds boric acid, a neutron-absorbing compound, for the purpose of slowing down the chain reaction.

Two days after the earthquake, it is the turn of reactor 3 to undergo a cooling failure. TEPCO also arranges for boron-enriched seawater to be brought in and performs various depressurisation sequences to maintain the system's integrity. Meanwhile, the information obtained from reactor 1 shows that the situation is deteriorating.

On Monday, 14 May, three days after the earthquake, a new explosion occurs, this time at the level of reactor 3. Eleven people are injured and many materials damaged, including those that enable the injection of water.

The situation continues to worsen. In the evening, TEPCO acknowledges the fact that a core meltdown is in progress in reactors 1, 2 and 3. In addition to this, a new explosion occurs on the morning of the fourth day. Reactor 2 has just suffered an identical fate to that of its 'neighbours', with one major difference, however: its containment building has been damaged. Unfortunately, the situation deteriorates further. On the same day, a fire breaks out in the fuel storage pool of reactor 4, followed a few hours later by an explosion and a massive release of ionising elements.

At this point, it becomes clear that the Fukushima power station has turned into a house of cards that's falling apart. All systems fail one by one, as radioactivity levels in the environment increase. On-site work now faces time restrictions as the radiation is too significant and only volunteers are employed there. The authorities advise the population residing within a thirty-kilometre radius to shut itself in and remain indoors, awaiting evacuation orders. On the site itself, dose rates of 400 mSv/h are measured in many places that thus become, according to legislation, prohibited areas, making the operators' work even more difficult.

Under the world's watchful eyes, the situation continues to be a source of concern. A fire erupts in reactor 4 and extends to radioactive materials, including caesium. The smoke thus released is highly charged with ionising particles. Staff can no longer approach certain areas, as the authorities hire helicopters to continue with the cooling operations. After a few hours, however, they are forced to put an end to their endeavours, as the dose rates are too high.

On the sixth day, the situation is very far from being stabilised... The containment building of reactor 3 has experienced significant damage. This is all the more worrying as the fuel supplying the reactor is, in this case, composed of MOX. A little further, in fuel storage pool number 4, there is almost no water left and the threat of a new outbreak of fire amidst the plant's most radioactive materials is looming on the horizon... Given the seriousness of the situation, international aid is gradually organised. Several countries such as Korea, France, and the United States offer to send men and equipment and give their own 'advice'. Abnormal levels of radioactivity are detected in the United States. A cloud carrying ionising elements travels repeatedly the entire world's breadth. On the Internet, both factual information and the wildest rumours begin to circulate.

On the site itself, meanwhile, the operator is doing everything in his power to stabilise the situation. And yet, after four weeks, reactors 1 and 3 are still in critical condition. In addition, highly contaminated water leaks are detected, as these effluents seep into the subsoil or stream into the sea. The authorities acknowledge the presence of rates millions of times higher than standard ones. Many attempts are made to try and contain this radioactivity (pumping, securing the area with sand bags, etc.), yielding mixed success. Whatever the case, intentional draining operations take place in parallel, as more or less contaminated liquids end up in the ocean…

At the beginning of the fifth week, the Japanese Nuclear Safety Agency categorises the Fukushima event as a level 7 accident on the INES scale, with the same degree of severity as the most serious nuclear disaster the world has ever seen, namely that of Chernobyl. Repair work and cooling attempts will continue for weeks. New explosions still occur, as unannounced leaks and intentional releasing continue. Contaminated or, in some cases, extremely contaminated water continues to seep into the subsoil and the Pacific Ocean. The slightest operation is a nightmare and unexpected occurrences a constant. Although the discharges are significant, polluting the air, ocean and surrounding land, most of the radioactivity remains confined. In early June, for example, the dose rate in the most damaged reactor reaches the considerable value of 250 Sv/h.

One month after the accident, the city of Fukushima, which lies at a sixty-kilometre distance from the plant, has not yet been evacuated, although rates of radioactivity up to 500 times higher than normal have been measured in certain locations. In the region, both surface waters (rivers, ponds, etc.) and soils are now contaminated. The distribution of radioactivity is, however, heterogeneous. Indeed, it depends mainly on the atmospheric conditions involved, including the direction and strength of the wind, obviously, but also the location and abundance of precipitations (which carry the ionising elements present in the air and create localised deposits).

The world is now well aware of the fact that the Fukushima accident caused widespread environmental contamination. In particular, it generated a radioactive cloud consisting mainly of iodine-131, xenon-133 and caesium-137. Although measurable throughout the northern hemisphere, it did not reach sufficient levels of radioactivity to be dangerous for humans (excluding in Japan). On the other hand, it resulted in significant pollution of both the region and the ocean. Many plants (tea, spinach, bamboo shoots, mushrooms) that were grown at a distance of up to 450 kilometres from the power station[9] thus showed higher levels of radioactivity than normal. Similarly, a number of marine species (cods, soles, congers, Alaskan pollocks) caught in north-eastern Japan have presented and still show caesium contamination in excess of the authorised limits.

In the end, the Fukushima disaster led to massive environmental pollution.[10] In Japan itself, it disrupted the lives of nearly 150,000 people, led to social unrest and triggered an economic depression.

AREAS AND POPULATION EVACUATED

On 11 March 2011 (the day of the accident), at 20:50, the authorities set up a two-kilometre forbidden zone around the plant. They subsequently proceed to widen it to three kilometres half an hour later and ask the population living within a radius of twenty kilometres to shut itself in at home. On 12 March, the evacuation zone is enlarged to ten kilometres and then, once more, to twenty kilometres that same evening. On the 15 of March, the containment zone is yet again increased, reaching a thirty-kilometre radius. It is

9 AN: Here is an example — 160 kilos of tea originating from the Shizuoka prefecture (450 kilometres from the plant) were seized at the Charles de Gaulle airport in Paris. Although not posing a serious health threat, the batch displayed a caesium contamination twice as high as the authorised dose in France.

10 AN: To gain a better understanding of the Japanese perspective on the events and their social and political consequences, see Takashi Imashiro's *Nuclear Wrath*, Akata Editions, 2012.

not until 25 March, i.e. two weeks after the events, that voluntary evacuations (depending on the means available to each person or family) are recommended by the authorities to all those residing within the twenty- to thirty-kilometre zone. Inside the latter, the population is also urged to prepare for possible forced evacuation at a later point. On 22 April, the authorities order the evacuation of all areas where contamination is likely to exceed a dose rate of 20 mSv/year (with some cities actually lying beyond the thirty-kilometre area). On 16 June and 21 July, one witnesses calls for further evacuation to cover new areas likely to result in exposure greater than 20 mSv/h.

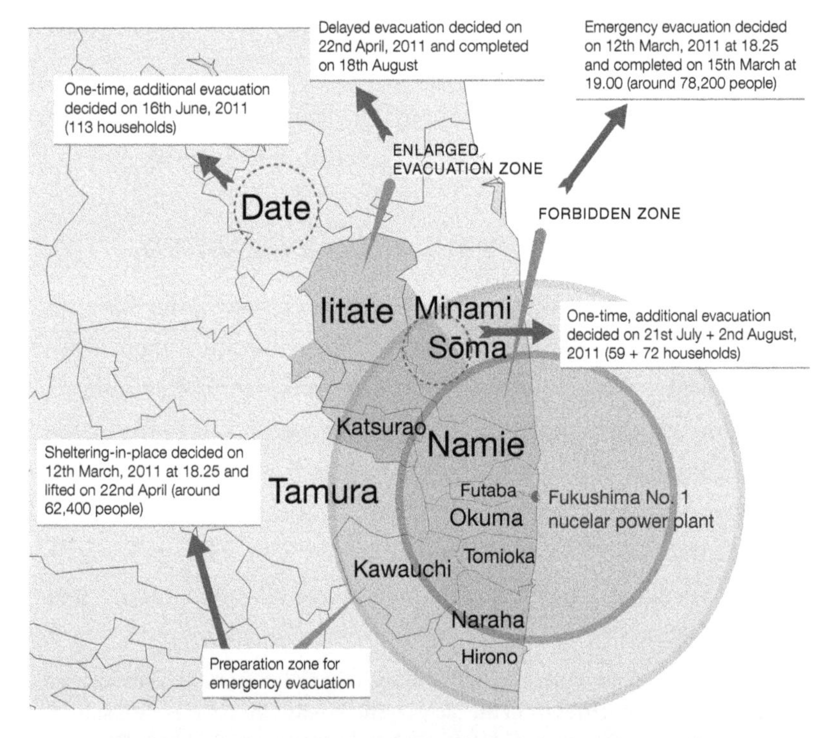

Figure 14. Post-Accident Management of Fukushima-Daiichi (Source: IRSN)

The drying up of leaks at the plant and the decontamination actions undertaken (the washing/cleaning of floors and buildings, the removal of soil layers, etc.) allowed the Japanese authorities to reduce access restrictions as of the month of April 2012. Gradually, some areas were reopened to traffic and others became habitable again.

The map below displays the equivalent of the exclusion zones set up for Fukushima and Chernobyl, superimposed on nuclear power plant sites in France[11]

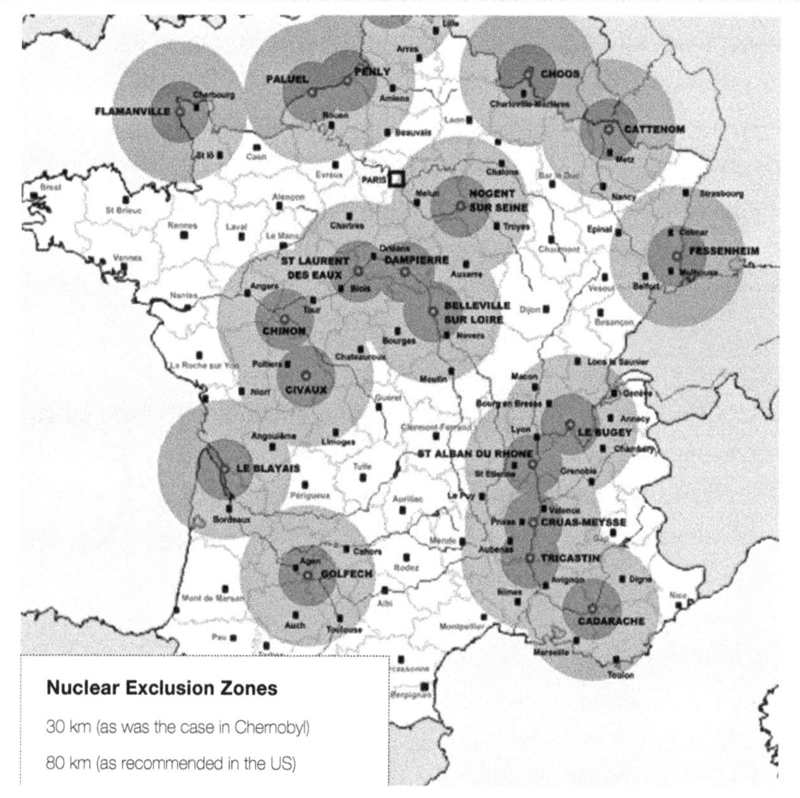

Figure 15. Nuclear Risk Map of France

11 AN: For Switzerland, the equivalent map can be accessed through the following link: http://www.sortirdunucleaire.ch/content/opening/images/intro-20151012-densite-de-popuplation.png.

It is very difficult to determine how much damage can be attributed to the earthquake. Has it caused cracks, damaged pipes, undermined safety systems? It would, at first glance, seem not to have been the main factor responsible for the disaster. Whatever the case, its action, combined with that of the tidal wave that followed, eventually led to a defect in the reactors' cooling process. Likewise, one cannot exclude the possibility that human errors may have accelerated the degradation process.

To understand the situation, it is useful to know that the chain reaction can be stopped through the control bars. However, the system is characterised by a significant sort of inertia and does not react in the same way that the ON/OFF button of your desk lamp does. It takes many days of cooling to bring down the temperature and eliminate any risk of potential runaway. In the case of Fukushima, reactor power dropped dramatically (as expected) when the automatic shutdown procedure was initiated as a result of the earthquake. Unfortunately, the shutdown of the cooling devices and failures of the emergency systems in the hours that followed no longer made it possible to release the residual energy. The temperature of the water inside the reactor thus rose. This water then began to turn into steam, gradually increasing pressure. This is why the operator authorised controlled gas releases (primarily made of water vapour, but also comprising small amounts of radioactive elements, despite the filters provided for this purpose) in order to maintain the system's integrity.

Although this kind of event is not at all good news, the situation is not yet hopeless. Even if the releases cause a slight increase in radioactivity, the tens of tonnes of nuclear fuel present in the reactor remain isolated from the outside and contained in the metal vessel (second barrier). Major concerns begin if, at this stage, no water supply nor cooling take place at reactor level. Under such conditions, the temperature would continue to rise, causing more frequent releases of steam, perhaps even accidental leaks. As the water is no longer renewed, a level decrease in the reactor becomes inevitable.

In the absence of contact with liquid water, this mechanism ends up dewatering the fuel assemblies (the reactor core). Ultimately, the latter find themselves 'in the open air', so to speak, without anything to cool them. At a temperature ranging from 700 to 900 °C, zirconium alloy claddings begin to slowly degrade and react to water. This reaction is exacerbated if the temperature continues to rise. Thus, at 1200 °C, the loss of watertightness impacting the first containment barrier becomes critical. In addition, the surrounding water vapour undergoes a decomposition process, leading to a release of hydrogen, a highly flammable gas… Everyone will have understood, by now, that this is the actual phenomenon that triggered the explosion: the hydrogen released with both gases and pressurised water steam was bound to ignite at the slightest spark.

Without any effective action to raise the water level, the situation continues to evolve unfavourably, as zirconium claddings deteriorate further. From 2500 °C onwards, the heat is so intense that even the ceramic fuel pellets begin to liquefy. This process is commonly known as the 'melting of the reactor core'. This melting can be total if the whole core melts, or partial if only the upper part is affected.

The magma of molten products (corium) resulting from this ultimate degradation can reach extreme temperatures, enabling it to pierce the metal vessel which constitutes the reactor's containment envelope (second barrier). Although the gases, vapours and liquid effluents that escape are heavily charged with ionising particles, they remain trapped by the concrete dome. Under these conditions, the level of radioactivity within the last bulwark can be considerably high. It is thus obvious that the slightest discharge out of this final containment building (third barrier) becomes a genuine source of concern.

Differences Between Chernobyl and Fukushima[12]

CHERNOBYL (26 APRIL 1986)	FUKUSHIMA (11 MARCH 2011)
Type of technology	**Type of technology**
Water-graphite reactor (RBMK)	Boiling water reactor (REB)
Circumstances	**Circumstances**
The operators' reckless actions, the wilful choice to ignore the alarms and the obsolete safety systems led to the reactor's sudden runaway. It exploded, destroying its containment building and causing a fire.	The loss of the cooling and emergency systems triggered an increase in temperature, which then led to the melting of the core in reactors 1, 2 and 3, as well as numerous discharges into the atmosphere, a leakage of liquids, and explosions resulting from the presence of hydrogen.
Impact on the installations	**Impact on the installations**
The reactor is split wide open and the containment building blown, as fuel is dispersed by the explosion and fire.	The containment chambers do not suffer major damage. The fuel remains confined and never reaches the outside.
Consequences	**Consequences**
Continuous discharges and fumes highly charged with radioactive elements spread in the atmosphere for approximately ten days.	Discontinuous atmospheric releases occur mainly during the first two weeks. Occasional liquid spills and sporadic gas/vapour leaks continue for months.
Evacuated population	**Evacuated population**
270,000 people	150,000 people

The following chart gives us an estimate of the total radioactivity released into the atmosphere:

12 AN: Source—Institute of Radioprotection and Nuclear Safety (IRSN): http://www.irsn.fr/FR/knowledge/Environment/expertises-incidents-accidents/comparaison-chernobyl-fukushima/Pages/1-repercussions-post-accidentelles-differentes.aspx?dId=5d0cc222-c748-41ea-bae7-33f47b490598&dwID=ebe35772-4442-413c-b628-068fde521abe.

	CHERNOBYL (AIEA, 2005)	FUKUSHIMA (IRSN ESTIMATION, 2011)
Noble gases (xenon…)	6,533 Pbq[13]	6,550 Pbq
Iodine	4,260 Pbq	408 Pbq
Tellurium	1,390 Pbq	145 Pbq
Caesium	168 Pbq	58 Pbq
Others (including strontium and plutonium)	1,227 Pbq	28 Pbq

- Except for noble gases, radioactive releases were much higher in the case of Chernobyl. This is logical, since the explosion released tonnes of fuel. Concerning the high levels of xenon measured in Fukushima, they stemmed from the ability of the elements belonging to this family to easily pass through the filters.

- The noble gases and tellurium present in both cases have a low half-life (a few days) and thus disappear quite quickly. Caesium, on the other hand, is characterised by slower decay, including caesium-137, which has a life span of about thirty years. Contamination spots whose distribution varies in accordance with wind conditions and precipitations will thus be formed.

- Elements such as plutonium are very heavy and tend to fall back quickly. They are particularly found in the areas surrounding the accident's location.

REMARK

In the case of Fukushima, a massive contamination of the ocean due to leakage and a discharge of contaminated water must be added. For caesium alone, the estimate is 27 Pbq.

13 AN: Pbq = petaBecquerel = 10^{15} becquerels.

CONCLUSION

It is logical to note that, as a result of the release and burning of tonnes of fuel, the explosion of the reactor core at Chernobyl caused greater contamination than the Fukushima accident ever did. If we consider caesium-137 deposits greater than 600,000 Bq/m^2, the extent of the impacted areas is 600 km^2 for Fukushima and 13,000 km^2 in the case of Chernobyl. We can also conclude that such an accident in Japan, one of our planet's richest and technologically most developed countries, pales before the one that occurred in Chernobyl, in a USSR whose economy was planned, inert and inefficient, and whose technologies were obsolete. Both cases, however, involved hesitation, slowness and misinformation—not to say lies— in the authorities' communication with the public.

Is the Nuclear Option Safe?

According to the IAEA, there was a worldwide total[14] of 442 operational nuclear reactors at the end of 2015 (with over 66 under construction), of which 99 were located in the United States (+5 under construction), 58 in France (+1), 48 in Japan, 34 in Russia (+8), 23 in China (+25), 23 in South Korea (+5), 21 in India (+1), 20 in Canada, 16 in the United Kingdom, 15 in the Ukraine (+2), 10 in Sweden, 9 in Spain, 9 in Germany, 7 in Belgium, 6 in the Czech Republic, 5 in Switzerland, 1 in Iran, 1 in Israel, etc.

Figure 16. Map of the World's Nuclear Power Plants

In addition to these civilian reactors, there are also those used to propel surface ships or submarines. Between 1954 and 2015, more than 780 nuclear reactors were built around the world to equip (war)ships.[15] Approximately 250 were still in service in 2016. Last but not least,

---

14 AN: http://www-pub.iaea.org/MTCD/Publications/PDF/rds2-35web-85937611.pdf.

15 AN: The first nuclear-powered ship was the American attack submarine USS Nautilus, which was commissioned in 1955 and is famous for having been the first submarine to complete a submerged transit of the North Pole.

there are also some 200[16] nuclear research reactors operating in more than 60 countries. It is therefore very important to determine the level of risk involved. However, the debates on the topic have been so passionate and publicised since the 1970s (in particular in the aftermath of the Three Mile Island, Chernobyl and Fukushima accidents) that it is difficult to give a definite, simple and clear answer to the question whether 'civilian nuclear power is safe'. In addition to these three accidents, the following list enumerates the most important events, indexed in order of occurrence:

- **1952:** a level-5 accident[17] (on the INES scale) impacted one of the reactors at the Chalk River Research Centre in Canada.
- **1969:** a level-4 accident[18] on the INES scale took place at the Lucens experimental plant in Switzerland.
- **1969:** a level-4 accident[19] (INES scale) happened at the Saint-Laurent-des-Eaux power station in France.
- **1979:** a level-5 accident[20] (INES scale) occurred at the Three Mile Island power plant in the United States.

---

16 AN: Only about forty of these have any significant power (more than 5 megawatts).

17 AN: Partial coolant loss.

18 AN: A cooling issue resulted in a partial melting of the core.

19 AN: Incorrect handling during the core loading of reactor 1 led to the melting of 50 kilos of uranium. http://www.lepoint.fr/societe/le-jour-ou-la-france-a-frole-le-pire-22-03-2011-1316269_23.php.

20 AN: The core of reactor 2 partially melted, causing the release of contaminants into the environment. http://www.nrc.gov/reading-rm/doc-collections/fact-sheets/3mile-isle.html.

- **1980:** another accident[21] (once again a level-4 event on the INES scale) took place at the Saint-Laurent-des-Eaux power station in France.

To these events, one must furthermore add the nuclear submarines that have sunk and whose reactor remains at the bottom of the oceans.

- **1963:** USS Thresher, off Cape Cod.
- **1968:** USS Scorpion, southwest of the Azores.
- **1968:** K-27, in the Kara Sea (a salvaging plan is being concocted).
- **1970:** K-8, in the Bay of Biscay.
- **1986:** K-219 off Bermuda.
- **1989:** K-278, in the Barents Sea.

Nevertheless, what is interesting to bear in mind is that the deadliest disasters related to the production of electricity have not been caused by nuclear power but by hydropower:

- **1959:** 423 deaths in France during the failure of the Malpasset[22] dam.
- **1963:** 1,900 to 2,500 casualties in Italy, during the Vajont-Longarone[23] dam burst.[24]

---

21 AN: The melting of 20 kilos of uranium in the core led to the latter's emergency stop and the discharge of plutonium into the Loire. http://www.lanouvellerepublique.fr/Loir-et-Cher/Actualite/Environnement/n/Contenus/Articles/2015/05/07/Mars-1980-du-plutonium-deverse-dans-la-Loire-2320446.

22 AN: https://fr.wikipedia.org/wiki/Barrage_de_Malpasset#La_rupture_du_barrage.

23 AN: https://en.wikipedia.org/wiki/Vajont_Dam#Early_signs_of_disaster.

24 TN: To my knowledge, as seemingly confirmed by the link above, it was an instance of overtopping rather than a burst.

- **1975:** 171,000 direct and indirect deaths following the dam failure in Banqiao, China.[25]
- **1979:** 1,800 to 25,000 people were killed when the Machchu-2 dam[26] burst in India.

Apart from this safety element, other aspects also come into play. In order to acquire a better general understanding of things, one should equally be interested in the very high costs of electricity when produced by other technologies as well as the resulting environmental destruction, as is the case with the extraction and use of coal (which does not, of course, mean that nuclear power has nothing but advantages).

In the end, nuclear energy comes across as rather safe compared to other technologies, but it could still have a huge impact on the environment in case of a major malfunction. Indeed, there is no such thing as zero risk and the danger varies from country to country. In addition, because of the high technicality required for the operation of an electricity-producing nuclear power plant, it all depends on personnel training and the ensuring of proper maintenance during the decades in which a plant is operational.

What About the Waste?

The storage and disposal of long-lived high and intermediate-level nuclear waste is carried out/meant to take place in deep geological layers. At first glance, this system seems safe with regard to the stability of such final storage places as coal or salt mines. Indeed, the latter only experience very little movement because of the very slow geological rhythms involved—a salt mine, for example, can remain geologically stable for a good hundred million years. Nevertheless, the example of the American '*Waste Isolation Pilot Plant*', which housed its first radioactive waste at a 650-metre depth in 1991, remains a matter

25 AN: https://en.wikipedia.org/wiki/Banqiao_Dam.

26 AN: https://en.wikipedia.org/wiki/1979_Machchhu_dam_failure.

of concern. Indeed, it took less than twenty-five years for radioactive leaks to be detected and the site was thus closed on 14 February 2014... It is therefore important to remain very vigilant when conducting preliminary studies and making use of a certain storage area, if only for the sake of our future generations.

Chinese Syndrome?

This term refers to insane rumours alleging that corium could somehow pass through the metal casing of the primary circuit and then traverse the several metres of concrete at the bottom of the containment building, before seeping into the ground and crossing the centre of the Earth to eventually reach the other side of the planet. Others add that such an event would trigger a chain reaction at the level of the Earth's mantle which would subsequently destroy entire continents. Of course, this kind of theory is completely whimsical. Even if corium could somehow pierce all these barriers and kept on seeping towards the centre of the Earth:

1. It would create no chain reaction in the mantle, which is comprised of inert and more or less molten minerals.

2. It could never come out on the other side of the planet, simply because gravity would make this impossible. It would simply mingle with the core, which is largely composed of molten metals. Just for the record, a certain part of our planet's internal heat comes from radioactivity![27]

3. The main risk would actually be embodied by the extreme contamination of the groundwater that it could encounter on its path.

In Fukushima, reactors 1, 2 and 3 experienced a melting of their respective core, allowing the corium to pierce the primary circuit

27 AN: For details, see the chapter on the topic of radioactivity.

envelope and probably sink into the concrete (yet no information on any sort of breach has been issued to date). Under these conditions, the level of radioactivity rose alarmingly within the containment building: barriers 1 and 2 were thus destroyed or left ineffective, with the concrete dome acting as the last bastion. The slightest liquid or gaseous leak would then result in massive environmental radioactive contamination.

Could a Nuclear Power Plant Explode Like an Atomic Bomb?

What nuclear power plants and nuclear bombs have in common is that they both take advantage of the extraordinary energy contained in the nucleus of atoms, making use of nuclear fission reactions. The nuclear reaction that takes place both at the heart of a nuclear power plant and in a uranium or plutonium-based atomic bomb is one of fission, meaning that it involves a chain reaction. Whenever atomic splitting occurs, a very large amount of energy is released.

In a nuclear bomb, the process occurs in an extremely short time. In a nuclear power plant, on the other hand, this reaction is controlled and the energy produced is thus distributed over time. Whether or not a runaway is triggered as part of the reaction depends on the proximity of the atoms themselves. In the absence of a sufficient number of uranium-235 nuclei, the neutrons produced will be able to escape and a chain reaction will not take place. There is therefore a critical mass of fissile atoms (235U) at which the chain reaction is carried out. In theory, this mass is 200 kilos, but by surrounding uranium-235 with reflectors that divert the emitted neutrons back towards the centre of the 'fuel', the value of this critical mass can be reduced to 15 kilos.

Since, in their natural state, uranium ores consist of only 0.7% uranium-235, this concentration is far too low to initiate a chain reaction and it is necessary to enrich uranium to increase the concentration of uranium-235. For military use in nuclear bombs, it requires a minimum of 80% uranium-235. The enrichment of uranium is a very complex, very long and very expensive process, because it is necessary to separate uranium-235 nuclei from those of uranium-238—which is why not all countries are capable of building a nuclear bomb, even though nuclear bomb technology is relatively simple to understand on paper. In nuclear power plants, so as to make sure that a runaway reaction does not occur, all necessary steps are taken to ensure that the neutrons produced by the decay of a nucleus do not encounter more than one single uranium-235 nucleus. For this purpose, fuel is used that has only been enriched by 3% to 5% and the neutrons are slowed down using a moderator (usually water or heavy water) so that they become more effective. In order to avoid a potential runaway, the reaction is controlled by means of two main mechanisms:

- control rods made of neutron absorbing materials (such as boron, a silver-indium-cadmium alloy, etc.) which are driven more or less into the reactor core.

- neutron-absorbing compounds that have been dissolved in water (boron in the form of boric acid, for instance), whose concentration can be controlled over time.

In actual fact, because of the low content of uranium-235 (between 3% and 5%), no explosion can occur as it does with atomic bombs. On the other hand, should the moderator disappear, the chain reaction would continue to produce larger amounts of energy capable of melting the reactor core. That's precisely what happened in Fukushima and Chernobyl. The explosion was not of the nuclear kind, but was, instead, due to overpressure or flammable gases. Considering the tonnes of radioactive fuels used, however, the threat to the environment is much more significant…

If a damaged nuclear reactor (in a power plant, submarine, etc.) came into contact with water, could it pollute all the oceans of the world?

The effects of radioactive pollution around Fukushima are dramatic, with, for example, 20 to 50 times higher thyroid cancer rates among under-aged people than before the accident,[28] and the impact on the fauna and flora is very real indeed.[29] However, the radionuclides that have leaked out of the Fukushima power station and reached the ocean will gradually become diluted in the colossal volume of the Pacific and eventually turn into a real yet tiny source of pollution. Several studies have confirmed the fact that fish are far more affected by industrial pollution (i.e. by heavy metals, plastics, and chemicals) than by radiation.[30] This, however, is by no means a reason to continue building near-shore nuclear power plants in a country that actually coined the very term 'tsunami' and is located in one of the world's most active seismic zones!

If a Chernobyl- or Fukushima-type of disaster were to occur in France, no matter how very unlikely such an event is, one may wonder what the planned evacuation measures would be like and whether they would suffice. A special intervention plan (PPI)[31] is in place for each of our nineteen nuclear power plants. This plan allows for the necessary emergency measures to be taken in the event of a major disaster in order to protect the population during the first twenty-four hours. The measures in question are the following: the evacuation and

28 AN: http://www.fukushima-blog.com/2015/10/fukushima-bilan-d-une-situation-sanitaire-inquietante.html.

29 AN: http://www.vivre-apres-fukushima.fr/tag/contamination/.

30 AN: http://fukushimainform.ca.

31 AN: Designed and drafted by the public authorities, the *Plan Particulier d'Intervention* (PPI) or Special Intervention Plan is part of the ORSEC departmental system. It interfaces with the emergency plans devised by the industrialists whose activities are at the very source of the risks in question. See: http://www.interieur.gouv.fr/Le-ministere/Securite-civile.

sheltering of people living within a five-to-ten-kilometre radius and the distribution of iodine tablets within this area. Populations living inside the PPI perimeter are regularly informed on the procedure to follow in the event of an accident, particularly through the distribution of brochures and the availability of a toll-free number.

IMPACT OF A POWER CUT ON NUCLEAR POWER PLANTS

What would happen in the event of prolonged power outage?

This could be caused by electromagnetic impulses, either as a result of nuclear explosions in the atmosphere or due to solar flares.

In 1859, a series of unusually intense solar flares (the 'Stewart' and 'Carrington' events) caused a magnetic storm that lasted for a week, leading to aurora borealis phenomena which stretched to almost tropical latitudes. It also induced powerful electric currents in telegraph networks, the only electrical networks that existed at the time.

Since it has already happened, the manifestation of a solar storm as intense as that of 1859 is not science fiction. Moreover, if such a storm occurred today, the effects would be much more serious.

Indeed, humanity has since become highly dependent on electrical energy, whether for food, heating, transportation, industry, or health. A magnetic storm of equal intensity to that of 1859 could result in the destruction of a very large share of the world's electrical transformers, electronic systems and means of communication, thus creating a global blackout. Such a global outage could last several months and plunge the whole world into an unprecedented crisis.

When it comes to the nuclear industry, 438 civil nuclear reactors[32] and 250 military nuclear reactors are currently in operation.

A nuclear power station is dependent on a permanent power supply. The latter is necessary not only to enable its operation, but also during the reactor's shutdown, to prevent the overheating of the core as a result of the enormous residual heat stemming from the decay of radioactive elements. Cooling must therefore be maintained for months or even years.[33]

Depending on the actual power of the operational 'CNPE', the tertiary circuit's water requirements range from 2 m^3/s for plants with cooling towers to 50 m^3/s when all the cooling water comes from an external source such as the sea, a lake or a river. Pumps thus play a vital role that ultimately enables the circulation of water in the three circuits (the primary, secondary and tertiary one) to ensure the cooling of the reactor. This process must be maintained for months following reactor shutdown. As for used fuel, it must then be stored in cooling pools for a period of three[34] to five years. Its 'complete' cooling takes ten times longer, i.e. about fifty years.[35]

A nuclear power plant does not stop the way a lamp does. It requires time and a constant supply of electricity throughout the period that follows. In addition to this, a nuclear reaction is run by computers, which also operate thanks to electricity. If a solar storm caused a lasting global blackout across the whole planet, would the backup generators that power the cooling circuit of nuclear power plants remain operational? Would most electronic systems

32 AN: 438 in operation, and 67 under construction in 2015: http://www.nei.org/Knowledge-Center/Nuclear-Statistics/World-Statistics/World-Nuclear-Power-Plants-in-Operation.

33 AN: http://www.laradioactivite.com/fr/site/pages/lerefroidissementducombustible.htm.

34 AN: https://www.edf.fr/groupe-edf/producteur-industriel/nucleaire/atouts/expertise-nucleaire/cycle-du-combustible-nucleaire-edf-present-sur-toutes-les-phases.

35 AN: http://www.laradioactivite.com/fr/site/pages/laradioactivitedumox.htm.

be burned as a result of electrical overload? Would technicians in those power plants be rendered 'blind' and unable to control the state of the reactor?

And even if, as we shall soon see in the case of Switzerland, many countries do take measures to protect nuclear installations, are they immune to a lack of long-term electricity supply caused by a sudden economic crisis, social chaos, or war?

The American Nuclear Regulatory Commission (NRC)[36] estimates that 50% of all core melting scenarios originate from a power outage in the reactor.[37]

Is such a scenario anticipated by the nuclear industry? In order to find out, we asked David Suchet of the ENSI (the Brugg-based Swiss Federal Nuclear Safety Inspectorate, which acts as the governmental body responsible for the safety of the Swiss nuclear power plants of Beznau 1 and 2, Gösgen, Leibstadt or Mühleberg) a series of questions.

Is a Carrington type of scenario provided for?

Mr. David Suchet: With regard to the scenario you mentioned, I would like to draw your attention to the fact that, in the event of a power shortage or outage, Swiss nuclear power plants would operate as if they were islands in the middle of the sea: even when separated from the network, they remain autonomous in terms of power supply. The safety systems have enough energy to ensure safe operation thanks to their ability to generate their own power and the presence of backup diesel generators.

The International Atomic Energy Agency (IAEA) is aware of the fact that a stable electricity network is one of the basic conditions for the operation of nuclear power plants. The latter thus produce power. They also need it in order to guarantee the cooling of the reactor. This electric current is also necessary for the

36 AN: http://www.nrc.gov.

37 AN: Helmut HIRSCH, *Nuclear Reactor Hazards Report*, p.121.

operation of measuring instruments and other important safety-related systems. In the event of a prolonged power outage, nuclear power plants would have to source their own energy, which they are able to do. They had to demonstrate this back in 2011, for example, as part of the European Union's so-called resistance test. The Federal Nuclear Safety Inspectorate (ENSI) concluded that thanks to the availability of multiple and diverse power supplies, the Loss of Offsite Power (LOOP) scenario was under control in all Swiss power stations.

Are nuclear power plants well-protected against solar storms?

D.S.: Yes, Swiss nuclear power plants are well-protected against solar storms. By contrast, transformers and high-voltage overhead lines are not. Equipping the Mühleberg and Beznau nuclear power plants with new emergency control room systems has allowed us to introduce additional measures within the framework of lightning protection, which also contributes to better protection against electromagnetic impulses. All the walls of the buildings that house the emergency control rooms thus include, within their frame structure, electrically interconnected grid networks with a fifteen-centimetre mesh aperture. The remote connections to the outside through optical fibres or copper cables were put in place using double shielding. In the Gösgen and Leibstadt nuclear power stations, these instructions had already been taken into account during their actual construction.

If, despite these measures, a power plant network failure still occurred, the reactor would be deactivated within a few seconds through automatic shutdown. This automatic shutdown leads to the immediate interruption of the chain fission reaction inside the reactor. However, even in the case of a shut-down power plant, it is still necessary to evacuate excess thermal energy from the reactor. This process is ensured by pumps that must be supplied with electricity. If a black-out were to take place, nuclear power plants are endowed with alternative systems for producing electricity, including diesel generators. This equipment is capable of cooling a nuclear power plant until the power distribution network has reverted to its restored function.

A prolonged failure of the ultra-high-voltage network is an extremely unlikely scenario. The *Swissgrid*[38] organisation is in charge of managing the high-voltage distribution network in Switzerland. When the ENSI asked Swissgrid about the possible effects of a solar storm on the electricity network, they received the following answer: "Solar storms could indeed affect transmission networks, since they induce continuous currents within the three-phase network, currents which contribute, above all, to the saturation of transformers, with the possibility of lasting damage. This is a well-known difficulty encountered in the case of long-distance transmission lines oriented from the north to the south and close to the poles, since the electromagnetic effects are particularly intense."

In recent years, a few publications have dealt with this phenomenon, particularly in the United States. The European Network of Transmission System Operators for Electricity (ENTSO-E)[39] has also begun to analyse the possible risks. Specific experiments have, thus far, only been conducted in connection to the Canadian and South African networks. Various approaches and solutions have been developed and their effects actively monitored by Swissgrid.

As regards Switzerland, Swissgrid believes that no special measures are necessary at present, as electricity transmission lines are relatively short in this country, and the effects will be reduced because of the relative remoteness of the poles. As part of the network's ongoing upgrading process, Swissgrid will now be paying

38 AN: http://www.swissgrid.ch/swissgrid/fr/home.html.

39 AN: www.entsoe.eu.

more attention to this topic, if only because of the north-south orientation of important lines planned for the coming years.

Solar storms are thus a known phenomenon which Switzerland is well-prepared for.[40]

What are the technical and organisational procedures for shutting down and cooling reactors?

D.S.: The cooling of the reactors must be ensured both during normal operation and in the event of a failure.

The fuelling of the reactor core is intended to operate in cycles. At the end of each cycle, the efficiency is no longer as high. The reactor is then stopped using conventional means.

Some malfunctions require the immediate cessation of nuclear energy production. In such cases, shut-off rods are inserted into the core of boiling water reactors. In pressurised water reactors, shutdown rod clusters slide into the core. The chain reaction is thus immediately stopped. So far, no failure of this system has been observed during power operations.

Despite this fact, the failure of the reactor emergency shutdown system cannot be ruled out in the case of light-water reactors. The latter are therefore equipped with a second, independent system involving the massive injection of boric acid into the reactor vessel. Its purpose is to allow one to intervene should the reactor's emergency shutdown device fail.[41]

Once the core's reaction has been interrupted, it is still necessary to evacuate the thermal energy released by the decay of isotopes within the nuclear fuel, a process which extends over a longer period. This residual heat then decreases rapidly over time.

---

40 AN: All the relevant information can be found here: http://www.ensi.ch/fr/2011/12/13/les-centrales-nucleaires-sont-armees-contre-les-tempetes-solaires/.

41 AN: http://www.ensi.ch/fr/2013/08/15/defense-en-profondeur-systemes-de-securite-et-objectifs-de-protection-1re-partie-713/.

However, it still represents a few kilowatts per fuel assembly, even after years.[42]

How long does it take to completely shut down and cool a reactor?

D.S.: A reactor can be shut down immediately if necessary. The cooling of the reactor must be maintained as long as the plant is still in operation. Insufficient cooling of the fuel assemblies or a complete cooling failure either in the reactor pressure vessel or in the spent fuel assembly's storage pool can lead to very serious consequences such as the bursting of the fuel rods or fuel meltdown. Cooling must therefore be ensured in all cases (both before and after the shutdown of a reactor).

Are there any auxiliary power generators that can run long enough?

D.S.: In the event of an external electricity network failure, the nuclear power plant is rerouted to enter a self-contained operation mode known as "islanding". This is done in accordance with the proper directives. In case of an incident, the decoupling of the Swiss power network into self-contained mode is performed in a fully automatic manner. Reactor power is thus decreased to the necessary level to enable an autonomous power supply. It then provides power to all devices, including security systems. The operation enabling an autonomous power supply can be maintained as long as necessary.

A diversified network-based electrical supply is ensured through a direct connection to a nearby hydroelectric plant.

Should the rerouting to a self-contained operation mode fail, the use of emergency diesel generators to supply the necessary electrical power would serve as a fallback solution. Each plant

42 AN: http://www.ensi.ch/fr/2013/08/22/defense-en-profondeur-systemes-de-securite-et-objectifs-de-protection-2e-partie-813/.

has its own specific generators. Thanks to these devices, a power station can operate in a state of autarky for several days, without requiring any additional energy.

In addition to this, emergency units are also available to ensure the power supply. Some are mobile and others fixed. They are protected against floods and earthquakes and can, if necessary, be used to implement accident management measures.

An exercise dealing chiefly with the Swiss electricity network's failure scenario took place from 3 to 21 November 2014. The focus was on controlling the consequences of a power failure and a pandemic. Topics such as mobility, health, supplies, management and logistics were addressed as a priority.[43]

The prospect of a complete power failure was also examined as part of the European Union's resistance test in the aftermath of the Fukushima disaster. The tests[44] proved that Swiss nuclear power plants can withstand major earthquakes, major floods, major climate disruptions, but also power network outages (both LOOP and Total SBO—Total Station Black Out). To be more specific with regard to these last two scenarios, the reactors of the power stations can operate in total autonomy for seventy-two hours on battery power. Spent Fuel Pools (SFPs) are also resistant to natural disasters and can provide adequate cooling for ninety-eight hours, without any kind of power supply.

Generators then take over and can be refuelled, as required, from secure storage centres provided for this very purpose and with access to both road and helicopter transport. Lastly, with the exception of the Mühleberg nuclear power plant, all emergency control station systems at Swiss nuclear power plants have their own ground water supply systems that serve as alternative

43 AN: As seen here—http://www.ensi.ch/fr/2014/11/07/les-centrales-nucleaires-suisses-autonomes-en-matiere-dalimentation-electrique/.

44 AN: Here is the link—http://static.ensi.ch/1326182677/swiss-national-report_eu-stress-test_20111231_final.pdf.

last-resort heat sinks in the event that the river can no longer be used. In Mühleberg itself, this alternative heat sink is currently being installed. On the international level, however, emergency control station systems are far from the norm and are only required in a handful of countries such as Germany. Complex crisis management procedures are in place for these scenarios and plant staff are regularly trained.

How does one ensure that the necessary staff is indeed present in the event of a very grave crisis?

D.S.: The staff present on site can resort to emergency control stations when needed.[45]

In addition to categorised safety systems, all Swiss nuclear power plants have a so-called emergency control station system. The latter is designed to meet the requirements of a crisis situation in which, due to external events, the shift team is unable to perform any actions. The emergency control station system is also there to guarantee the functions of conventional security systems in a redundant and diversified manner. Its role is to ensure, in the event of a request, the safe shutdown of the installation and the evacuation of residual heat. No personnel intervention is required during the first ten hours of an incident.

The emergency control station system is completely independent of conventional security systems. It has its own power supply, as well as its own ventilation and cooling systems. It is especially protected against earthquakes, falling aircraft, floods and any third-party involvement.

All nuclear power plants in Switzerland have ready-to-use auxiliary emergency rooms located within the station itself. In addition to on-site emergency management posts, intervention facilities outside the nuclear power plant's perimeter must be made

45 AN: http://www.ensi.ch/fr/2013/08/29/defense-en-profondeur-systeme-de-poste-de-commande-durgence-913/.

available, enabling response staff to work safely and effectively under aggravated conditions. Another point is that only a minimal number of staff members must be kept on the premises to manage such an emergency.

The ENSI therefore requires operators to prepare an external emergency response centre. The usefulness of such a centre and the planning of staff movement within this location must be tested by an adequate drill in 2016.

The equipment necessary for the emergency staff during a crisis must be kept in a suitable place. One must ensure that all intervention equipment remains accessible in case of an event. This also applies to the radiation protection equipment and foodstuffs meant for the various intervening teams. Indeed, the latter initially find themselves alone, acting under carefully defined conditions.

The ENSI requires operators to ensure that the means of intervention necessary for emergency protection, including radiation protection equipment, are stored in such a way that they remain readily available. In their response strategy, operators must document the manner in which this goal is achieved.

An emergency in a nuclear power plant does not require only protection of the population; the staff, too, must be shielded. The ENSI thus also emphasises the importance of inhaled air. Operators must offer the intervening teams the necessary guarantees that both air radioactivity and the concentration of toxic respiratory substances such as carbon dioxide are monitored. Similarly, they must provide safe premises during the entirety of the operation.

So as to ensure the effective management and coordination of an emergency situation, communication tools are essential. Operators must therefore be able to propose a back-up solution that works under deteriorating conditions.

The presence of a minimal number of qualified personnel members is crucial for the management of an emergency. In addition to readily available protective equipment, the ENSI thus also requires the presence of radiation protection specialists.

In its desire to test intervention strategies, the ENSI envisioned the following extreme scenario: a major accident initiated by a natural event and leading to disruptions in the long-term infrastructure as well as to an unfiltered release of radioactive substances into the environment. In this scenario, the networks and external power supply become unavailable for several days and the internal power supply, within the installation itself, temporarily inoperative.

Immediately after the reactor accidents that occurred at the Japanese nuclear power plant of Fukushima on 11 March 2011, the ENSI initiated a new safety audit of Swiss nuclear power plants. Following this study, four ENSI decisions[46] were published. The first three (dated 18 March, 1 April and 5 May, 2011) required immediate action and additional audits.

These concerned the creation of an external emergency warehouse, a warehouse common to all Swiss nuclear power plants and comprising the supplies and equipment necessary for the management of emergency situations and the repair of spent fuel storage pools. These additional audits were also meant to examine the design of Swiss nuclear power plants in relation to earthquakes and/or submergence resulting from external sources.

Last but not least, orders were given to test the cooling water supply of the safety and auxiliary systems, as well as that of fuel assembly storage tanks.

In parallel to these operator-related tests, the ENSI also carried out targeted inspections. In 2011, they focused on storage tank safety devices, protection against flood effects, and the adequacy of filtered pressure relief systems for containment buildings. In 2012, these inspections focused on the strategies and responses in

46 AN: The topic can be explored here—http://www.ensi.ch/fr/2015/03/04/lifsn-exige-des-centrales-nucleaires-des-postes-dintervention-externes/.

the event of a power failure, external event evaluation processes, and the emergency response rooms of power stations. In 2013, the radiation protection equipment available at each nuclear power plant was inspected. Its presence and proper functioning are essential in case of a serious failure.

The results of the tests conducted by the ENSI[47] confirmed the fact that Swiss nuclear power plants are characterised by a high level of protection against the results of earthquakes, floods and the combination of these two phenomena. They also validated the measures taken in connection to power-supply failure and heat sinks. Each analysed default scenario appears to be under control when considering all currently envisaged risk assumptions. The basic safety-related legal requirements (i.e. reactivity management, the cooling of fuel assemblies and the containment of radioactive substances) are thus all fulfilled. With a view to further improving safety, however, the ENSI has formulated a series of new requests regarding significant retrofits such as the presence of an auxiliary heat sink unaffected by the consequences of earthquakes and floods.

How does one manage the operation of fuel cooling pools and how long does it take to render the stored fuel inert?

D.S.: As a general rule, a fuel element (a set of assemblies with multiple fuel rods, each containing several hundred cylindrical pellets of fuel, usually uranium oxide) remains in a reactor for a period ranging from four to seven years. It must then be replaced and discharged from the reactor vessel. When removed, these spent fuel elements are still extremely hot, but significantly less so than if the reactor were still in operation. They must therefore be cooled in an intermediate repository for years. This phase takes place in special water-filled pools located in the nuclear power plants themselves. What we are referring to here are storage pools.

47 AN: http://www.ensi.ch/fr/document/plan-daction-fukushima-2015/.

Used fuel is usually kept there for up to five years, as it waits to be dry-packed for transport. In the case of the Gösgen plant, the assemblies are stored in a warehouse-pool.

Once a year, a portion of the reactor vessel's fuel elements is transferred to the storage pool. This step takes place exclusively under water, which, on the one hand, acts as an anti-radiation screen and, on the other, enables the cooling of assemblies. The fuel elements thus remain permanently covered by several meters of water both during transfer and pool storage.

In order to prevent any increase in the storage tank's temperature, the residual heat produced by these fuel elements must be continuously evacuated. Reliable cooling is therefore a necessity. It does not matter if the nuclear reactor is operational or not. Even after the plant has been shut down, spent assemblies must be cooled for several years in separately operated storage pools.

After several years, the fuel will have lost enough of its activity and its temperature sufficiently decreased for it to be unloaded from the pool. It is then transferred to special armoured containers, which, this time, are cooled by the ambient air.

Each nuclear power plant has at least one storage basin for water-cooled assemblies on its premises. For the dry storage of older fuel elements, which have declined significantly, Swiss nuclear power plants make use of two intermediary storage facilities. One is located near the Beznau power station and is used for storing power plant assemblies. The other is the ZWILAG intermediary storage facility in Würenlingen, which stores spent fuel assemblies and highly radioactive waste originating from all Swiss nuclear power plants and research institutes.

Insufficient cooling of fuel elements or an utter cooling failure either in the reactor pressure vessel or in the storage tanks can lead to serious consequences. Such a situation may for example result in the bursting of fuel rod claddings or the latter's melting. This is precisely what happened in the Fukushima nuclear power plant.

Following the Fukushima accident, several questions[48] regarding storage tanks arose and the ENSI thus required Swiss plant operators to test the safety of their storage pools.

Operators were expected to respond immediately to the following questions:

1. Is the supplying of the safety and emergency systems' cooling source ensured by a diversified source that is resistant to earthquakes, floods and contaminations (an additional supply by means of a water spring)?
2. Are storage pools outside primary containment adequately protected against both external and internal adverse effects?
3. Is the cooling of storage pools a particularly well-protected safety function? Can it be powered and managed through the "bunkerised" backup system?

Having studied the reports, the ENSI identified various weaknesses in the storage basins which will now be improved.

Have the necessary electrical systems (including the generators, transformers, etc.) been reinforced for potential Carrington-type or EMP-type49 scenarios?

D.S.: Primary reactor protection systems are located behind the primary and secondary containment buildings. The latter do therefore offer electromagnetic protection. In addition, independent backup systems are located inside "bunkerised" installations. They meet strict requirements for both lightning protection and electromagnetic compatibility.

---

48 AN: http://www.ensi.ch/fr/2011/08/25/comment-les-elements-combustibles-sont-ils-stockes/.

49 TN: Typically abbreviated as nuclear EMP or NEMP, a (nuclear) electromagnetic pulse is a burst of electromagnetic radiation triggered by a nuclear explosion.

And what about France?

We attempted to ask the French authorities responsible for our nuclear energy these questions, but did not receive any official response to our queries, unfortunately. We did, however, manage to obtain the testimony of an EDF executive who wished to remain anonymous.

Why is it that we were unable to get any answer?

Mr. X: I was employed by EDF until recently, including in a nuclear power plant. I am not surprised that you have failed to receive an answer. Despite the fact that, since 2008, communication channels have been opened at a local level, there is a prevalent tradition of secrecy in France.

Above all else, one is instructed to communicate as little as possible the moment the topic relates to security issues. This is still the case today, despite the internal scandals that have been taking place… Recently, an unauthorised person was arrested and found to be in possession of the detailed plans of a certain power plant… And due to the latent threat of terrorism that we have been facing since the 1990s, the authorities are always very suspicious.

What are the risks today?

Mr. X: The first risk is, in my opinion, an economic one. Nuclear power in France is comprised of three major parts: EDF and the power plant operation domain; Areva and the sphere of EPR reactor construction,[50] uranium exploitation and the reprocessing of waste; and the CEA[51] in the research field. Following the closure of several CEA centres across France, Areva proceeded, for reasons of centralisation, to downsize operations by over 5,000 positions in 2015. In EDF's case, the management plans to make 4,000

50 AN: EDF has recently claimed world leadership in this sector.

51 AN: The *Commissariat à l'énergie atomique et aux énergies alternatives* or French Alternative Energies and Atomic Energy Commission.

people redundant by 2018. This development will impact working hours and therefore staff quality as well. It will additionally affect external service providers, who will be compelled to plane down their margins or the quality of their services.

What would happen if an emergency shut-off were implemented at a power station?

Mr. X: In France, the procedure is different from Switzerland, especially because the technologies on which our reactors and security systems are based are not identical. First of all, in the event of an emergency shutdown (whatever the cause may be), there are redundancy systems that enable the reactors to keep cooling down while allowing the plant's vital elements to operate. Elements that prevent the phenomenon of atomic fragmentation from taking place are added to the pools. Using a large red emergency-stop press button, the activation is initiated from the control room itself, where three people are permanently on duty. There is no command or remote-control centre to my knowledge.

These systems are not connected to any batteries but, instead, to generators that enjoy a fifteen-day fuel autonomy. If the plant is no longer supplied with electricity from outside, the reactor and fuel can continue to be cooled during this period of time.

And after that?

Mr. X: It depends on the degradation of the surrounding area. As long as we can safely channel fuel because we actually control the territory, we can do so indefinitely. In addition to this, there is EDF's Nuclear Rapid Action Force (FARN), which can provide reinforcements to any power plant. But there must be fuel for this to happen, considering the reduction of the army's strategic stocks… On site, diesel generators are checked several times a year, regularly dismantled and very carefully reassembled. The work quality is of a very high standard there!

You mentioned the prospect of terrorist threats. Without revealing any security plan details, what can you tell us about them?

Mr. X: In France, there are Specialised Gendarmerie Protection Platoons (PSPG) that are permanently assigned to nuclear sites. They are well-trained and know what to do in case of a conventional armed attack, and can, if necessary, hold the ground until the arrival of the National Gendarmerie Intervention Group (GIGN). In my opinion, serious security lacks should instead be sought at the level of the actual selection and auditing of personnel given access to the sites. One of my colleagues, for instance, lied in his CV and was given access to positions far beyond his skills. It should be noted that every year, our government services carry out more than 100,000 administrative inquiries into the 73,000 employees, and a further 23,000 into the service providers working in our nineteen nuclear power plants. On average, 700 of them are then denied access. But because of the decrease in resources, ethics checks are no longer carried out as accurately as before—if they are carried out at all, that is.

Why is that?

Mr. X: It is all due to a lack of means and will. As the markets in question are European, one is more forgiving in order to lower costs. One proceeds to hire Moldovans and Romanians, who have not been given the same training as in France. For economic reasons and in a desire to increase shareholder margins, one brings in people who do not have the required training level—and the ultimate result is that the employees, especially the security officers among them, do not always have the necessary skill level to perform the tasks assigned to them.

Some sites leak like a sieve. Not from a physical point of view, since our CNPEs are well-protected by security gates and systems (which have, incidentally, been tested by the GIGN); the leakage

I am referring to concerns the monitoring of personnel and vehicle entries and exits. This aspect rests, in fact, in the hands of private security companies which, owing to their lack of time and resources, do not always implement efficient checks. No incoming delivery truck is thoroughly searched, for instance. Another fact: there are one-way security gates that remain unsupervised and through which two people can pass simultaneously, one after another… The staff could thus aid or be forced to bring in/out both ill-intentioned people and sensitive material.

Recently, during a meeting of the High Committee for Transparency and Information on Nuclear Security (HCTISN), it was acknowledged that employees were sometimes denied access to power plants, particularly for reasons of radicalisation.[52]

I would, however, like to reassure you on one level: an act of genuine sabotage impacting a reactor cannot be easily committed, because the reactor is kept behind thick armoured doors, as is the "K building" that stores fuel. Furthermore, these places are well-guarded. On the other hand, the transformers, the buildings that house generators, and EL (electricity) buildings are easier to access, and there, an attack or act of sabotage could indeed put a power plant out of service and have one hell of an impact on the country's electrical network. In addition, although some of our CEA sites are well-monitored and protected, others such as the Fontenay-aux-Roses site are secured by nothing more than a simple fence… It might thus be easy for any sufficiently motivated person to enter the premises and steal sensitive material or radioactive sources.

52 AN: http://www.lejdd.fr/Societe/Le-nucleaire-n-echappe-pas-aux-derives-islamistes-760935.

And what can citizens do?

Mr. X: I suggest they get in touch with their town halls to obtain weekly updated reports on the current state of our power stations and to acquire iodine from them.

Last but not least, citizens must remain attentive in case the alarm is raised—although ever since they tried to trick people into believing that the "Chernobyl radioactive cloud" could somehow stop at our borders, I, for one, am unsure whether we can put our faith in the authorities and the information that is spread.

6. Atomic Bombs

Now I am become Death, the destroyer of worlds.[1]

— American physicist ROBERT OPPENHEIMER (1904–1967)

Greetings Professor Falken.
Hello.
A strange game.
The only winning game is not to play.
How about a nice game of chess?

— *Wargames*, 1983

On 16 July 1945, humanity enters the nuclear age. On that day, at the Alamogordo Air Force Base in the state of New Mexico, the *Trinity* test is successful. It involves the first explosion of an A-bomb, commonly known as an atomic bomb or nuclear bomb. The test is the result of an arms race, a race from which the United States emerged victorious against the USSR thanks to the Manhattan project. This immense technological-industrial effort was led by physicist Robert Oppenheimer, surrounded by a team that included Nobel

1 AN: During the first nuclear explosion test on 16 July 1945 in Alamogordo, New Mexico, it was this phrase, originally found in the Bhagavad-Gita, that Robert Oppenheimer, the scientific director of the Manhattan Project, thought of.

Prize-winning physicists such as Niels Bohr, James Chadwick, Enrico Fermi and Isidor Isaac Rabi.

The idea of an atomic bomb was born in the 1910s, when physicists such as Otto Hahn and Enrico Fermi considered the use of atomic energy for military and energetic purposes. The first nuclear fission was carried out in December 1938 at the Kaiser Wilhelm Institute in Berlin. As part of its *Projekt Uranium*, however, Germany's Third Reich failed to develop a bomb, despite the construction of several experimental atomic reactors.

Following the success of the *Trinity* test, a quick decision was made to use such a bomb against Japan, which was still at war with the United States and the Allies.

Hiroshima

On 6 August 1945, at 02:45 a.m., the Boeing B-29 Superfortress bomber baptised 'Enola Gay' by its commander (in honour of his own mother) took off from Tinian Island in the Mariana Islands and headed for Japan. It was accompanied by six other aircraft of the same type whose purpose was to make a reconnaissance, bring back weather records and take photographs. As for the Enola Gay B-29 bomber, it carried in its bosom an atomic bomb weighing approximately four tons and nicknamed *Little Boy*.

After travelling approximately 3,200 km in the course of six hours, the bomber finds itself at a high altitude (9,500m) above Hiroshima. The clear, cloudless weather conditions allow it to confirm its target. At 08:15 (local time), Captain Tibbets, who is flying the aircraft, has the airdrop initiated. The bomb drops for forty-three seconds before exploding at a 580-metre distance from the ground. A fireball with a radius of about 400 metres is formed, emitting powerful lumino-thermal radiation which, as suggested by its name, spreads at the speed of light. Although very brief, the flash causes the death of thousands of people. With the gravity of the burns depending on the distance and surface of the exposed body, the victims would agonise for a period ranging from a few minutes to several hours. In addition, many fires are created by this intense heat flow, extending across a radius of several kilometres around the centre of the explosion.

A few seconds later, a surge wave carried by winds that reach a speed of 300 to 800 km/h pulverises the buildings, crushing many survivors that had remained sheltered behind the wooden walls of their homes. For those that had miraculously survived, the ordeal is not over yet, as masses of burning air come flowing in from all sides. Fires spread throughout the city. Most electronic components[2] are damaged by the electromagnetic pulse resulting from the explosion. An atomic mushroom made of dust obscures the sky. After a few hours, the cloud falls in the form of rain. Thick black drops[3] thus descend upon Hiroshima. Filled with radioactive ash, they would contaminate the city and the surrounding areas, stretching across a thirty-kilometre radius and along a north-west axis.

Consequences

In the case of Hiroshima, *Little Boy* had an estimated power equalling 15 kt of TNT (i.e. 15,000 tonnes). The diameter of the generated fireball was 400 metres long and the temperature at its centre several million degrees, with an estimated 4000 °C at ground level. Third degree burns were possible up to a distance of eight kilometres. As for the atomic mushroom, it reached an approximate height of seven kilometres. Buildings located within a one-kilometre radius from the epicentre — or a 1.5-kilometre radius in the case of wooden buildings — were destroyed.

2 AN: Today, this effect would definitely be even more disabling, since electronic devices play a major role in our society (and are found in our cars, radios, mobile phones, computers, etc.).

3 AN: Known as 'black rain' in English.

By comparison, the most powerful explosion ever recorded on our planet's surface (the *Tsar Bomba*, a 50-Mt[4] hydrogen bomb tested by the USSR on 30 October 1961 in northern Siberia) spawned a fireball eight kilometres in diameter and a cloud that reached a sixty-four-kilometre height, causing burns and building destruction up to a distance of 100 kilometres. It was both visible and felt up to 1,000 kilometres away, triggering a 5.2-magnitude earthquake on the Richter scale, an earthquake that would have reached a 7.1-magnitude if the explosion had occurred underground.

According to the Hiroshima Peace Memorial, the number of victims totalled 140,000. In the years that followed, however, the figure increased significantly, mainly because of the cancer and various complications resulting from radioactive contamination.

The attack on Hiroshima was followed by another on 9 August 1945 against the city of Nagasaki, causing between 40,000 and 70,000 immediate deaths and more than 140,000 in the next five years. One day after the Nagasaki attack, Japan agreed to surrender unconditionally.

Reacting to this event, Joseph Stalin, the leader of the USSR, would go on to say that if the Americans could destroy a city in an instant using a single bomb, he could do it in one day with his artillery. There is, however, a difference between destroying a city with a single bomb and using ten thousand bombs or millions of shells to achieve this. Everyone at the time sensed that this was not just a new, powerful bomb but, indeed, a destroyer of civilisations.

The Atom's Destructive Power

An atomic bomb uses the power of the atom to release a colossal amount of energy in a very short time. Its destruction capacity is beyond comparison with conventional (chemical) explosives and can be expressed through TNT-equivalence.[5] Its power is considerably vari-

4 TN: A Megatonne is a unit that totals 1 million (106) tonnes.

5 AN: TNT = Trinitrotoluene. It serves as a reference when comparing the power of explosives with each other.

able: the smallest (Davy Crockett, USA) 'only' totalled 20 tonnes of TNT; the bomb dropped on Hiroshima had a power of 15 kilotonnes or kt (i.e. the equivalent of 15,000 tonnes of TNT); the power generated by the most fearsome bomb of all, the *Tsar Bomba* (USSR), was equivalent to 50,000,000 tonnes of TNT (50 Mt). Between the two extremes, an entire range has been developed by the different nuclear powers, knowing that the more precise a missile or bomb is, the lower the power required to destroy a particular objective. Some examples are shown in the table below:

			POWER (TNT EQUIVALENT)	
HELD BY	**NAME**	**TYPE**	**TONNES**	**KILOTONNES (KT)**
CHINA	Dong-Feng 5	Intercontinental ballistic missile	5,000,000	5,000
USA	Davy Crockett	Tactical nuclear recoilless gun projectile	20	0.020
	Little boy (Hiroshima)	Air-dropped bomb	15,000	15
	W-80-1	Cruise missile warhead	150,000	150
	W-87	Minuteman III intercontinental missile warhead	300,000	300
FRANCE	TNA[6]	Medium-range air-to-surface missile warhead	300,000	300
RUSSIA	Topol	SS-25 missile warhead	800,000	800
	Tsar Bomba	Air-dropped bomb	50,000,000	50,000

The explosion of a nuclear bomb has different effects, namely:

6 AN: *Tête Nucléaire Aéroportée* or Airborne Nuclear Warhead.

- **Lumino-thermal effects**

  35% of the energy is dissipated in this form. The radiation generated by the fireball spreads at the speed of light. Although its duration was less than three seconds in the case of Hiroshima, this time-span increases with the power of the weapon itself (fifteen seconds for a 600-kt bomb, for example). Depending on the distance, the intensity of the heat flow and light flux can cause death, lead to temporary or permanent blindness, trigger third-degree burns and set any combustible element ablaze.

- **Mechanical effects**

  These represent about 50% of the energy dissipated. They are manifested by a surge wave that destroys both buildings and infrastructure, as well as by violent winds that ravage everything. From the human perspective, three categories of effects are thus generated:

1. The primary effect: the overpressure that damages the eardrums, lungs and other internal organs.

2. The secondary effect: high pressures that cause a 'crushing' type of damage due to wall collapse; powerful winds causing various lesions and injuries by means of propelled objects.

3. Tertiary effect: people themselves are thrown against hard or raised surfaces before being rolled across the ground.

- **Radioactive effects**

  Their share is estimated at 15% of the dissipated energy. They are divided into two types:

1. The initial radiation (5% of the energy), consisting mainly of gamma and neutron radiation emitted during the first minute after the explosion. Generally, the lethal zone resulting from this effect is smaller in size than that of the lumino-thermal flash and surge

wave; meaning that the people that are meant to succumb to the effects of the initial radiation will have already perished, having either been carbonised or met their fate as a result of various deep injuries. The greater the power of the bomb, the more this is the case. It is, however, clear that significant exposure to gamma and neutron radiation is an aggravating factor in the case of seriously injured survivors.

2. Residual radiation (10% of the energy) is responsible for the contamination phenomenon. It is composed of unconsumed nuclear material and any unstable products resulting from the chain reaction. Whenever the explosion occurs close to the ground, a radioactive dust and ash effect is added to the phenomenon.

Note number 1: The radioactive contamination caused by the explosion of an atomic bomb is mainly dependent on three factors:

1. Its power and the quantity of nuclear fuel contained in it.

2. The type of element that causes the chain reaction. A uranium or plutonium weapon generates more contamination than an H-bomb, for instance.

3. The altitude of the explosion.

The dust and ashes sucked into the atomic mushroom become radioactive through an activation mechanism (neutron capture). The closer the bomb is to the surface at the time of its use, the more significant this phenomenon becomes. The generated cloud then causes a radioactive fallout that pollutes the air, soil and water. In the case of Hiroshima, the explosion occurred at a height of 580 metres. Knowing that the power of the weapon was 15 kt, an altitude of 160 metres would have been necessary to obtain a significant contribution of radioactivity through the above-mentioned mechanism.

Note number 2: It is important to understand that the Chernobyl and Fukushima accidents have had more serious environmental consequences than the attacks on Hiroshima and Nagasaki. This is mainly due to the large amounts of radioactive material released for extended periods of time. Although it is inferior, the fallout resulting from an atomic explosion is, however, very real indeed. Even if it does not cause immediate lethal effects, such residual radiation will contaminate both the area and the people. The danger is even greater if individuals inhale or ingest these radioactive substances. In such a case, the risk of various kinds of cancer occurring years later is increased. Adequate measures should therefore be taken by anyone who finds himself amidst a fallout so as to avoid external or, in the worst case, internal contamination.

Note number 3: A nuclear explosion can also produce an electromagnetic pulse with the ability to destroy any unprotected electronic equipment. This effect is all the more potent when the bomb is powerful enough (more than 200 kt, generally speaking) and the explosion occurs at altitude (optimally at heights greater than thirty kilometres). The scope is very significant, as damage can be observed in entire regions or even whole countries. Special atomic weapons have been developed to achieve this type of impulse, including hydrogen bombs. The EMP topic will be discussed in more detail in Chapter 5 (Nuclear Attacks).

TYPES OF ATOMIC BOMB

There are several types of atomic bomb:

- **The fission bomb** (= nuclear bomb or A-bomb)

As the name suggests, this kind of bomb operates on the principle of nuclear fission. It requires uranium-235 (as was the case of the *Little Boy* dropped on Hiroshima) or plutonium-239 (which was true of the *Fat Man* bomb, used against Nagasaki).

- **The fusion bomb** (= thermonuclear bomb or H-bomb)

This kind of bomb involves a fusion reaction of small atoms such as deuterium and tritium, all of which are hydrogen isotopes (hence the name 'H-bomb'). In order to achieve the pressure and temperature necessary to initiate the fusion reaction, a small fission bomb is required.

- **The neutron bomb** (= reinforced radiation bomb or N-bomb)

This is a special fusion bomb whose goal is to produce a maximal number of neutrons and cause minimal conventional damage and fallout. Meant to kill individuals and 'spare' equipment, this type of weapon seems to have gone out of use today.

The detonation of the Tsar Bomb on 30 October 1961.

~

Fiction

Jerome and Laurent walked ahead, under the cover of a copse of old beech trees. Over several metres, they plunged into the thick vegetation before pushing the last undergrowth aside and sitting down on an old stone bench, at the end of their breath.

Although they had only climbed a hundred meters, it was enough. The gendarmerie would never come looking for them there. The path did lead up the hill, but no one ever took it these days. Tourists no longer came to admire the exceptional view of the plain and the great city that lay at the foot of the elevation; and understandably so — for people had other priorities now that the war had begun… In addition to this, the restaurant that stood at the top of the ridge had been looted six months earlier and set ablaze. The two friends would not be disturbed there.

— 'Amazing to what lengths one has to go to get some peace and quiet,' said Jerome.

— 'What are you complaining about? The weather's perfect. Not too hot, not too cold. And not a single cloud in sight!'

— 'Still! We'll have to find a hideout somewhere in town. I'm sick and tired of climbing here every day.'

— 'If you'd rather get caught by our friends the gendarmes, then by all means, feel free!'

Jerome sighed in response. He took a pouch out of his backpack and, unpacking the contents, went on to say:

— 'Seems the Biston brothers have bitten the dust.'

— 'Oh, had no idea. Who told you?'

— 'A neighbour told my mother.'

— 'You shouldn't go home, it's not safe. You're a deserter, remember? You'll get caught one of these days.'

—'Anyway, the Russian troops will be there soon, it would seem. And this time, the Ricans will not be there to help us. They're too busy dealing with the Chinese.'

—'Mind you, we did ask for it. We were the ones who started this bloody war.'

The two friends spent the next few minutes without uttering a word, with only the rustling leaves breaking the silence.

—'Why don't you go ahead and roll yourself a joint,' Laurent suddenly said. 'Let's not get all melancholic now, shall we?'

—'Yeah, you're right.'

—'What have you got today?'

—'Not a clue. The shit is so expensive nowadays. So, I took a new thing. The guy told me it's based on plants that witches used back in the Middle Ages.'

—'If you ask me, you've been had.'

Jerome finished rolling his joint and lit it quietly. He took a big puff, then a second one, before handing it to his comrade.

Laurent needed no encouragement to grab it. He was just about to hand the cone-shaped cigarette back to his friend when he exclaimed:

—'Shit! What's that light?'

—'Argh! It's burning my eyes!'

—'God damn it! It bloody burns!'

The phenomenon disappeared as quickly as it had appeared. The trees that protected them had all lost their leaves and were now black in colour. The two men's skin had turned red, as if they had got sunburnt.

—'Wow,' said Laurent. 'Your stuff is a killer. Never seen anything like it!'

In response, Jerome, merely raised a finger and pointed towards the plain, his face twisted by fear.

—'But that's downright hallucinogenic!' continued Laurent, turning his head. 'You might have to use less next time.'

A huge mushroom of dust rose silently towards the sky. On the ground, a grey wave crushed everything in its path, advancing at a vertiginous speed.

— 'That's not good,' Jerome blurted.

— 'Personally, it reminds me of a movie.'

— 'Exactly, we have to get out of here.'

— 'Hey! What's that noise?' Laurent suddenly shouted. 'The whole place is shaking!'

— 'Quick, get down!'

Jerome did not get a chance to finish his sentence. The grey wave had reached them. He barely had time to feel himself being lifted, hurled, and sent rolling across the ground. In actual fact, his lungs had already exploded as a result of the overpressure, and blood was flowing out of his ears. And yet, he was not aware of any of it — not of the branch that had become embedded in his back, nor of the fact that his spine had just snapped.

In truth, everything had happened too fast. The pain and fear he felt were the last information his brain ever received.

Ten months earlier, Jerome and Laurent had chosen desertion, but on that day, the war had caught up with them.

Unfortunately, they were not the only ones in this case. The plain now resembled a pancake that had been left on a burner for far too long; as for the city, it looked like a pile of ashes.

Even if, a mere few minutes earlier, tens of thousands of people had still been busy with their daily activities. They had traded, helped each other, worked, and sometimes quarrelled. And yet they had lived, driven by the hope of seeing better days.

Everything was over now.

Unaware of the danger that threatened them, a crowd of innocent people perished in a matter of seconds. But who had thrown the bomb? Not that it mattered… History would be written by those that emerged victorious from the current war. It would not be the first time that someone used the 'common good' argument to justify atrocities...

Those that had survived, on the other hand, had other priorities. Mostly wounded or burnt, thirsty and shocked, they would now have to escape from this hell. Even if they had lost everything, their survival instinct would drive them forward as long as they had as little as an ounce of strength left.

The majority kept turning around to gaze in horror at the atomic mushroom that was rising into the sky. As if it never intended to stop, it rose for dozens of minutes, forming a gigantic shadow that would mark forever the minds of those poor wretches. But what would happen when it fell in the form of rain? Many of the survivors surely knew that under no circumstance should one drink this water, which had now become poison. But how many of them actually knew that they had to protect themselves from it as well? How many realised that this contamination would end up in rivers, plants and soil and remain there for days, months or even years?

Naturally, this story is a tale of fiction centred around an imaginary event. However, as all of you will have understood, it can be likened to the two very real nuclear explosions that targeted the civilian populations of Hiroshima and Nagasaki.

7. Myths

Our doctrine is to use nuclear weapons as soon as we consider it necessary, so as to protect our forces and achieve our goals.

— US Secretary of Defence ROBERT MCNAMARA (1916–2009)

Successful books such as Neville Chute's *On the Beach* and many other artistic works set in a post-apocalyptic world have instilled into popular culture many myths regarding nuclear weapons. One of them is that their use is allegedly synonymous with the end of the world.

A nuclear accident scenario, and even a global thermonuclear war, however tragic it would be, however colossal and inestimable its consequences, would not signify the end of human life on earth. Even among 'survivalists' or other *preppers*, the topic of preparing for nuclear warfare is often overlooked — indeed, it is thought to be 'useless' because it would be 'too hard' or 'too horrible'.

It is therefore necessary to deconstruct the myths connected with nuclear energy in order to enable a clear understanding of reality with the least possible prejudice and emotion.

Myth: The radiation resulting from a nuclear war would poison the air and the entire environment. Everyone would die!

The facts: When a nuclear explosion occurs, especially if it is close to the ground, thousands of tonnes of material are irradiated. This contaminated material is demolished by the explosion (the temperature, the blast) and transported into the atmosphere by the ascending effect of explosion-heated air. This typical mushroom-shaped cloud causes these particles to spread through the atmosphere and then disperse and fall, depending on the wind.

The particles which are heavier and therefore more loaded with radioactive nuclides fall back more rapidly. They create highly radioactive deposits with occasionally well-defined borders and located near the site of the explosion. As for the finest particles, they are not to be overlooked either. Indeed, they remain dangerous because they can easily be ingested or inhaled while remaining invisible to the naked eye. Falling more slowly, they settle on larger and broader areas, usually causing less contamination yet remaining very real.

During a nuclear explosion, the people who are quick to take refuge in a shelter and stay there for a few days are unlikely to enter into contact with these radioactive particles, especially if the shelter is equipped with an air filter. Indeed, most of these particles will have settled during the first forty-eight hours. Over time, dust deposits loaded with radionuclides will be carried along by water and their effects will fade, except in places where, due to an accumulation effect (sludges, mudbanks, ponds, etc.), their concentration will grow and remain dangerous.

Back in 1986, during the explosion of reactor 4 in Chernobyl, the large number of casualties was mainly due to the slow evacuation of the areas located near the accident site. For several days, the inhabitants inhaled an air laden with radioactive dust that reached a level 400 times higher than the 'acceptable' limit.

Fortunately, the danger of radiation decreases over time: this is due to the half-life of radioactive elements, as previously indicated. Since this decrease is exponential, the danger quickly fades. During the Cold War, civil defence programmes estimated that during a nuclear war, it was relatively safe for people to come out of their shelters after two weeks. Nevertheless, it is more than recommended to avoid places located downwind of any targets that have been affected by nuclear warheads, including large cities or strategic sites (missile silos,[1] etc.).

The study of the effects of the atomic bombing of Hiroshima and Nagasaki shows that only a small part of the population that absorbed radiation doses died or suffered long-term deleterious effects. A 1980 simulation[2] conducted by the British government showed that a Soviet attack against the United Kingdom using 130 nuclear warheads would kill 53% of the population and cause a 12% rate of severe injuries. These forecasts are not far removed from those made by famous German physicist Albert Einstein, who stated: 'I do not believe that civilisation will be wiped out in a war fought with the atomic bomb, but perhaps two-thirds of the people of the Earth will be killed'. Let us rejoice, then!

Myth: Radiation penetrates everything and there is no way to escape its effects.

The facts: The degree to which the impact of gamma rays is attenuated depends on the type of screen used. The denser and thicker the constituent material, the more effective it is. However, any substance is useful to some extent: several meters of earth can stop most gamma rays. Concerning the neutrons emitted during a nuclear explosion, the attenuation phenomenon is identical, with the difference that it is preferable to use light materials such as water.

Myth: In an explosion like the ones in Hiroshima and Nagasaki, all buildings are destroyed and everyone is killed by the shock wave, radiation or fire.

The facts: In Nagasaki, many people survived without sustaining injuries because they were hiding in shelters built to protect the

---

1 TN: A silo is an underground chamber where a guided missile is kept, ready to be fired.

2 AN: https://en.wikipedia.org/wiki/Square_Leg.

population against conventional air strikes. Some of these survivors were in shelters within 500 metres of *ground zero*.[3] Indeed, many family shelters that had simply been built into the earth were not destroyed, while the buildings located on the surface were swept away or consumed by the flames.

Myth: A nuclear war would set cities on fire and create 'firestorms' that would consume all the oxygen and kill all inhabitants, including those hiding in shelters.

The facts: On a clear day and in dry weather conditions, the heat generated by a nuclear explosion is such that it instantly ignites all flammable materials (rugs, [fitted] carpets, tapestries, paper, dry wood, newspapers, dry grass, bushes, curtains, etc.), including the skin of human beings and certain animals. These effects are very powerful near the explosion site and diminish with the distance. Under cloudy or humid weather conditions, they are slightly reduced, as air humidity will absorb some of the heat. Although burns can still be severe beyond this 'ignition' zone (second-degree burns, specifically), the thermal flash is no longer sufficient to ignite materials. Furthermore, nuclear tests carried out on buildings specifically built for the purpose of studying the effects of explosions have shown that the blast which follows the thermal effect often extinguishes the fires. Nevertheless, should it still ravage cities, the experience of the worst incendiary bombings carried out by the American air force against the cities of Dresden (13 February 1945) and Tokyo (9 May 1945) has taught us that even in the case of real firestorms, it takes more than that to completely drain all oxygen from the air. Certainly, the heat of the fires can become such that the air is rendered unbreathable and the steel of metallic structures melts, but there is always some oxygen in the atmosphere.

Unlike old cities with lots of flammable materials such as Paris, London or Strasbourg, more modern cities (New York, Dubai, etc.),

3 AN: The point on the earth's surface directly below an exploding nuclear bomb.

Nuclear warfare and cinema: screenshots from
Stanley Kubrick's *Dr. Strangelove Gold: How I Learned to Stop Worrying and Love the Bomb.* (1964)

which are all made of concrete, steel and glass, do not comprise enough combustible materials for this phenomenon to occur. In addition, if the density is low, as is the case in areas of low-rise housing, such 'firestorms' are unlikely to take place.

Lastly, let us note that soil offers excellent thermal insulation and, therefore, any underground shelter represents a decent source of protection against heatwaves.

Myth: Food and water will become radioactive and the population will die even if enough food and water is available to the survivors.

The facts: It is important to understand the difference between irradiation and contamination. If foods were traversed (irradiated) by gamma rays, they would probably remain edible. (Moreover, most of the vegetables and fresh products that we consume are irradiated in order to be sterilised.) In the case of contamination, if radionuclide-laden particles do not come into contact with food, there is no risk. Water and food that are kept dust-free (in cans, sealed bags, etc.) will not be contaminated. The simple paring of fruit and vegetables and filtering of water reduces the risk considerably. Be careful, though: plants grown in contaminated soils are not suitable for consumption. For more information, see the chapter on food preparation below.

Myth: Most newborns will be malformed and the genetic heritage of future generations so affected that the human population will decline and disappear.

The facts: According to medical studies conducted on the effects of radiation on the populations of Hiroshima and Nagasaki,[4] abnormal births among these populations were not significantly higher than among those not exposed to radiation. In contrast, there was a higher number of abnormally formed foetuses causing miscarriage, as well as a slight increase in the cancer rate among survivors of the explosion. Radiation thus has a real and tragic effect, but does not change the human population to the point of making it disappear.

Myth: Because of the destruction of the atmosphere's ozone layer, which would allow the passage of too much solar ultraviolet radiation, humans and animals would be left blind.

---

4 AN: *A Thirty-Year Study of the Survivors of Hiroshima and Nagasaki*, National Academy of Sciences, 1977.

The Facts: Nuclear explosions release a large amount of nitrogen oxide into the stratosphere, which is a gas that destroys the ozone layer. However, according to the calculations of scientists[5] involved in atmospheric nuclear tests carried out in both the USA and the USSR between 1952 and 1962, these amounts are too low to result in any significant destruction and to allow the amount of ultraviolet radiation that passes through the layer ozone to noticeably increase.

In addition to this, since modern nuclear weapons have much less power than those of the 1950s and 1960s, the effects on the ozone layer would be smaller or even inexistent. On the contrary, the dust projected into the atmosphere could, by means of 'smog'-type reactions (particles + humidity), generate additional protection against ultraviolet rays in the troposphere[6] itself. The survivors of a nuclear war would at least have the consolation of being able to expose themselves to the sun without danger!

Myth: A 'nuclear winter' making any and all survival impossible would follow a nuclear war. The smoke of the fires generated by the war and the dust projected into the atmosphere would engulf the earth in a dark layer that would remain impervious to the sun's rays. The result would be darkness and biting cold. Harvests would freeze, even in the tropics. Famine would rage and humanity would not survive.

The Facts: The 'nuclear winter' theory that frightened so many people was born in 1982 in the mind of German chemist Paul Crutzenet and then publicised by pacifistic scientists such as the very famous Carl

5 AN: M. H. Foley and M. A. Ruderman, *Stratospheric NO from Past Nuclear Explosions*, in *Journal of Geophysics*, Res. 78, 4441–4450; Julius S. Chang et Donald J. Wuebbles, *Atmospheric Nuclear Tests of the 1950s and 1960s: A possible test of Ozone depletion Theories*, in *JournalofGeophysicalResearch*84, 1979.

6 AN: The lowest part of the atmosphere, stretching from the earth's surface to an altitude ranging between six and ten kilometres.

Sagan,[7] as well as by the entire Soviet scientific community, advocating the dismantling of all nuclear weapons in the process… In 1986, an extensive scientific study[8] estimated that in the event of a global nuclear war, climate effects would not be as severe as the above-mentioned theory predicts. As part of the scenario of such a war, the temperature of the temperate zones of the Northern Hemisphere would drop in summer by a total of twenty degrees compared to the average, but only for a few days. Furthermore, Soviet studies that preceded the nuclear winter theory showed that the amount of dust propelled into the atmosphere would be much lower than what would be required to actually darken the sky.

Apart from immediate destruction and radiation, the real danger of a nuclear war lies in the disruption of infrastructures and supply systems, especially urban ones, in relation to all that concerns water, food, medicines, spare parts, petrol, etc. If the global economy were paralysed, supermarkets would be left without supplies, pharmacies and hospitals without medicine, and people without water or electricity. Sewer systems would collapse and the consequences for the millions—or billions, globally speaking—of city dwellers would be catastrophic.

Myth: A war between nuclear powers would be suicidal, and none of them would thus ever take the risk of triggering one.

The Facts: It would be great if this myth were true, and history has proven that from 1945 until now, the deterrent effect of such weapons has meant that no major attack has ever taken place against a country possessing a nuclear weapon.

---

7 AN: Especially in an article dated 23 December 1983 and published in *Science: Nuclear Winter, Global Consequences of Multiple Nuclear Explosions*, R. P. Turco, O. B. Toon, T. P. Ackerman, J. B. Pollack, et C. Sagan.

8 AN: Starley L. Thompson et Stephen H. Schneider, *Nuclear Winter Reappraised*, Foreign Affairs, 1986.

Alas… Declassified documents dating back to the 1950s have shown that the US military was willing to take the risk of launching a surprise attack against the USSR, Eastern Europe and China.[9] In the 1990s and 2000s, the United States[10] and Israel[11] developed atomic 'micro-bombs' intended for use against so-called 'hard' targets such as underground bunkers and other command and communication objectives, or nuclear research centres buried dozens of metres in the earth. These weapons were almost used against Iraq and the Iranian nuclear programme. Political powers have, for the time being, been reluctant to use them, probably because of their fear of a negative impact on public opinion.

Myth: It is possible to find out which country has manufactured the radioactive materials of an atomic bomb by simply studying the radiation resulting from the explosion.

The facts: No, it's not. Although one can conceivably determine the nature of the bomb (fission or fusion) during the explosion itself, it is impossible to identify the source of the radioactive materials that constitute it. On the other hand, if a sample of the radioelements of an unexploded bomb could be recovered, it would be possible to identify the producing country and, in most cases, even identify the production site. This can be particularly useful in cases involving unexploded 'home-made' nuclear bombs of terrorist origin, as scientists can trace it back to the original producer.

Myth: The use of depleted uranium ammunition in conflicts will render entire regions unfit for life.

---

9 AN: http://nsarchive.gwu.edu/nukevault/ebb538-Cold-War-Nuclear-Target-List-Declassified-First-Ever.

10 AN: http://www.informationclearinghouse.info/article17206.htm.

11 AN: http://www.reuters.com/article/us-nuclear-iran-israel-nukes-idUSTRE-62P1LH20100326.

The facts: The answer is a complex one, because uranium is a natural element in the environment and may be present either in trace amounts or in larger quantities, depending on the region in question. Also, depleted uranium is 40% less radioactive than natural uranium. Nevertheless, there are potential health risks from the point of view of radiation toxicity and chemo-toxicity. When depleted uranium ammunition or armour-piercing shells are used, some of the radioisotope spreads in the form of dust and contaminates the area. And that is where the real danger lies. The main means of exposure threatening nearby populations are therefore ingestion and inhalation, both of which would lead to internal contamination.

Fortunately, this material is far less dangerous than the isotopes involved in atomic bomb-type nuclear incidents or nuclear power plant accidents. What is more, traces of depleted uranium are scattered over time by precipitations, the wind, and so on in the form of dust. In areas where a significant number of such shells have been fired, however, the risk of water contamination cannot be ruled out.

All things considered, the irradiation involved is negligible and only internal contamination could turn out to be a problem. Exploring the destroyed body of a tank struck by a depleted uranium shell, for instance, does not generate any health problems. As a precaution, though, it is best to wear a dust mask, or even a pair of disposable gloves. In contrast, chronic damage to the liver and lungs may occur in the case of *regularly* exposed populations. This is especially true for children who play inside armoured vehicle wrecks, stir up the dust and put their hands in their mouths…

BIOLOGICAL RISKS

Epidemiologists believe that smallpox has killed an estimated one billion people during the last one hundred years of activity on Earth.

— American writer RICHARD PRESTON, *The Demon in the Freezer*, 2003.

[We are] determined, for the sake of all mankind, to exclude completely the possibility of bacteriological (biological) agents and toxins being used as weapons. [We are] convinced that such use would be repugnant to the conscience of mankind and that no effort should be spared to minimise this risk.

— Preamble to the Convention on the Prohibition of the Development, Production and Stockpiling of Bacteriological (Biological) and Toxin Weapons and on Their Destruction, (drafted and opened for signature in)[1] 1972.

---

1 TN: I have added the words in brackets because the content only entered into force on 26 March 1975.

1. Living Organisms

Nothing in biology makes any sense
except in the light of evolution.

— Theodosius Dobjansky, Ukrainian biologist (1900–1975)

General Facts

Having emerged almost 3.8 billion years ago in the form of tiny and primitive entities, never since has life ceased to evolve, adapt and multiply, thus offering the spectacle of breath-taking beauty that we can contemplate today.

Titanic cataclysms have sometimes brutally impeded its progress, but life has always regained the upper hand. Over time, it has filled gaps and reconquered lost domains, just as water invades every corner of the space in which it finds itself.

At present, the diversity characterising the living world far exceeds what ordinary people could ever imagine. Few of us thus have any idea of the considerable number of species that have so far been described. Here are some examples:

- 5,500 among mammals
- 10,000 among birds
- 31,000 among fish

- 100,000 among spiders
- 280,000 among plants
- 1,000,000 among insects

Yet the list compiled to date is only a small portion of the living world that surrounds us. Many species have not even been listed, much less described in detail. This is especially true of insects and micro-organisms. 99% of all bacteria, for instance, are still unknown to us.

Now, at the beginning of the twenty-first century, some scientists[1] estimate that there are anywhere between ten and twelve million species, of which only 1.9 million have been described. If we consider that between 16,000 and 18,000 are identified and added to the index each year, it will take more than half a millennium at this rate to identify almost all of them. And the current assessments would first have to be entirely correct and not minimalistic. According to zoologist Terry Erwin, for example, **there are supposedly more than fifty million animal species on Earth, including no less than thirty million insects.**

And that's not all! All the species listed thus far are only one-thousandth of what our planet has borne to date.

Although the number of species is impressive, the number of 'individual lifeforms' is even more bewildering. Did you know, for instance, that our gut contains up to fifty million bacteria per cm^2? And that there are nearly 1,000 billion micro-organisms present on our skin, i.e. about 140 times more than the world's human population?

These numbers can make you dizzy, yet our Earth is potentially at the dawn of a new mass extinction!

Indeed, the species extinction rate is said to be 50 to 560 times[2] greater than what nature has thus far experienced (excluding in

---

1 AN: A.D. Chapman, *Numbers of Living Species in Australia and the World*, 2nd edition, 2009.

2 AN: Source — CNRS. Other studies are even more alarming.

times of cataclysm). According to the International Union for the Conservation of Nature (IUCN), 22,413 of the 76,199 species that have been studied are threatened with extinction. As for the 2005 report presented by the *Millennium Ecosystem Assessment* (a group of international scientists), it mentions the disappearance of 12% of all birds, 25% of all mammals and 32% of all amphibians by 2100.

Although these numbers are subject to discussion (and could either be higher or lower), the general consensus is that the sixth extinction of species is already in motion…

However, it does not seem to be the case that a meteorite has struck the Earth in recent times or that a super-volcano has erupted and spewed ashes for weeks. No, the real reason is much less spectacular and much more pernicious than that: what is responsible for the extinction in progress is man himself (and his activities). The consequences of our actions are numerous and have a negative impact on the environment, and even directly on living beings. The examples include:

- Alteration or destruction of various species' habitats;
- Over-exploitation of resources;
- Water, soil and air pollution;
- The introduction of invasive species that decimate indigenous populations;
- Global warming.

Classification

Throughout his existence, and depending on his knowledge, man has attempted to inventory all that surrounded him, particularly living beings. Several types of classifications have followed one another over time and each region of the world has witnessed the emergence of popular names referring to animals and plants. The words 'chauve-souris' and 'pomme de terre'[3] are thus known to most French speakers. Such designations sometimes result from similarities, but often remain imprecise and vary from country to country. For example, the cork oak has a distinct name depending on the different locations:

France: chêne-liège	Portugal: sobreiro
Catalonia: suro	Italy: sughera
Provence: suvé	Germany: korkeiche
Gascony: corcier	England: cork oak
Spain: alcornoque, alzinasurera	North Africa: fernan

In addition, these common names (= vernacular names) generally include a group of species rather than only one. The term 'ants', for example, represents the entire family and actually includes some 12,000 species!

Among scientists, the need for a common language and the necessity to meet well-defined criteria has led to the development of classifications.

Traditional/Classic Classification

Traditional classification derives from that of Linnaeus,[4] which was initiated in 1730 and published between 1735 and 1758. It remained in use and evolved for many years before slowly falling into disuse from 1950 onwards, when phylogenetic classification was born. Although less used by scientists, it continues to be resorted to by the general public due to its simplicity and the visible logic of its distribution system (everyone can recognise a fish, a bird, etc.). This traditional model began with two kingdoms (that of animals and plants) and

3 TN: 'Bat' and 'potato' respectively.

4 AN: A Swedish naturalist, Linnaeus (1707–1778) came up with the notion of a binomial nomenclature (i.e. the use of two names to characterise a species).

then incorporated a third one once microbes (including protists) were discovered. Woese's version (1977) comprises six kingdoms:[5]

1. **Bacteria** (= unicellular prokaryotes[6])
   Bacteria consist of a single cell without a nucleus. They are so small that a microscope is necessary to distinguish them—hence the term *microbes*[7] sometimes used to designate them. They generally measure anywhere between 0.1 and 50 micrometres (one millionth of a metre). They may be curved or elongated (bacilli), spherical (cocci), or spiral (spirilla). Some could be pathogenic (i.e. causing disease), like the plague bacillus, but the great majority is not. They are found virtually everywhere in the environment and are, in some cases, used by humans (e.g. lactic acid bacteria, which play a key role in the production of yogurts).

2. **Archaea** (= unicellular prokaryotes)
   Bacteria-like archaea have long been considered to be special bacteria living in very extreme environments (extremophiles). This kingdom has existed for billions of years and has colonised the bowels of the Earth, the abyss (i.e. the very deep areas of the ocean), acid lakes, polar ice, etc.

3. **Protists** (= unicellular eukaryotes[8])
   These are organisms consisting of a single cell with its own nucleus (unlike bacteria and archaea). The latter can be of different kinds

5 AN: Other versions have up to eight kingdoms or added levels of hierarchy (such as domains).

6 AN: Prokaryotes are living entities whose cellular structure is devoid of any sort of nucleus.

7 AN: Microbes do not really constitute a distinct group in the scientific sense of the term. All they are is a heterogeneous set of microorganisms (requiring a microscope to become visible) that includes bacteria, yeasts and some parasites.

8 AN: Eukaryotes are defined as (unicellular or multicellular) living entities whose cells contain both a nucleus and mitochondria (tiny 'powerhouses' that supply cells with energy).

and fulfils many functions necessary for life. It may, for example, comprise complex organelles such as flagella (filaments that enable the mobility of the cell). Protists have conquered all kinds of environments and adapted to them. Some are, furthermore, parasites that could be very dangerous to humans (e.g. amoebas). They are usually microscopic: the smallest ones are the size of large bacteria, but most measure between 10 and 50 thousandths of a millimetre. There are, however, some species that are visible to the naked eye, with some even reaching the size of a coin.

4. **Fungi** (= unicellular or multicellular eukaryotes with poorly differentiated tissues and without chlorophyll)
   Fungi are a group characterised by large variety; they range from microscopic species (such as yeast) to organisms weighing several kilos. They are distinguished from plants by their inability to perform photosynthesis: they cannot sustain themselves autonomously and must consume either decaying or living organic matter.

5. **Plants** (= eukaryotes that carry out photosynthesis)
   What makes plants (including algae) different from animals is their ability to carry out photosynthesis; they thus do not need to ingest other living things. This kingdom includes both plants without roots or stems (thallophytes) and others with roots and stems (cormophytes) and a set of leafy sprigs.

6. **Animals** (= eukaryotes comprising neither chlorophyll nor a cellulosic cell wall)
   This kingdom brings together all animals, meaning organisms composed of several cells and needing to ingest food (by eating other living beings).

They represent a highly varied group, some of whose members have completely disappeared and are only found in a fossilised state (e.g. dinosaurs).

Within this traditional classification, the living world can be hierarchised using major categories stemming from one another. Seven levels are thus defined, in accordance with the following principle:

WORLD OF THE LIVING

Kingdom → Phylum → Class → Order → Family → Genus → Species

The following chart displays three examples:

	MAN	CORK OAK	THE PLAGUE BACILLUS
KINGDOM	Animal	Plant	Bacteria
PHYLUM/ DIVISION	Vertebrates	Magnoliophyta	Proteobacteria
CLASS	Mammals	Magnoliopsida	Gammaprotobacteria
ORDER	Primates	Fagales	Enterobacteriales
FAMILY	Hominids	Fagaceae	Enterobacteriaceae
GENUS	Homo	Quercus	Yersinia
SPECIES	H. sapiens	Q. suber	Y. pestis

Phylogenetic Classification

Proposed back in 1950 by Hennig[9] and regularly revised, this type of classification has gradually been replacing the traditional one. It makes use of the progress of genetics and refers to a model based on evolution, i.e. on the fact that all species stem from a common ancestor and have differentiated themselves over time. Henceforth, three domains occupy the top of the hierarchy, replacing the six kingdoms of traditional classification:

9 AN: Emil Hans Willi Hennig (20 April 1913–5 November 1976) was a German biologist.

1. **Eubacteria** — these are single-celled and non-nucleated organisms which have a cell wall consisting of peptidoglycan.

2. **Archaea** — these are also single-celled and non-nucleated organisms. They have a cell wall comprised of specific lipids. They are very similar to eubacteria and, from an ecological point of view, are often (but not always) extremophiles.

3. **Eukaryotes** — they can be either unicellular or multicellular. Their genetic material is enclosed in a nucleus delimited by a membrane; they are endowed with mitochondria and their DNA is distributed in chromosomes. They are characterised by reproduction of the sexual type.

Whereas the traditional classification has seven levels of hierarchy that lead to a given species, the phylogenetic model uses a branching system, with the number of branches depending on the evolutionary 'distance' from a common ancestor. To reach man from the domain of the eukaryotes, it is thus necessary to go through twenty-six branches.

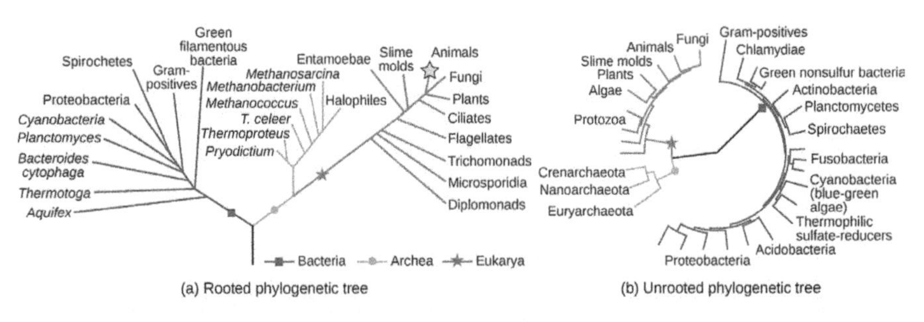

Figure 17. The Phylogenetic Tree of Life[10]

Some groups, such as mammals, have been preserved in the phylogenetic model. Others, on the other hand, have disappeared from it, including reptiles and fish (as seen below). Similarly, some species have been classified into a different group, due to kinship ties which

10 TN: Source: https://courses.lumenlearning.com/wmopen-biology2/chapter/phylogenies-and-the-history-of-life/.

are, in fact, closer than they appear. Elephants, for instance, which, as everyone knows, are four-legged terrestrial animals with tusks, now belong to the same group as manatees, which are aquatic animals with flippers.

TWO EXAMPLES: REPTILES AND FISH

The case of reptiles:

According to traditional classification, reptiles are terrestrial vertebrates of variable temperature, usually covered with scales. They include crocodiles, snakes, lizards, turtles etc., comprising a total of over 7,000 species. Phylogenetic classification, however, shows that crocodiles are closer—in terms of evolution—to birds than to lizards or snakes… A black caiman thus has more in common with a pigeon than with a Nile monitor![11] The very concept of reptiles is therefore obsolete.

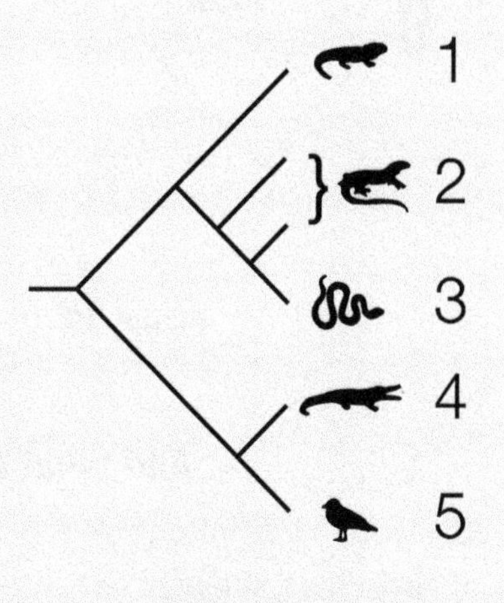

Figure 18.
The Classification of Reptiles

The case of fish:

According to traditional classification, fish are aquatic vertebrate animals with gills, fins and bodies mostly covered with scales. Yet just as the previous example, this concept is no longer applicable. Indeed, some fish such as coelacanths[12] are, from an evolutionary perspective, closer to humans than to any trout…

---

11 TN: The Nile monitor ('Varanus niloticus') is a large-sized member of the monitor lizard family (Varanidae) encountered over much of Africa.

12 TN: With some of its species considered critically endangered, coelacanths are an order of fish that was long regarded by scientists as a 'living fossil', because it was believed to be the only survivor of a *taxon* (a group of one or more than

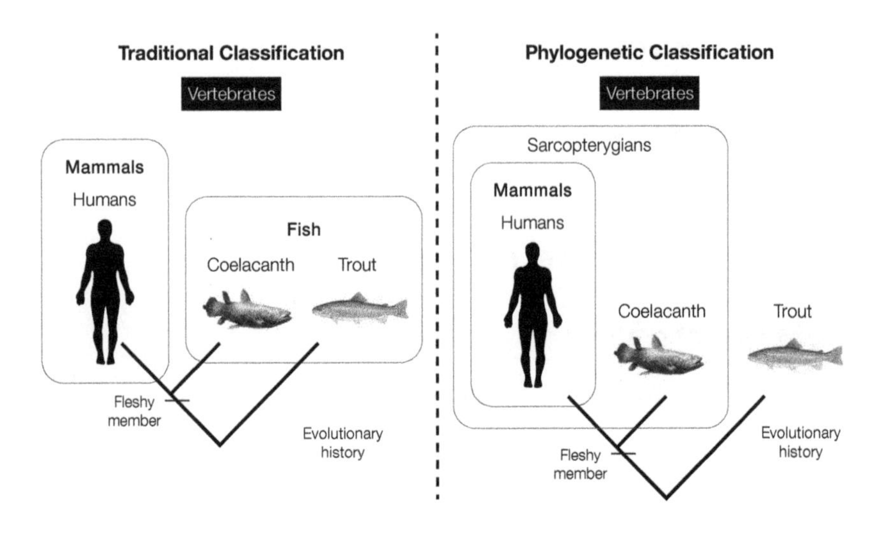

Figure 19. The Classification of Fish

And What About Viruses?

The word 'virus' means 'poison' in Latin. The debate about the nature of viruses is a complex one and much ink has been spilled over it. To date, they are generally regarded rather as mere associations of molecules than as living organisms. They do not fit, therefore, into either of the previously detailed models, but have their own classifications based on characteristics such as the type of nucleic acid involved, their morphology, their replication methods, etc.

---

one population of organisms constituting one single unit) that had otherwise been encountered solely in fossilised form. Recent studies, however, have highlighted a greater diversity in coelacanth body shapes than anyone had ever thought possible and brought to light a certain similarity to humans on an evolutionary level.

2. The Biological Threat

Bioterrorism is a real threat to our country. It's a threat to every nation that loves freedom. Terrorist groups seek biological weapons; we know some rogue states already have them... It's important that we confront these real threats to our country and prepare for future emergencies.

— US President George W. Bush, 28 April 2004

As we have just seen, Life has managed to colonise almost all of our planet. Still, not everything is as peaceful as it seems. It is all, in fact, characterised by a precarious sort of balance, by a perpetual struggle in which escape and predation constantly clash and the competition for food and the most favourable position identify the victor. Evolution is ultimately a constant and endless adaptation process whose sole goal is to ensure one's survival!

Sometimes, this harmony reaches a breaking point when certain species try to gain the upper hand over others. Life, however, has more than one trick up its sleeve and usually manages to maintain self-regulation. If a given species experiences excessive growth, lack of food forces it to reduce its numbers, or deadly epidemics spread and decimate its population. Nevertheless, it does seem to be the case that ever since the dawn of man, these different regulatory mechanisms have reached their limits.

And yet, history has shown that our species is not immune to these plagues. Admittedly, the dangers themselves have changed since prehistoric times. Large predators are no longer a threat to most of our society, and our enemies are henceforth invisible to the naked eye. Although tiny, they are no less terrifying, and can sometimes cause the deaths of millions of people. The following (non-exhaustive) list offers a few striking examples:

The Black Death of the fourteenth century (1346–1350) killed more than one out of three people across both France (seven million out of seventeen) and all of Europe. It emerged during the exodus of ships fleeing the besieged city of Caffa (Crimea), where the epidemic was wreaking havoc. It then spread to surrounding countries through rodents (rats, gerbils) whose fleas carried the bacteria responsible for the disease.

Smallpox once decimated the Inca people in Latin America and, at a later point, the American Indian population. Whether through voluntary actions (the donation of infected blankets) or mere natural transmission, the disease, which had until then been unknown in that part of the world, spread quickly across the continent. It contributed greatly to the collapse of these civilisations, annihilating up to 90% of some tribes and causing far more deaths than any conflict with 'white' invaders.

The Spanish flu of 1918–1920 resulted in the deaths of at least thirty million people,[1] thus surpassing even the Great War, despite the fact that the latter was responsible for nearly nineteen million fatal casualties (both military and civilian).

1 AN: Twenty to forty million according to the *Institut Pasteur*. Other sources put the figure at seventy-five million.

In addition to these 'one-off' pandemics,[2] there are many endemic or recurrent diseases annually responsible for the deaths of hundreds of thousands of people:

Malaria (paludism) is widespread in the tropics. In 2013 alone, the World Health Organisation (WHO) estimated the number of malaria cases at 198 million and the total number of men, women and children who had died from it at 582,000.

AIDS (Acquired Immunodeficiency Syndrome) is the consequence of a collapse of one's immune defences under the impact of a virus. Contaminated individuals become very vulnerable to diseases (even commonplace ones) and can actually die from these infections. Since the emergence of AIDS in 1981, thirty-nine million people[3] have died as a result of this syndrome.

Typhus, dengue, cholera, and yellow fever are further examples to add to the list. This list of diseases is, in fact, a very long one, and the number of deaths for which they are responsible is impossible to keep track of. Since the dawn of time, they have wreaked havoc upon indigenous peoples, and the past arrival of expeditionary forces, prisoners and settlers has only amplified the phenomenon — the number of victims is thus simply colossal. Furthermore, additional threats have surfaced since the end of the twentieth century. Called 'emerging diseases', they are linked to the evolution (i.e. the mutation or recombination) of biological agents such as certain viruses or parasites. On a regular basis, they make the headlines, and everyone has already heard their names at least once: SARS (Severe Acute Respiratory Syndrome), bird flu (including H5N1), haemorrhagic fevers (such as Ebola), etc. To this day, authorities fear the outbreak of a total pandemic caused particularly by one of these viruses and having acquired unusually strong

2 AN: A pandemic is an epidemic that spreads across continents.

3 AN: Source — UNAIDS (a United Nations specialised agency that combats AIDS).

virulence or contagion factors, as this would thus trigger a significant fatality rate among humans. Despite the rapid progress of medicine in recent decades, science remains powerless against this kind of plague.

It would seem, then, that life continues to adapt, as if it were trying to restore the balance upset by human beings. Microorganisms thus develop certain resistances and mutate. In reality, what they are doing is constantly adjusting to their environment and targets. Given that we are barely familiar with 1% of all viruses and bacteria, and that they continue to evolve for the sole purpose of survival, it seems likely that, sooner or later, a major pandemic will strike at our civilisation, causing tens of millions of deaths and inconceivable chaos.

On numerous occasions already, we have narrowly evaded disaster. And yet this does not seem to have frightened our society, which continues to play the part of the sorcerer's apprentice. Some scientists seek to create or modify microorganisms in order to make them far deadlier. Some (US and Russian) laboratories have, for instance, kept the smallpox virus, which would otherwise have completely disappeared by now. Others still are striving to awaken microbes such as the bacterium responsible for the Black Death, or even viruses dating back to another age and preserved in ice for thousands of years… It should be added that these biological agents — whether organisms or toxins — are not always developed to act directly and specifically upon humans. Alternative strategies may, for example, involve the targeting of livestock or crops to trigger economic crises or famines…

As if all this were not enough, terrorist groups are increasingly interested in biological weapons, whether in the form of poisons (toxins) or deadly diseases that can easily be transmitted. In this last instance (the dissemination of agents capable of multiplying), the developers would lose all control over the actual evolution of the epidemic. This kind of threat remains plausible only if terrorists are driven by the will to deeply harm society or provoke a combined state of panic and general destabilisation, ***and*** if they have no fear of sacrificing themselves, or do not understand the true scope of their own actions.

Types of Biological Agents

As we have just seen, the biological threat is quite varied indeed, both with regard to the type of agents involved (viruses, bacteria, toxins[4]) and to their different possible origins (natural causes, laboratory accidents, terrorist acts, military operations,[5] etc.) It is interesting to note that, in actual fact, these biological agents consist of two large groups:

1. Biological agents that are alive or capable of proliferating (viruses, bacteria, etc.)
2. Toxins that behave more like poisons and cannot replicate.

The main agents can be divided in accordance with the following model:

VIRUSES	Infectious microorganisms with a well-defined structure that exclusively parasitise living cells; they have a specific type of nucleic acid and reproduce using their own genetic material.
BACTERIA	Unicellular micro-organisms representing an autonomous kingdom that is neither animal nor plant. They come in various forms and can live as saprophytes (soil, water, living organisms) or as human, animal, and plant parasites.
FUNGI	Biological organisms devoid of chlorophyll; they are either immobile or with limited mobility. They consist of nucleated cells and therefore belong to the eukaryotic domain.
PARASITES[6]	Protists or animals capable of parasitising other organisms.
TOXINS	Toxic substances produced by a living organism (a bacterium, poisonous mushroom, arthropod, amphibian, venomous snake, etc.).

4 AN: A toxin is a poison secreted by living organisms.

5 AN: 162 countries have thus far ratified or acceded to the Convention on the Prohibition of Biological Weapons. Its effectiveness is, however, limited because it has not provided any control device following the United States' refusal to impose such measures.

6 AN: As previously seen, parasites do not represent a group used in modern classifications (and actually include members of various groups). There are several definitions, with the current one excluding plants and fungi, which have their own category in this chart. The actual term 'parasites' has the advantage of being very explicit in defining beings that live at the expense of other organisms.

Bio-Agent Classifications

Several attempts have been made to classify human and animal pathogens according to the risks they present to health:

EUROPEAN COMMUNITY

In 2000, the European Parliament and the Council published Directive 2000/54/EC on the protection of workers from risks related to exposure to biological agents at work: http://www.eurogip.fr/images/documents/3526/Directive%20200054EC.pdf.

This Directive refers to biological agents instead of micro-organisms, and specifies four risk groups based on the infection risk level.

It has been implemented by the Members States and several lists of human pathogens have been produced by the different countries.

GROUP 1

- Group 1 biological agents are unlikely to cause disease in humans. This applies to the vast majority of microorganisms.

GROUP 2

- Group 2 biological agents can cause human *disease.*
- They pose a *danger* to workers.
- Their spread through the community is unlikely.
- There is *prophylaxis* or *effective treatment.*

This group is composed of:

  - 130 species of bacteria (including those responsible for cholera, legionellosis, tetanus, botulism, gangrene, etc.).
  - 65 virus species (including those responsible for herpes, measles, mumps, hepatitis A, etc.).
  - 60 species of parasites (including tapeworms and those responsible for leishmaniasis, toxoplasmosis, etc.).

- 21 species of fungi (including those responsible for candidiasis, aspergillosis, etc.).

GROUP 3

- Group 3 biological agents can lead to *serious illness* in humans.
- They pose a *serious danger* to workers.
- Their spread in the *community* is possible.
- *Prophylaxis* or *effective treatment* is available.

This group consists of:

- 28 species of bacteria (including those responsible for leprosy, tuberculosis, typhus, dysentery, plague, etc.).
- 57 virus species (including those responsible for dengue, hepatitis B, C and G, yellow fever, rabies, chikungunya, etc.).
- 10 species of parasites (including those responsible for malaria or paludism, the sleeping sickness transmitted by the tsetse fly, etc.).
- 6 species of fungi (including the source of histoplasmosis, which is characterised by symptoms similar to those of tuberculosis).

GROUP 4

- Group 4 biological agents cause *serious diseases* in humans.
- They pose a *serious danger* to workers.
- *High risk of spread* throughout the community.
- There is *no prophylaxis or effective treatment.*

It consists ONLY of viruses (11 species), including those responsible for haemorrhagic fevers (Ebola, Marburg, Lassa), smallpox, etc.

1. *Arenaviridae*: Lassa virus
2. *Arenaviridae*: Guanarito virus
3. *Arenaviridae*: Junin virus
4. *Arenaviridae*: Sabia virus

5. *Arenaviridae*: Machupo virus
6. Equine morbillivirus
7. *Nairovirus*: haemorrhagic virus of Crimean-Congo fever
8. *Filoviridae*: Ebola virus
9. *Filoviridae*: Marburg virus
10. *Poxyviridae*: smallpox virus (*variola major* and *minor*). An effective vaccine is available.
11. *Poxyviridae*: cottonpox virus. An effective vaccine is available.

UNITED KINGDOM

The Advisory Committee on Dangerous Pathogens (ACDP) provides an Approved List of Biological Agents (2004): http://www.hse.gov.uk/pubns/misc208.pdf.

In the Approved List, each biological agent listed is assigned to a hazard group according to its human infection risk level (only agents in Groups 2, 3 and 4 are listed).

Hazard Group 1 agents are considered not to pose a risk to human health.

Hazard Group 4 agents present the greatest risk.

The Hazard group definition is similar to the European Directive 2000/54/EC.

UNITED STATES

The first attempt to establish a classification for pathogenic micro-organisms in the USA was made by the United States Public Health Service. In 1969 and 1974, they published a description of four classes of etiological agents (bacteria, fungi and viruses) ranking them from those that pose no or minimal hazard (class 1) to those responsible for very serious diseases (class 4).

The current classification provided by the National Institutes of Health (NIH) uses Risk Groups (Rgs) instead of Risk Classes, and agents are classified according to their relative pathogenicity for healthy adult humans: http://osp.od.nih.gov/sites/default/files/NIH_Guidelines.html.

- **Risk Group 1 (RG1):** Agents that are not associated with disease in healthy adult humans.
- **Risk Group 2 (RG2):** Agents that are associated with human diseases that are rarely serious and for which preventive or therapeutic interventions are often available.
- **Risk Group 3 (RG3):** Agents that are associated with serious or lethal human disease for which preventive or therapeutic interventions may be available (high individual danger, but low community risk).
- **Risk Group 4 (RG4):** Agents that are likely to cause serious or lethal human diseases for which preventive or therapeutic interventions are not usually available (high individual and community risk).

WORLD HEALTH ORGANISATION (WHO)

The WHO defines four risk groups based on the relative hazard of infective micro-organisms to laboratory workers, the community, livestock and the environment.

- **Risk Group I:** low individual and community risk.
  A microorganism that is unlikely to cause any human disease or animal disease of veterinary significance.
- **Risk Group II:** moderate individual risk, limited community risk.
  A pathogen that can cause human or animal disease but is unlikely to be a serious hazard to laboratory workers, the community, livestock, or the environment. Laboratory exposure may cause serious infection, but effective treatment and preventive measures are available and the risk of any spread is limited.

- **Risk Group III:** high individual risk, low community risk.
  A pathogen that usually causes serious human disease, but does not ordinarily spread from one infected individual to another.

- **Risk Group IV:** high individual and community risk.
  A pathogen that usually results in serious human or animal disease and may be readily transmitted from one individual to another, directly or indirectly.

Although no list of infective agents has been provided, the WHO recommends that each country draw up its own classification (by risk group) of the agents encountered in that country, based on the following factors:

1. The agent's pathogenicity.

2. The agent's mode of transmission and host range. These may be influenced by existing levels of immunity, the density and movement of the host population, the presence of appropriate vectors, and the standards of environmental hygiene.

3. The availability of effective preventive measures. Such measures may include: prophylaxis by means of vaccination or antisera; sanitary measures such as food and water hygiene; the management of animal reservoirs or arthropod vectors; the movement of people or animals; and the importation of infected animals or animal products.

4. The availability of effective treatment, including passive immunisation and post-exposure vaccination, antibiotics, and chemotherapeutic agents, while taking into consideration the possibility of resistant strain emergence.

Military or Terrorism-Related Biological Agents

The biological risk agents that may be used for military or terrorist purposes in order to cause damage to humans, domestic animals or crops must have different 'qualities'. The ability to cause serious illness is an essential yet insufficient aspect if the ultimate goal is to use them as a biological weapon. Eleven criteria were thus defined by Theodore Rosebury in 1949 to characterise a 'suitable' military biological agent.

The authors of the book '*NRBC-E, savoir pour agir*'[7] have made the following transcript:

1. A low infection threshold;

2. Strong virulence with the ability to cause an acute, life-threatening or incapacitating disease;

3. Stable pathogenicity during manufacture, storage and transportation;

4. A short incubation period;

5. Low contagiousness to avoid any boomerang effect against the attacker;

6. The absence of a vaccine or natural immunity against this agent in the target population, coupled with the existence of possible protection against this agent for the troops of the aggressor country (vaccine, antibiotic therapy, protective clothing, masks);

7. Resistance to traditionally used antibiotics;

8. The ability to withstand aerosolisation;

7 TN: *CBRN-E, from Knowledge to Action*, published by Xavier Montauban editions (2nd edition).

9. The ability to withstand the environment during dispersal (heat, light, desiccation, explosion), long enough to infect the target population;

10. Ease of transportation and the ability to survive during battlefield storage and dispersal;

11. Large-scale production at low cost.

A terrorist organisation would focus on acquiring an agent that would meet the following conditions:

1. Effectiveness in terms of morbidity or mortality;

2. Ease of obtention and production;

3. Absence/difficulty of treatment;

4. Ease of transmission and contagion.

For these extremist groups, the 'benefits' of such an agent would be:

1. An often easy and inexpensive production;

2. A delay in the onset of symptoms that reduces the chances of the source being detected;

3. A major psychological impact that generates reactions of terror, panic and isolation, sometimes even the disruption of the functioning of statal structures.

The *Center for Disease Control and Prevention* (CDC) is an American (US) agency that enjoys worldwide recognition for its resources, work and expertise in the field of biology, as well as in the management of crises associated with this risk. This entity surfaces regularly in disaster films centred around events of such nature. The CDC classifies the agents that can be used in bioterrorist actions into three categories (A, B and C).

Classification of Biological Agents that Could Be Used for Aggressive Purposes

CATEGORY A	CATEGORY B	CATEGORY C
▪ Easily disseminated or transmissible from one person to another. ▪ High mortality rate, high impact on public health. ▪ Likely to generate panic and social disorganisation. ▪ Requires special preparatory measures by the competent authorities.	▪ Characterised by a moderate dissemination capacity. ▪ Moderate morbidity rate, low mortality rate. ▪ Requires authorities to improve their diagnostic capabilities and increase their monitoring.	Emerging agents used due to their: ▪ Availability; ▪ Ease of production and dissemination; ▪ High morbidity and mortality rate potential. Major impact on health.
Bacteria responsible for: ▪ Anthrax (=> *Bacillus anthracis*) ▪ Plague (=> *Yersinia pestis*) ▪ Tularaemia (=> *Francisella tularensis*)	**Bacteria responsible for:** ▪ Brucellosis (=> *Brucella spp.*) ▪ Cholera (=> *Vibrio cholerae*) ▪ Dysentery (=> *Shigella dysenteriae*) ▪ Food poisoning (=> *Escherichia coli* O 157: H7) ▪ Q fever (=> *Coxiella burnettii*) ▪ Melioidosis (=> *Burkholderia Pseudomallei*) ▪ Glanders (=> *Burkholderia mallei*) ▪ Ornithosis (=> *Chlamydia psittaci*) ▪ Typhoid/Salmonellosis (=> *Salmonella spp.*) ▪ Typhus (=> *Rickettsia prowazecki*)	**Bacteria responsible for:** ▪ Multidrug-resistant tuberculosis (=> multidrug-resistant *Mycobacterium tuberculosis*)

CATEGORY A	CATEGORY B	CATEGORY C
Viruses responsible for: ▪ Smallpox (=> *Variola major*) ▪ Haemorrhagic fevers: => Filoviridae (Marburg, Ebola) => Arenaviridae (Lassa, Junin, Machupo)	**Viruses responsible for:** ▪ Equine encephalitis (=> Alphavirus)	**Viruses responsible for:** ▪ Animal haemorrhagic fevers that may accidentally affect humans (=> e.g. hantavirus) ▪ Neurological and respiratory diseases that can accidentally affect humans (=> e.g. Nipah virus) ▪ Chikungunya, etc. ▪ Coronavirus for acute respiratory syndrome (MERS and SARS)
Parasites/Fungi -	**Parasites/Fungi** ▪ Cryptosporidiosis (=> *Cryptosporidium parvum*)	**Parasites/Fungi** -
Toxins ▪ Botulinum toxin (=> which stems from the bacterium *Clostridium botulinum*)	**Toxins** ▪ Ricin (=> from the castor oil plant) ▪ Enterotoxin B (=> from the bacterium *Staphylococcus aureus*) ▪ Epsilon toxin (=> from the bacterium *Clostridium perfringens*)	**Toxins** -

3. Examples of Biological Agents

Could it not be contrived to send the small pox among those disaffected tribes of Indians?

— Jeffery Amherst, 1st Baron of Montreal (1717–1797)

Biological agents pose a serious threat because they are relatively accessible and can spread both quickly and without initial detection. They can infect/poison their targets in many ways:

- Through inhalation — the agent is used in an aerosol spray, as a fine powder, etc. It can be spread using a spraying method (aerosol sprays, garden sprayers, agricultural aircraft, and so on) or through common mechanical action. This was the case of the envelopes containing anthrax bacilli spores, the mere opening of which allowed a quantity sufficient to inoculate anthrax[1] to become airborne.

- Through ingestion — the most obvious way is the absorption of contaminated food and/or water. However, the fact of putting

1 AN: The term 'anthrax' in English is equivalent to the French '*maladie du charbon*' and is used here as such. In French, however, 'anthrax' has a completely different meaning, since it involves boils or abscesses due to a bacterium called *staphylococcus*.

one's 'dirty' hands into one's mouth usually produces the very same result.

- Through dermal means—the skin provides effective protection against many biological agents, but some of them (such as the tularaemia agent) can still get through. In addition, it is only on rare occasions that the skin is entirely healthy, due to the usual presence of micro-cuts, pimples, etc. Injections also fall into this category, regardless of whether they are intentional (as in the case of ricin in the Bulgarian umbrella[2] incident) or whether they involve vectors such as insects (e.g. fleas whose bites transmit plague bacilli). This last method was tested in real life by unit 731 (a biological research entity belonging to the Japanese army)[3] during World War II. It involved the scattering, by means of planes, of bags of both wheat and rice containing fleas that bore the *Yersinia pestis* bacterium. These 'tests' resulted in a series of plague epidemics and the deaths of thousands of innocent people.

It is interesting to note that unlike chemical toxins that can act quickly, biological agents require a more or less significant latency period.

The biological agents that we shall describe here can be of two kinds: 'living' ones (i.e. bacteria and viruses) and 'non-living' ones (toxins).

Bacteria

As already mentioned, only 1% of these microorganisms are currently known to us. Bacteria are present in all types of environments and some species have the potential to cause diseases in plants, animals and humans. In order to do this, they have special abilities such as:

2 AN: In 1978, a former Soviet agent, Giorgi Markov, was injected with a tiny ricin-bearing ball through the tip of an umbrella. He died four days later.

3 AN: https://en.wikipedia.org/wiki/Unit_731.

- adhering to host cells;
- proliferating very quickly;
- producing toxins;
- constructing a protective envelope (a capsule);
- evading the body's defences;
- hiding in the cells of their host.

Depending on these characteristics, as well as on the number of bacteria involved and the infected person's resistance, a certain amount of time is required for the first symptoms to surface (this is known as the incubation period). The following chart offers some examples:

AGENT	DISEASE	INFECTIVE DOSE	INCUBATION PERIOD	HUMAN-TO-HUMAN TRANSMISSION
Bacillus anthracis	Anthrax	8,000–50,000 spores	2–6 days	No
Yersinia pestis	Plague	100–500 bacteria	2–3 days	+ + (pulmonary)
Francisella tularensis	Tularaemia	10–50 bacteria	1–10 days	No
Brucella	Brucellosis	10–50 bacteria	5–60 days	No
Coxiella burnetii	Q fever	1–10 bacteria	10–40 days	No

If one wished to eradicate (i.e. totally eliminate) these diseases, one would have to fulfill three conditions:

1. To ensure that no one is infected anymore;
2. To guarantee the absence of any and all disease reservoirs (animals, fleas, etc.);

3. To make sure that all the bacteria responsible for it in the external environment are destroyed.

Unfortunately, the last two points are problematic. Indeed, it does seem difficult to kill all living beings as potential reservoirs, and most bacteria are likely to survive for a long time in the environment: so some will always remain out there, in the wild, ready to infect an unlucky host. This resistance in an external environment varies in accordance with the species, and some agents make use of highly effective means.

The best-known example is probably the anthrax bacillus (*Bacillus anthracis*), which has the ability to produce a spore. This is a small 'seed' containing all the material needed to recreate a new 'individual'. If the bacterium ends up being destroyed, this ball of concentrate survives further and, as soon as the external environment becomes favourable, reconstitutes a complete organism. In the case of the *Bacillus anthracis*, this form of subsistence has amazing properties because it can persist in the soil for entire decades, is invulnerable to both cold and fire and is impervious to ultraviolet light and most disinfectants…

Throughout their history, humans have thus been at the mercy of many bacterial epidemics. Nevertheless, enormous progress was made in the struggle against this type of biological agent in 1928, when Sir Alexander Fleming discovered penicillin, the first in a long series of antibiotics. These substances act as antibacterials and can destroy or block the growth of bacteria. The event was a true medical revolution, allowing medicine to create other families of such compounds over the years. As a result, leprosy, plague and most bacterial diseases suffered a sharp decline.

Nature, however, had not said its last word! And thus, in 1940, four years after the introduction of sulfonamides (the first family of marketed antibiotics), some bacteria showed signs of resistance. In the years that followed, whenever new compounds were used, the phenomenon repeated. This is all the more worrying because it is becoming increasingly difficult to find families of innovative molecules. The extensive use of such active substances by humans, in fact, helps to create resistant strains. The latter consequently survive the treatments and continue to multiply; and all their descendants retain this insensitivity to a given type of antibiotic. When this phenomenon is repeated, it leads to the emergence of multi-resistant strains.

The following files are based mainly on WHO (World Health Organization) publications and on the book entitled 'CBRN-E Risks, from Knowledge to Action' (Xavier Montauban Editions).

THE PLAGUE (*YERSINIA PESTIS*) FACT SHEET

Class 3 Agent (Possible Bioterroristic Use)

Yersinia pestis is the most contagious and most virulent bacterial disease that humans have ever known and has been responsible for several pandemics in the past. Today, there are still some residual hotbeds, mainly in Africa (Madagascar, the Democratic Republic of the Congo, etc.) and in Asia (India, the countries of the former Union, etc.). Their distribution is superimposed on that of the rodents that serve as reservoirs.

Note: *Yersinia pestis* is unable to withstand a harsh environment. The bacterium remains infectious for approximately one hour in an aerosol and is sensitive to ultraviolet light, high temperatures, as well as most disinfectants. Nevertheless, it can survive for several months in moist soil and away from light.

Factors that make the bacterium pathogenic

Y. pestis has the ability to multiply rapidly and produce a protective cap. In addition, it synthesises substances that will limit the host's immune response and allow it to evade phagocytosis (i.e. digestion by certain types of white blood cells).

In the absence of treatment, the disease is fatal in 30 to 60% of cases.

Transmission

The different modes of transmission are the following: flea bites[4] in the case of bubonic plague, and the respiratory tract for pulmonary plague (as the subject expels large amounts of bacteria when coughing).

Symptoms

Infected persons usually develop flu-like symptoms after a three-to-seven-day incubation period. These include a sudden onset of fever, headache, body aches, weakness, vomiting and nausea. Six to eight hours after the emergence of the first symptoms, very painful lumps (or buboes)[5] begin to form under the skin. They grow larger and become darker as infection takes over the tissues. The swelling of the lymph nodes (which are responsible for lymph filtration) is very painful, especially in the neck, groin and armpits, so much so that there have been instances of comatose patients writhing in pain as they agonised.

There are three kinds of plague:

The bubonic plague (or Black Death of medieval Europe) is the most common one. The bacterium is injected into the body as a result of a flea bite. It then migrates to the nearest lymph nodes and proliferates, causing a sensitive abscess (= bubo). The disease can evolve into a septicaemic or pulmonary form.

Septicaemic plague occurs when the bacteria reproduce and enter the blood without first invading the ganglia (i.e. without forming buboes), overwhelming the host's immune defences.

---

4 AN: One single infected flea bite can inject up to 24,000 plague bacilli into the victim's blood or lymphatic system.

5 TN: Technically speaking, buboes are actually swollen and inflamed *lymph nodes* typically present in one's groin or armpit.

It is accompanied by necrosis of the extremities and, in the absence of treatment, leads to death. It can result from flea bites or even direct contact with infected materials.

The pulmonary form is the most contagious one and generates a very high mortality rate. The first symptoms (fever, shivers, etc.) appear after an incubation period of one to four days, and death occurs two to four days later. This type of plague stems from either airborne transmission or the evolution of a bubo. A sneeze or coughing fit is sometimes enough to allow it to pass from one person to another, and bacteria invade the bronchial system. Pneumonia sets in when the fluid invades the lungs, depriving the most distant organs of oxygen. As for the incubation period of pneumonic plague, it is brief and rarely takes more than a few days. The symptoms appear very suddenly and are often difficult to distinguish from those of other infectious diseases. A wrong or late diagnosis could thus be fatal.

Treatment

The administration of effective antibiotics should be done as soon as possible, ideally within twenty-four hours of the initial symptoms (one should thus consult a doctor). The main active substances used for this purpose are: streptomycin (reference treatment), gentamicin, chloramphenicol and tetracyclines. The treatment lasts ten days, except in the case of chloramphenicol, when one has been afflicted with plague meningitis—then it's twenty-one days.

During a plague epidemic, flea-eliminating disinfection and deratting are necessary in order to break the cycle of transmission.

Note: Although there is a vaccine, it is not available to the general public. It is reserved for high-risk groups (professionally exposed persons, some members of the military). In addition, it only offers low protection and for a limited amount of time.

THE ANTHRAX (*BACILLUS ANTHRACIS*) FACT SHEET

Class 3 Agent (Possible Bioterroristic Use)

Although this disease mainly impacts herbivores, it can also infect other mammals (including humans), some birds, etc. There are no reported cases of human-to-human transmission.

Factors that render the bacterium pathogenic

Bacillus anthracis can synthesise a protective capsule and produces two toxins composed of three proteins and causing oedema and necrosis. In addition, it is able to sporulate (i.e. create a spore) to withstand harsh environments.

Transmission

Transmission occurs through spores, especially as a result of skin contact with contaminated material or infected animals. In the event of a malicious action, the digestive and respiratory tracts are prime targets. The effect of such a pathogen would be a most fearsome one, because the mortality rate is as much as 20% when ingested. Airborne, it is even more dangerous than that, with a mortality rate close to 90%. It would, in fact, suffice to use fifty kilos and disseminate it all (by plane) using a relatively uniform spray to impact the entirety of an urban centre inhabited by 500,000 people.

Symptoms

These vary depending on the entry route:

Cutaneous anthrax: After an incubation period of a few days, a lesion is formed at the point of inoculation. In 80% of all cases, it evolves into a painless pimple covered with a black crust before healing without complications. In the remaining

20%, the wound spreads, results in oedema and a feverish condition which, if left untreated, could lead to death.

Gastrointestinal anthrax: This form of anthrax stems from the ingestion of spore-containing food. The diagnosis is difficult, because the symptoms are similar to conventional acute gastroenteritis. After an incubation period of one to seven days, nausea and vomiting appear and are quickly followed by bloody diarrhoea. Effective and timely antibiotic treatment is imperative to avoid life-threatening sepsis.

Inhalation anthrax: This is the most dangerous form of all and was used in the 2001 anthrax spore envelope attacks in the United States. After an incubation period of four to six days, the victim experiences both fever and fatigue and is afflicted with a dry cough. Two or three days later, their state of health suddenly deteriorates, with the emergence of respiratory distress and possible vomiting. In 95% of all cases, the victim has but a few hours left to live...

Any case involving a *Bacillus anthracis* infection should be reported to the authorities.

Treatment

The administration of effective antibiotics is imperative as soon as possible (one should thus consult a doctor). This process can be initiated intravenously and then relayed orally without undue delay. The main treatments are based on a mixture of different active ingredients: Ciprofloxacin, Ofloxacin, Doxycycline or Amoxicycline for a period of sixty days.

Note: Although there is a vaccine, it is reserved for high-risk groups (professionally exposed persons, some members of the military).

TULARAEMIA (*FRANCISELLA TULARENSIS*) FACT SHEET

Class 3 Agent (Possible Bioterroristic Use)

Tularaemia is a disease that is mainly encountered in wooded areas of the northern hemisphere. It affects lagomorphs (hares, etc.) and rodents (squirrels, voles, etc.), which act as its main reservoir. It can be accidentally transmitted to other vertebrate or invertebrate species.

Note: What is characteristic of *F. tularensis* is its ability to survive a few days in corpses and several months in contaminated soil or water.

Factors that render the bacterium pathogenic

F. tularensis is a small oval bacterium that can produce a protective capsule and has the ability to rapidly proliferate. Its life is intracellular (meaning that it lives in other cells). In addition, it has the ability to resist while in macrophages (which are white blood cells that are supposed to digest it) and can even reproduce within them.

Transmission

The transmission of tularaemia occurs mainly through the skin:

- through direct contact with infected animals (*F. tularensis* can pass through healthy skin, but the presence of skin lesions promotes its penetration);
- through contact with contaminated urine or excrement (i.e. indirectly);
- through infected insect bites (including those of ticks);

Airborne transmission (the inhalation of contaminated particles or aerosols) is also possible.

The presence of very few bacteria (ranging from ten to fifty) is enough to infect a subject.

Symptoms

After an average incubation period of three to five days, the host usually exhibits flu-like symptoms — fever, fatigue, headache, nausea and occasional vomiting.

There are different types and possible developments of tularaemia, the main ones being:

Ulceroglandular tularaemia: This is the most common one, encompassing 80% of all cases. It is characterised by a local lesion (ulceration) at the bacterium's point of penetration. *F. tularensis* can then proliferate and migrate towards the lymph nodes. The development is mostly favourable, but the convalescence period extends over several weeks.

Septicaemic tularaemia: This is rare and only includes 2% of all cases. It can occur in the event of massive contamination or in an immunocompromised host.

Pulmonary tularaemia: it results either from airborne contamination or from the evolution of one of the two previously mentioned forms. It is accompanied by a dry cough, and in the absence of treatment, its mortality rate reaches 30%.

Treatment

Treatment[6] is based on the administration of antibiotics, usually orally and for a period of fourteen days:

Ciprofloxacin (500 mg twice daily), Ofloxacin (400 mg twice daily), or Levofloxacin (500 mg once daily).

In cases of severe pulmonary tularaemia, Gentamicin may be combined with the 5 mg/kg daily dose of Ciprofloxacin.

Note: There is a vaccine, but its effectiveness is limited both in relation to the protection it provides and with regard to the duration.

---

6 AN: *CBRN-E Risks, from Knowledge to Action* (Montauban, 2nd edition).

Viruses

Unlike bacteria, viruses cannot proliferate of their own accord. It is thus necessary for them to enter a cell, access the existing 'machinery' and divert it to serve their interests. The latter then begins to produce significant amounts of virus, which will subsequently be released and, in turn, enter a new cell, thus continuing the infection cycle.

Another major difference compared to bacteria is that antibiotics have no effect on viruses. Indeed, in most cases, there is no genuinely effective treatment to date for people that are already infected.

Some of these biological agents' potential to be transmitted from person to person makes them particularly dangerous, since they may end up becoming the source of real pandemics. This is all the more the case because the virus dose required to infect a subject is generally very low.

AGENT	DISEASE	INFECTIVE VIRUS DOSE	INCUBATION PERIOD	HUMAN-TO-HUMAN TRANSMISSION
Orthopoxvirus	Smallpox	10–100	7–17 days	+ + +
Haemorrhagic fever virus	Haemorrhagic fevers	1–10	4–21 days	+ + +

It is interesting to note that there are viruses capable of infecting just about any living being: animals, plants, bacteria, etc.

Man is, of course, no exception to the rule, since these agents can be responsible for many human diseases, some of which are serious or even fatal. Some examples are presented in the chart below.

Viruses consist of genetic material (DNA or RNA) protected by a shell composed of proteins (= capsid). In some cases, they also have an additional envelope. Their varied and more or less complex form can be used as a classification system. However, the most frequently used model is the Baltimore classification, which is based on the virus' genomic type:

			EXAMPLES OF VIRUS FAMILIES	EXAMPLES OF DISEASE
DNA VIRUSES	Group I	Double-stranded DNA viruses	*Poxviridae*	=> Smallpox
	Group II	Single-stranded DNA viruses	*Parvoviridae*	=> Few diseases affecting humans
RNA VIRUSES	Group III	Double-stranded RNA viruses	*Reoviridae*	=> Infantile gastroenteritis/ Colorado tick fever
	Group IV	Single-stranded positive sense RNA viruses	*Coronaviradea* *Flaviviridae* *Togaviridae*	=> SARS => yellow fever, dengue, Zika, etc. => rubella, chikungunya
	Group V	Single-stranded negative sense RNA viruses	*Rhabdoviridae* *Orthomyxoviridae* *Filoviridae*	=> rabies => flu => Haemorrhagic fevers (Ebola, Marburg)
	Group VI	Positive-sense single-stranded RNA viruses that replicate through a DNA intermediate	*Retroviridae*	=> AIDS
	Group VII	Double-stranded DNA Pararetroviruses[7]	*Hepadnaviridae*	=> hepatitis B

A fundamental point is that viruses can mutate quite easily, or even recombine (merge) with other viral strains. The consequences are highly variable:

7 TN: Despite their name, these are *not* considered DNA viruses (Baltimore classification class I), but rather reverse transcribing viruses, since their replication occurs through an RNA intermediate.

- Generally speaking, if the biological agent changes, the protection acquired by our body (through vaccines, previous contact) decreases.[8] Indeed, our natural defences thus find it more difficult to find and destroy the invader.
- Although some modifications do not bring any 'added value' to the virus, others will increase its pathogenicity (better penetration into host cells, a more intense replication, etc.), thus making it more dangerous.
- A recombination generates more significant random changes than simple mutations ever could. Under such conditions, a virus can acquire the ability to infect other species, as well as additional characteristics that can increase its pathogenicity or its ability to be transmitted from person to person.

Emerging Viral Diseases

The emergence of a viral disease is the appearance of a disease that has never existed either in the world or in a given region. Such diseases can be caused by completely new viruses (such as AIDS) or mutated/recombined ones (SARS, for instance), or even be transmitted from animals to human beings (e.g. bird flu). Although this phenomenon has existed since the dawn of time, many recent examples can be given:

- **1967:** Marburg haemorrhagic fever.
- **1969:** Lassa haemorrhagic fever.
- **1976:** Ebola haemorrhagic fever.
- **1976:** Hantaan haemorrhagic fever.

8 AN: This is one of the reasons why one is told to regularly undergo vaccination against the flu virus.

- **1981:** AIDS (caused by HIV).
- **1997 (resurgence in 2003–2004):** Avian influenza due to the A-H5N1 viruses.
- **2003:** Severe Acute Respiratory Syndrome-associated CoronaVirus (SARS-Cov).
- **2009:** Avian influenza due to A-H1N1 viruses.
- **2013:** Middle East Respiratory Syndrome CoronaVirus (MERS-CoV).
- **2014:** Ebola haemorrhagic fever.
- **SARS (Severe Acute Respiratory Syndrome):**
  SARS is an infection of the lungs caused by a coronavirus. After an incubation period of two to ten days, high fever appears, often accompanied by chills and headaches and paired with coughing and breathing difficulties. Detected in China back in 2002, SARS turned into an epidemic in 2003, infecting up to 8,000 people and killing 800 of them. Airborne or direct contact-based human-to-human transmission is possible. Bats are said to have been the original reservoir of the virus, which may have been transmitted to humans by the animals themselves or through a small wild carnivorous animal usually consumed in southern China.
- **Bird flu:**
  This disease is a viral flu that affects birds. Often asymptomatic in wildlife, it can lead to real hecatombs among farm animals such as chickens. The cause is the spread of an Influenza A virus. In general, it does not infect humans, but some strains (such as H5N1)[9] do have this ability. They present a significant mortality

9 AN: H and N refer to two antigens, namely hemagglutinin and neuraminidase. The attached number relates to the subtype and allows one to define the

rate approaching 60% (around 400 deaths since the 2003 resurgence). The symptoms are flu-like (fever, body aches, a sore throat, etc.) and can evolve into respiratory distress and superinfections. Cases of human-to-human transmission are very rare. Virtually all of those impacted by the disease had contact with birds. What scientists fear most is that the virus could mutate or exchange genes with other influenza strains, such as those that normally infect humans. This could indeed render human-to-human transmission possible and pave the way for a terrible pandemic.

- **MERS (Middle East Respiratory Syndrome):**
  Discovered in 2012 in Saudi Arabia, this virus is a coronavirus (just like SARS). It led to an epidemic in the Gulf countries in 2013–2014, with sporadic cases surfacing mainly in Europe and the Maghreb. The mortality rate is estimated at 30%. Since the first cases emerged in 2012, it has been responsible for the deaths of more than 350 people.[10] Just like a common flu, the MERS virus can involve human-to-human transmission (saliva, touching, and so on), but less easily. Camels are thought to have been its natural reservoir. The symptoms are flu-like (fever, fatigue, body aches, coughing, etc.) and can develop into respiratory distress (acute pneumonia) and sometimes trigger kidney complications.

  Note: Just like the H5N1 virus, the MERS virus has as animal reservoir. Cases of human contamination surface on a regular basis, even when there is no epidemic in progress.

- **The Zika virus:**
  There are still many unknown aspects surrounding the Zika virus, often abbreviated to ZIKV in English. It is an arbovirus belonging

---

pathogenicity and affinity with the various types. 149 combinations are possible—16 haemagglutinins and 9 neuraminidases. The H1Nx, H2Nx, H3Nx, or HxN1, HxN2 or H7N9 viruses are likely to infect humans.

10 AN: A figure declared by the World Health Organisation in February 2015.

to the *Flaviviridae* family and the *Flavivirus* genus. This virus takes its name from the Zika Forest in Uganda —where it was originally identified in 1947—and is transmitted through the bites of mosquitoes of the *Aedes* genus, just like chikungunya, dengue or the West Nile virus. Zika was diagnosed in tropical areas of Africa and Asia for entire decades, but did not seem to cause symptoms more severe than mild fevers and rashes, and 80% of all infected individuals did not develop any symptoms at all.

In 2007, a Zika outbreak on Yap Island in the Pacific infected 75% of the local population, but the symptoms remained mild. In November 2015, a health alert was issued in Brazil concerning a sudden outbreak of nearly 4,000 cases of infant microcephaly,[11] while a total of only 150 cases had been recorded during the previous five years. This development, however, has coincided with the presence of the Zika virus in Brazil since May 2015, and some of the women whose children are afflicted with microcephaly have tested positive for it. To date, the correlation remains uncertain. Through research, one is now trying to determine whether this virus has always caused such microcephalic diseases (without them being reported) or has mutated and developed a new pathology, or whether there are other factors involved.

There is currently no vaccine against Zika, and the CDC advises pregnant women to avoid staying in countries where the presence of the virus has been confirmed. As for those living there, the advice is to avoid mosquito bites by using anti-mosquito products, wearing long sleeves/leg-covering garments, and removing stagnant water, which serves as a breeding ground for mosquitoes.

11 AN: Microcephaly is an abnormality in the growth of the skull and can result in very significant mental retardation. The causes are mainly nutritional (including alcohol consumption during pregnancy) and genetic.

SMALLPOX FACT SHEET

Class 4 Biological Risk Agent

The smallpox virus is a double-stranded DNA (Deoxyribonucleic Acid) virus belonging to the *Poxiviradea* family and the *Orthopoxvirus* genus.

The disease caused by this agent is one of the worst calamities of biological origin that humanity has ever experienced. The number of fatal casualties has been in the tens of millions and all continents have been impacted by it.

However, following an effective vaccination[12] campaign, smallpox was eradicated and, in 1980, declared as such by the WHO. Indeed, since man is the only reservoir and the virus can only resist for a few weeks in an external environment, it became obvious that since there were no patients left, the disease itself had completely disappeared. The worst biological plague known to man had just vanished forever!

That's if we ignore the fact that two laboratories (in the United States and Russia) have kept some strains of the virus under the pretext of studying it...

Note: The virus can survive for weeks in a humid environment, but is sensitive to light and most antiseptics and disinfectants.

Factors that render the virus pathogenic

Once inside a cell, the agent responsible for smallpox will replicate and produce proteins that will weaken the host's immune response (defence) in accordance with three main strategies:

- The inhibition of infected cell recognition.

12 AN: The term 'vaccine' comes from the Latin word *vacca* (cow) and was coined by British doctor Edward Jenner, who discovered in 1796 that inoculating cowpox (whose effects are mild from the human perspective) enabled the body to fight against human smallpox (*Variola major*).

- The inhibition of one's innate antiviral response.
- The creation of decoys interfering with the immune system.

Transmission

Smallpox is an eruptive[13] and contagious disease. Transmission occurs through contact with a patient or contaminated object (clothes, etc.) or through the inhalation of droplets emitted by an infected person.

However, an individual is only infectious from the onset of fever (i.e. two to three days before the rash) until scabs are left on the skin.

Symptoms

The incubation phase lasts from ten to fourteen days and is followed by the sudden emergence of influenza symptoms—high fever, chills, headaches, lower back pain and possible vomiting.

A rash appears three days later on both the face and the torso, before spreading across the entire body in less than thirty-six hours. Unlike with chicken pox, the skin lesions all appear at the same stage and are also found on one's palms and soles. If the evolution is favourable, the pustules turn into scabs that fall off after about three weeks. However, in 35–50% of cases, the disease leads to the death of unvaccinated individuals (and this percentage is significantly higher in populations that have never been in contact with smallpox, as was the case in the Americas).

Treatment

As with most viral diseases, the treatment is symptomatic. Nevertheless, some antivirals such as ribavirin and cidofovir may be helpful to a limited extent.

Note: Vaccination is effective and offers long-term protection. However, since the latter has some rare, albeit major, unwanted

13 AN: Meaning one that results in vesicles reminiscent of chickenpox.

side effects affecting about ten per million people and the disease has disappeared off the face of the Earth, it has not been carried out in France since 1980. (Anyone born after this date is thus unprotected.) In addition to this, vaccination is not recommended in the case of immunocompromised individuals.

Since 2003, a number of specialised service staff members have been vaccinated in order to establish a prevention and control mechanism should even the smallest case of smallpox re-emerge in the world. For more information, see the Smallpox Plan developed in 2006 by the French State (and currently under revision).[14]

EBOLA VIRUS (HAEMORRHAGIC FEVER) FACT SHEET

Class 4 Biological Risk Agent

Ebola is a single-stranded RNA (ribonucleic acid) virus belonging to the *Filoviradea* family.

Having surfaced for the first time in 1976 in Sudan and Zaire (now the Democratic Republic of the Congo), it is responsible for haemorrhagic fevers and was named after the river that flows through Zaire.

To date, five virus species have been identified:

- **Bundibugyo** (BDBV),
- **Zaire** (EBOV),
- **Reston** (RESTV),
- **Sudan** (SUDV),
- **Taï Forest** (TAFV).

14 AN: http://www.sante.gouv.fr/IMG/pdf/plan_variole_2006-2.pdf.

Genetic material can fluctuate to a fairly great extent from one strain to another (up to 40% difference). Also, the various strains have a different pathogenicity with regard to humans and primates and generate a mortality rate ranging from 25% to 90% (with Reston having a low case fatality rate and Zaire being the most dangerous one).

Ebola has already caused several outbreaks on the African continent, including the one in:

- **1976** (150 deaths in Sudan and 280 in Zaire)
- **1995** (254 deaths in Zaire)
- **2000** (224 deaths in Uganda)
- **2003** (128 deaths in the Congo)
- **2007** (187 deaths in the Democratic Republic of Congo/former Zaire)

The last epidemic, which broke out in 2014, was due to the Zaire strain. It has a case fatality rate of about 50% and has caused an unprecedented disaster in West Africa, leading to the deaths of more than 10,000 people: 4,408 in Liberia, 3,831 in Sierra Leone and 2,333 in Guinea.[15]

The natural reservoir of the Ebola virus seems to be a certain frugivorous bat, which is considered to be its healthy carrier.

Factors that render the virus pathogenic

The exact factors of virulence and pathogenicity characterising Ebola are still poorly understood. Whatever the case, the result is a partial suppression of the host's immune response (defence) and severe inflammation, as well as vascular damage and coagulation disorders.

15 AN: Figures given by the WHO on 5th April, 2015.

Ebola also diverts the 'machinery' of infected cells so that they produce a glycoprotein which promotes the virus' penetration into other targets.

Transmission

Man is most often infected through contact with bats/infected monkeys or by ingesting 'bush meat'. When subjects display signs of the disease, however, human-to-human transmission is made easier through bodily fluids such as blood, saliva, sweat and even tears. The careless handling of cadavers (i.e. the use of adapted equipment and procedures) or mere contact with a relative exhibiting signs of a haemorrhagic fever is enough to become infected.

Symptoms

The incubation period usually lasts between five and twelve days, but can sometimes reach a maximum of twenty-one days. As is the case with many viral diseases, the first symptoms are influenza-like (a sudden emergence of fever, headaches, body aches, etc.). Then comes vomiting, diarrhoea, kidney and liver disorders, and internal and external haemorrhaging.

Treatment

Treatment is limited to intensive symptomatic care, focusing only on fever, pain and dehydration. Indeed, although recent tests have yielded encouraging results, there is currently no treatment or vaccine approved by the health authorities.

To reduce the spread of the epidemic, some anti-infectious precautions must be followed:

- wash your hands regularly;
- isolate the sick;
- avoid contact with infected liquids or contaminated objects;

- use gloves, masks, glasses, boiler suits, and boots to protect yourself.

Non-Living Biological Agents: Toxins

As we have already seen, toxins are 'poisons' produced by a large variety of living organisms such as bacteria (which are responsible for botulism, tetanus, etc.), plants (including ricin, etc.), fungi (rye ergot fungus, etc.), and so on. Some of them are considered the most dangerous substances in the world. Using adequate means, for instance, it would be enough to obtain:

- about 0.5 grammes of botulinum toxin to eliminate the Greater Paris population (6.7 million inhabitants);
- about 0.6 grammes of botulinum toxin to kill the population of Switzerland (more than 8 million inhabitants);
- less than 5 grammes (the equivalent of a sugar cube) to eliminate the entire French population (66 million).

These figures are, of course, purely theoretical, since the optimal dose should be distributed to reach each person under optimal conditions and be administered intravenously. Nevertheless, the numbers do give an idea of this 'poison's' destructive power. The following chart shows some examples of toxins:

TOXIN	EFFECTS	LETHAL* OR INCAPACITATING** DOSE	INCUBATION PERIOD
Botulinum toxin (mainly produced by the bacterium *Clostridium botulinum*)	Nausea, vomiting, abdominal cramps, difficulty swallowing, constipation, weakness and general paralysis beginning at the head and neck (the cranial nerves) and moving down to reach the muscles of the respiratory system. The pupils are dilated or fixed with a lowering of the upper eyelids.	LD_{50} = 3 ng/kg (airborne route) = 1 ng/kg (when injected) = 1 µg/kg (when ingested)	8–48 hours
Ricin (produced by the *Ricinuscommunis* or *Ricinussanguineus* plant)	Inhalation causes respiratory afflictions such as necrosis and pulmonary oedema; ingestion causes necrosis and gastrointestinal bleeding followed by various general effects and death.	LD_{50} = 1 mg/kg (when ingested) = 5 à 10 µg/kg (when inhaled or injected)	3–6 hours
Staphylococcal enterotoxin B (produced by the bacterium *Staphylococcus aureus*)	Through inhalation: sudden onset of fever and chills, headache, myalgia, non-productive cough leading to acute pneumonia and pleuritic pain. Through ingestion: severe abdominal cramps, vomiting, diarrhoea. Severe exposure will lead to septic shock and death.	LD_{50} = 27 µg/kg (when inhaled) ID_{50} = 0,4 µg/kg	3–12 hours

TOXIN	EFFECTS	LETHAL* OR INCAPACITATING** DOSE	INCUBATION PERIOD
Saxitoxin (produced by microscopic algae)	Sudden onset of numbness in the face and extremities, slurred speech, shortness of breath, dizziness.	LD_{50} (in the case of mice) = 10 µg/kg (when ingested) = 2 µg/kg (when inhaled)	10 minutes to a few hours

* LD50 = lethal dose in 50% of exposed subjects

** ID50 = incapacitating dose in 50% of exposed subjects

Unlike living biological agents, toxins are unable to replicate. In addition, there is no human-to-human transmission, in the sense that a toxin-afflicted person will not 'pass on' his 'illness' to someone else. On the other hand, if people touch contaminated materials, objects or people (the carriers of the substance), they may end up being 'poisoned' through various mechanisms: by breathing in airborne toxins, by ingesting them in their food or simply by putting their fingers in their mouths. From a 'terrorist' point of view, toxins offer several obvious advantages:

- They are easy to produce;
- They are easy to store;
- They have devastating effects.

BOTULINUM TOXIN (PRODUCED BY *CLOSTRIDIUM BOTULINUM*) FACT SHEET

This toxin is the cause of botulism, which is a disorder of the nervous system, and it is therefore a neurotoxin that can affect both humans and animals.

It is produced by certain bacteria of the *clostridium* genus, i.e. by agents that are common in the environment and are able to sporulate (thus surviving a long time in this form). It is the *C.*

botulinum species, however, that is the main cause of most of the poisoning.

Botulism is a reportable disease in France.

There are different types of toxins: A, B, C, D, E, F and G. The first three (A, B, C) are the most common in Western Europe.

Botulinum toxin is considered ***the most powerful poison in the world****. For example,* ***it is 5,000 times deadlier than ricin and 140,000 times more lethal than curare.***

Transmission

Botulism is not transmitted from human to human.

In most cases, it involves food poisoning that results from the ingestion of food or beverages contaminated with the toxin-producing *C. botulinum* bacterium.

Injury-related botulism is less common and is similar to tetanus contamination.

With regard to terrorist actions, aerosol usage can also be considered.

Note: There is a particular type of botulism called infantile botulism. It is contracted through the absorption of food contaminated with the spores or vegetative forms of the bacterium. In this case, the intestinal flora of young children (which is incomplete) is unable to prevent the proliferation of *clostridium* in the digestive system and the latter thus multiplies while producing toxins.

Symptoms

The first symptoms will appear a few hours or days after contamination. Their speed of occurrence depends on the absorbed dose and the route of entry used (inhalation, for instance, results in a longer delay). They consist in:

- ocular damage (accommodation issues, pupil deformity, photophobia, etc.);
- dry mouth and eyes, difficulty swallowing;

- digestive disorders (nausea, vomiting, diarrhoea, etc.).

Next, flaccid paralysis sets in, little by little, sometimes even causing death as a result of respiratory failure.

Treatment

Treatment is mainly symptomatic and essentially comprises respiratory and cardiac assistance for a period ranging from a few days to a few weeks.

The early administration of antibodies specific to the type of toxin in question allow one to limit both the severity of the intoxication and its lethality.

Vaccination

Although illegal in France, vaccination is possible in some countries (such as the United States). It is, however, reserved for risk categories and only provides temporary protection (the vaccine must be renewed every year).

RICIN FACT SHEET

Ricin is produced by the *Ricinus Communis* or *Ricinus Sanguineus* plant. This toxin stems from a castor plant seed, a common ornamental plant from which an oil of the same name is also extracted.

It induces irreversible inhibition of protein synthesis through the inactivation of ribosomes. In other words, the machinery of the cells is broken, inevitably causing the death of affected cells.

The toxin is characterised by high environmental resistance, but remains sensitive to bleach and high temperatures (and is thus destroyed when faced with a temperature of 80°C for 10 minutes or 50°C for one hour).

Note: The toxin is not actually present in castor oil.

In the industry field, the residues of the seeds from which oil has been extracted are heated in order to inactivate the toxin and are then transformed into fertiliser or cattle feed.

Transmission

In the majority of cases, the poisoning in question results from the ingestion of seeds (with one to three seeds sufficient to kill a child, and three to six for an adult).

In a context of malicious or terrorist use, however, other ways, apart from ingestion, are also possible: both the injection (Bulgarian umbrella) and aerosol use of the toxin should thus be considered.

Symptoms

In case of massive ingestion, the first symptoms may appear a few hours later, including nausea, vomiting, intestinal pain, diarrhoea, etc. The victim is thus at risk of dehydration, as well as renal and hepatic insufficiency. Somnolence and a loss of one's sense of direction accompany these attacks. Death can occur at any time between thirty-six hours and twelve days.

Injection results in similar effects with potential necrosis at the point of entry.

If inhaled, it triggers an irritation of the airways, coughing, bronchial congestion, pulmonary oedema, and other symptoms. This evolves into respiratory distress and hypoxaemia (a lack of oxygen in the blood), causing death in the space of two or three days.

Treatment

Treatment is merely symptomatic. At present, there is no specific medication or vaccine against ricin poisoning.

Research is ongoing both in the field of vaccination and into the discovery of active substances that can counteract the toxin.

Dark Winter: Simulation of a Terrorist Attack

'*Dark Winter*' is the code name of a bio-terrorist attack simulation exercise. This exercise took place on 22 and 23 June 2001 in the United States and was overseen by the *Johns Hopkins Center for Civilian Biodefense Strategies*, in collaboration with the *Center for Strategic and International Studies*, the *Analytic Services Institute for Homeland Security*, and the *Oklahoma National Memorial Institute for the Prevention of Terrorism*. The objective was to assess the capabilities and shortcomings of a coordinated United States government response to a scenario in which a terrorist organisation spread the smallpox virus, and to determine any improvements to be made.

The scenario began with three localised attacks in Oklahoma City shopping malls, resulting in the infection of 3,000 people,[16] followed by others in the cities of Philadelphia and Atlanta. It then simulated the spread of smallpox virus across the country via air and land transport, its subsequent rapid progression among the population, and the caused panic effect. What interested the organisers was an evaluation of the government agencies' capacity to manage the medical, organisational and coordination aspects between the different agencies.

Smallpox virus was chosen because it is well-known and well-documented both historically and medically. What is more, its mortality rate is between 30% and 50% and there is no effective treatment. The only protection is prevention through vaccination. However, since this was discontinued back in the 1980s, it is highly likely that only few people are still immune today and the population is therefore particularly sensitive to the virus. In addition, it was previously part of

---

16 AN: According to William Patrick, one of the scientific leaders of the American offensive biological warfare (BW) programme, one gramme of smallpox is enough to infect 100 people by means of an aerosol attack. Henderson DA, Ingelsby TV, Bartlett JG, et al., *Smallpox as a Biological Weapon: Medical and Public Health Management*, Working Group on Civilian Biodefense, JAMA, 1999, 281, 2127–37.

the biological warfare arsenal of many countries and is still preserved in some of the world's laboratories…

The results of the exercise[17] showed that such an attack would lead to a rapid progression of the virus and result in a very large number of victims, as well as in the inability of government agencies to limit the spread of the virus, either through vaccination or the isolation of populations. The consequences were a loss of public confidence in the American authorities, panic and looting, disorganised escapes from cities, and a saturation of hospital infrastructure. The government was thus compelled to resort to strong-arm methods to regain control of the situation.[18] The exercise also showed that the various government agencies (civil protection, military forces, public health services, and national and local governments) had great difficulty in both communicating with one another and working together. Furthermore, politicians only had a limited understanding of the nature of the danger and were thus unable to coordinate their actions with those of the private sector (hospitals, pharmacies, ambulances, transport, etc.).

The exercise revealed that the local political authorities wanted to impose measures of populational containment and isolation; the closing of borders; and the halting of road, rail, air and public transport-related traffic. On the other hand, the national authorities wanted to gain nation-wide control through a rapid intervention on the part of the armed forces (National Guard) and the federal state security authorities. The number of available vaccines was considered insufficient, and inconsistencies in distribution priorities were noted, as some authorities wanted to reserve the vaccines first and foremost for the police and caregivers, whereas others insisted on putting the population located in the immediate vicinity of contagious cases first.

---

17 AN: *Dark Winter—About the Exercise*, http://www.upmchealthsecurity.org/our-work/events/2001_darkwinter/.

18 AN: O'Toole, Tara; Michael, Mair; Inglesby, Thomas V., *Shining Light on 'Dark Winter'*, 2002; http://cid.oxfordjournals.org/content/34/7/972.full#sec-8, ClinicalInfectiousDiseases, 34 (7), 972–983.

The simulation also showed that the hospital system would quickly be overwhelmed — both by real cases and 'false positives', i.e. by people who thought they were infected but were actually not — and that the contagion would then spread to hospital and clinic staff, quickly crippling any and all ability to provide effective care. In some cases, quarantine was imposed on some localities, but this also caused a disruption in the supplying of medicines, food and other essential commodities. These draconian measures thus failed to improve the situation, since any sense of lack and imposed rationing only serve to increase both anxiety and panic among the population.

In the end, the simulation concluded that the authorities would have succeeded neither in curbing the panic effect through effective media communication, nor in making the right decisions to stem the contagion. This exercise, however, was considered a success because its objective was not so much to 'succeed' as to make authorities aware of the risks posed by a pandemic, thus enabling them to improve future measures and procedures. One of the first consequences lay in increasing the number of vaccine doses available in the United States, a number that went from twelve million in 2001 to 300 million doses in 2010 for a total population of 320 million.

According to a report published by the French Senate,[19] France has, by way of comparison, 72 million doses of the smallpox vaccine that could be used in the event of a threat. These are regularly tested by the AFSSAPS[20, 21] so as to ensure their effectiveness. The issue of their renewal will arise as soon as the tests indicate a loss of efficiency. These stocks could also be renewed if a new generation of vaccine, equally effective but with fewer side effects, were granted marketing

19 AN: Official bulletin dated 13th October, 2005, page 2658.

20 AN: *Agence Française de Sécurité Sanitaire des Produits de Santé* (the French Agency for the Sanitary Safety of Health Products), http://ansm.sante.fr.

21 TN: This agency's tasks/duties were superseded by the *Agence Nationale de Sécurité du Médicament et des Produits de Santé* (National Agency of Medicinal and Health Product Safety) on 1st May, 2012.

authorisation. In 2007, as part of its bioterrorist threat measures, the French authorities published a guide entitled '*Plague, Anthrax and Tularaemia Guide*', which is freely downloadable online.[22]

22 AN: http://www.sante.gouv.fr/IMG/pdf/Guide_Peste_-_Charbon_-_Tularemie_PCT.pdf.

4. Biological Weapons

Whoever sows sparingly will also reap sparingly, and whoever sows generously will also reap generously.

— Epistle of Saint Paul, New Testament

We have already covered the different categories of biological threats:

- **Bacteria:**
  Diseases whose origin is bacterial can usually be treated with antibiotics. In the event of a severe infection, however, appropriate treatment must be administered as soon as possible. The appearance of (natural or, in the case of organisms that have been genetically modified for military purposes, artificial) resistance complicates and delays care, as it takes time to find a new active substance to use against the bacteria.

- **Viruses:**
  Viruses are entities halfway between the inert and the living. It is absolutely necessary for them to enter a cell to be able to multiply. Insensitive to antibiotics and having the ability to mutate and acquire additional aptitudes, they represent the greatest threat for pandemic to date. The surge in Ebola cases that have been ravaging West Africa since 2014 is the latest example.

- **Toxins:**

  Easy to produce and store, toxins can be used as poisons. They do not cause a pandemic, but their formidable 'efficiency' could result in the deaths of thousands of individuals through the use of tiny amounts of substances.

  Several factors thus make the biological threat particularly pernicious and difficult to counter:

  - The effects are time-delayed (= latency or incubation period) and allow the contamination of a large number of people before the first measures can be initiated.
  - In contrast to chemical or radiological agents, it is very complicated to use alert systems with real-time detection against them.
  - Identification, especially in the case of new agents, leads to additional delays.
  - In some cases (most viruses, toxins such as ricin, etc.), there is no real, effective treatment.
  - The human-to-human transmission of certain diseases can lead to chaos at a local level and generate a risk of a major pandemic worthy of disaster films.

All in all, the use of biological agents is a particularly credible threat and the consequences could be extremely serious. The world of life is so vast and complex that it is only a matter of time before the natural evolution of one of these entities (mutation, recombination, resistance, etc.) leads to a new pandemic. Indeed, recent history (Ebola, H_5N_1, SARS, MERS…) offers us irrefutable proof of this. In addition, scientific and military research can lead to the creation of new agents with exacerbated lethal capacities. This is all the more serious since:

- **Accidents are frequent:**
    - Instead of routine influenza samples, the CDC mistakenly sent the H5N1 virus to the Ministry of Agriculture (United States, 2014);
    - Sixty-two CDC employees were possibly exposed to anthrax during a sample transfer between laboratories (United States, 2014);
    - More than 350 samples containing the biological agents of smallpox, dengue fever, Q fever, etc. were discovered in an abandoned laboratory (United States, 2014);
    - On 30 March 1979, an anthrax leak in a bacteriological weapons production plant in Sverdlovsk, Russia caused the deaths of 105 people over several months;[1]
    - In 2015, the state of Louisiana experienced contamination by a potentially deadly bacterium (*Burkholderia pseudomallei*), which had presumably 'escaped' from a research centre.

- **Malicious or terrorist actions do occur:**
    - The salmonella poisoning attack in a US restaurant (1984);
    - The mailing of envelopes containing anthrax (United States, 2001);[2]
    - The sending of letters containing ricin (United States, 2013).[3]

Should the dissemination of a militarised agent ever occur, it is likely that mankind would be faced with most terrifying pandemic in all of human history…

1 TN: Other sources specify a different date (2 April 1979) with a death toll of around 100 casualties.

2 AN: https://fr.wikipedia.org/wiki/Enveloppes_contaminées_au_bacille_du_charbon.

3 AN: http://www.lemonde.fr/ameriques/article/2013/04/17/un-senateur-americain-recoit-une-lettre-contenant-de-la-ricine_3160968_3222.html.

Some Examples of Biological Weapons Research

Even if some pathogens are relatively easy to obtain,[4] bacteriological weapons are not just missiles that can be loaded and then fired at will against a target. The most virulent culture developed in a test tube will never be operational unless it is subjected to a complex process that will turn it into a stable, predictable, and storable agent that can then be disseminated with precision and with great certainty of the result. In a sense, one could say that the real weapon is actually embodied by the manufacturing technique, which is even more difficult to develop than the actual pathogens themselves.

- **In the USSR:**
  Biopreparat was the main agency for the research and development of biological weapons in the Soviet Union. This agency, officially a civilian one but actually dependent on the 15th Directorate of the Red Army, served as a cover for a vast network of secret laboratories, employing more than 30,000 researchers dedicated to the study and improvement of pathogens intended for use in conflicts.[5] There were more than eighteen separate and large-scale specialised research laboratories, such as the Stepnogorsk Scientific and Technical Institute for Microbiology (North Kazakhstan); the Institute of Ultra-Pure Biochemical Preparations of Leningrad, conducting specialised research into the plague bacillus; the Vector State Research Centre of Virology and Biotechnology, which specialised in smallpox research; the Institute for Engineering Immunology in Lyubuchany; the Institute of Applied

4 AN: It is easy to acquire viruses and bacteria in thousands of organic banks around the world. Transnational trade between laboratories and pathogen research centres is quite free. The Zika virus, for instance, can be purchased here: https://www.atcc.org/Products/All/VR-84.aspx#history.

5 AN: Including 'Laboratory 12', where the ricin used by the Bulgarian services and their famous 'Bulgarian umbrella' was developed.

Biochemistry in Omutninsk; the pathogen production plants of Kirov, Zagorsk (both of which specialised in smallpox production) and Berdsk; and Military Compound 19 in Sverdlovsk, which carried out specialist research into anthrax.

Using often innocuous names, this organisation developed militarised strains of smallpox, bubonic plague, anthrax, equine encephalitis, tularaemia, influenza, brucellosis, and glanders from 1970 to 1990. The latter[6] is believed to have been used against the Mujahedeen in Afghanistan between 1982 and 1984. The militarisation of the Marburg and Machupo viruses was still in development in 1992, before the laboratories' activities were classified as a national defence secret by the new Russia. Rumours have emerged that *Biopreparat* may have attempted to develop hybrids of both equine encephalitis and smallpox (*Veepox*) and Ebola and smallpox (*Ebolapox*[7]), but this has not yet been confirmed.

- **In the USA:**

Following research initiated in 1940 by the United Kingdom and Canada into the production and dissemination of pathogenic biological agents, the United States took over these programmes in 1942 with the establishment of a secret unit headed by George W. Merck, the president of the major pharmaceutical company Merck & Co.

These efforts continued during the Cold War in research centres located in Fort Detrick, Maryland, and the Pine Bluff Arsenal in Arkansas, where research was undertaken into glanders, brucellosis, cholera, dysentery, plague, typhus, etc. A decision was also made to develop biological weapons (bioherbicides and mycoherbicides) as well as chemical ones (defoliants and herbicides) for

6 AN: Glanders (*Burkholderia mallei*) was used by the Germans against Romania during World War I to make the horses of the Romanian army sick.

7 AN: http://allnewspipeline.com/Ebola_Pox_Ultimate_Doomsday_Virus.php.

the purpose of destroying enemy crops[8] (wheat, rice, etc.) or the vegetation of certain military operation theatres. Biological weapons have also been developed to target animals, including cattle, pigs, sheep, and fish. The underlying idea was that the destruction of an enemy's agriculture could, from a strategic perspective, force him to end the war. The agents that were thus developed could be spread by planes or via cluster bombs in the targeted agricultural and fish-farming areas. In 1969 and 1970, when, under the Nixon administration, the United States officially decided to put a stop to its biological research programme, the majority of its arsenal — which was unaffected by these measures — was composed of these enterotoxins and mycotoxins. Cuba claims it has suffered numerous attacks on both its crops and livestock, including the dispersal of palm thrips (*Thrips palmi*), an insect carrying a virus (of the *Tospovirus* type) that attacks vegetable crops.

Although herbicides are technically considered chemical weapons, they are often developed in biological weapons programmes.

During the 1940s and 1950s, the armed forces of the United States and United Kingdom collaborated on the development of herbicides and defoliants for use in conflict. The British were the first to use them in Malaysia to destroy the crops and foliage of communist guerrilla zones. As a result of this experiment, the United States set up its own programme, which led to the use of *Agent Orange*[9] in an operation codenamed '*Ranch Hand*' in both Vietnam and Laos (1961–1971).

The US biological research programme was officially terminated by the Nixon administration between 1969 and 1973, the facilities destroyed and the expertise and staff redirected towards the study of protection against biological agents within the framework of the United States Army Medical Research and Materiel

8 AN: Particularly in 1951, with the development of 'agricultural' agents intended to destroy Soviet wheat crops and the rice fields of Communist China.

9 AN: https://en.wikipedia.org/wiki/Agent_Orange.

Command (USAMRMC). However, it is highly likely that research into offensive pathogens is still ongoing in secret laboratories or perhaps even in collusion with private companies.

- **In South Africa:**
  South Africa conducted a very dynamic research programme from the late 1970s until the end of the apartheid regime. Headed by Doctor Wouter Basson, this programme, known as *Project Coast*,[10] aimed to develop the country's defensive and offensive capabilities. The offensive part included research into combat gases and biological pathogens to poison the leaders of terrorist organisations linked to the African National Congress (ANC) and possibly also to poison and sterilise the black population. During the war in Rhodesia, cholera bacilli were disseminated in communist guerrilla zones and spores of anthrax were delivered to the Rhodesian Armed Forces.[11] The effects were relatively limited: in 1979, eighty-two guerrilla members succumbed and thousands more fell ill, which did not, however, influence the conflict's outcome.

LABORATORIES AND RESEARCH CENTRES

Research into biological agents is carried out in specially equipped laboratories with different levels of safety depending on the danger posed by the microorganisms being studied. There are four levels (three in the former USSR) of security ranging from least to most secure. Level 4 is officially known as 'BSL4', which, in English, stands for *Biosafety Level 4* (in French, the abbreviation NSB4, i.e. *Niveau de Sécurité Biologique 4*,[12] is used instead). This level corresponds to an area with pressurised laboratories; successive airlock access systems and locker rooms; secure and exclusive access to

10 AN: https://en.wikipedia.org/wiki/Project_Coast.

11 AN: http://www.pbs.org/wgbh/pages/frontline/shows/plague/sa.

12 TN: Biological Safety Level 4.

staff in fully impermeable and pressurised protective clothing; showers and decontamination measures at the exit; containment means in the event of contamination; and, last but not least, appropriate bio-organism neutralisation means... All of this to prevent these very dangerous pathogens from leaking out of these laboratories and spreading. Many countries have their own official level 4 research centres, and nations such as the United States, Russia, China, the United Kingdom and France, i.e. those with a proven capacity to develop bacteriological weapons, are among the 173 signatories to the **Biological Weapons Convention** (BWC), which prohibits the development of such weapons. Other countries are suspected of secretly conducting (offensive or defensive) research. These include Iran, Israel, North Korea, South Korea, Pakistan, Taiwan, Egypt, Vietnam, Laos, and so on. Other countries such as Libya or South Africa once had very active research programmes, but these have now been terminated. Saddam Hussein's Iraq was an example of a country conducting research in this area, particularly through the development of anthrax for military purposes. Fortunately, Iraq did not use these weapons until its labs were dismantled and stocks destroyed in the 1990s and 2000s. The main confirmed level 4 research centres are the following:

South Africa

- National Institute for Communicable Diseases Special Pathogens Unit, Sandringham, http://www.nicd.ac.za.

Germany

- Bernhard Nocht Institute for Tropical Medicine, Hamburg, http://www15.bni-hamburg.de.
- Friedrich Loeffler Institute, Federal Research Institute for Animal Health, Riems Island, Greifswald, http://www.fli.bund.de.

- Institute of Virology, Philipps University of Marburg, Marburg, http://www.uni-marburg.de.
- Robert Koch Institute, Berlin, http://www.rki.de.
- Institute of Microbiology of the Federal Armed Forces, Munich, http://dl.acm.org.

Australia

- Australian Animal Health Laboratory, Newcomb, http://www.csiro.au/places/AAHL.html.
- National High Security Laboratory / Victorian Infectious Disease Reference Laboratory (VIDRF), Glen Iris, http://www.vidrl.org.au.
- Virology Laboratory of the Queensland Department of Health. Coopers Plains.

Canada

- National Microbiology Laboratory, Winnipeg, Manitoba, http://www.nml-lnm.gc.ca.

China

- Wuhan Institute of Virology of the Chinese Academy of Sciences, Wuhan, http://english.whiov.cas.cn.

United States

- Centers for Disease Control and Prevention (CDC), Atlanta, Georgia, http://www.cdc.gov.
- Division of Consolidated Laboratory Services, Virginia, https://dgs.virginia.gov/division-of-consolidated-laboratory-services.
- Galveston National Laboratory, Galveston, Texas, http://www.utmb.edu/gnl.
- Georgia State University, Atlanta, Georgia, USA, http://www.gsu.edu.
- Integrated Research Facility, Bethesda, Maryland, http://orf.od.nih.gov.

- Middle Atlantic Regional Center for Biodefense and Emerging Infectious Diseases Research, Baltimore, Maryland, http://marce.vbi.vt.edu.
- National Bio and Agro-Defense Facility (NBA), Manhattan, Kansas, https://www.dhs.gov/science-and-technology/national-bio-and-agro-defense-facility.
- National Biodefense Analysis and Countermeasures Center (NBACC), Frederick, Maryland, http://www.nbacc.net.
- National Emerging Infectious Disease Laboratory (NEIDL), Boston, Massachusetts, http://web.bu.edu/dbin/neidl/en.
- National Institute of Allergy and Infectious Disease (NIAID), Rocky Mountain Laboratories, Hamilton, Montana. http://www3.niaid.nih.gov/about/organization/dir/rml.
- Southwest Foundation for Biomedical Research, San Antonio, Texas, SA, http://www.sfbr.org.
- US Army Medical Research Institute of Infectious Diseases (USAMRIID) Frederick, Maryland, http://www.usamriid.army.mil.

France

- Laboratoire de la DGA (Direction Générale de l'Armement), Vert-le-petit, Essonne.
- INSERM, P4 Jean Mérieux, Gerland, Lyon.

Gabon

- Centre International de Recherches Médicales de Franceville (CIRMF), Franceville.

India

- All India Institute of Medical Sciences, New Delhi, Delhi, http://www.aiims.edu.
- High Security Animal Disease Laboratory, Madhya Pradesh.

Italy

- Ospedale universitario Luigi Sacco, Milan, http://www.hsacco.it.
- Istituto Nazionale Malattie Infettive, Rome, http://www.inmi.it.

Japan

- Institute of Physical and Chemical Research (RIKEN), Koyadai, Tsukuba-shi, Ibaraki, http://www.riken.go.jp/engn.
- National Institute for Infectious Diseases (NIID), Tokyo, https://www.niaid.nih.gov.

Netherlands

- Netherlands National Institute for Public Health and the Environment, Bilthoven, http://www.rivm.nl/en.

Czech Republic

- Centre of Biological Protection, Prague, http://www.army.cz/en.

United Kingdom

- Centre for Emergency Preparedness and Response, Health Protection Agency, Defence Science and Technology Laboratory, Salisbury, Wiltshire, http://www.dstl.gov.uk.
- Health Protection Agency—Centre for Infections (HPA-CFI), London, http://www.hpa.org.uk/HPA/AboutTheHPA/WhoWeAre/CentreForInfections.
- National Institute for Medical Research (NIMR), London, http://www.nimr.mrc.ac.uk.

Russia

- State Research Centre for Virology and Biotechnology VECTOR, Koltsovo, Novosibirsk, http://www.vector.nsc.ru.

- Institute of Microbiology, Kirov.
- Virological Centre of the Institute of Microbiology, Sergiyev Posad, Moscow.

Singapore

- Defense Science Organization (DSO), http://www.dso.org.sg.

Sweden

- Swedish Institute for Infectious Disease Control, Solna, http://www.smittskyddsinstitutet.se/in-english/.

Switzerland

- Institute of Virology and Immunoprophylaxis, Mittelhausern, http://www.bvet.admin.ch/ivi/?lang=en.
- Laboratory of Spiez, Spiez.
- High Containment Laboratory DDPS, Bern, https://www.vbs.admin.ch.

Taiwan

- Kwen-yang Laboratory Center of Disease Control.
- Preventive Medical Institute of ROC Ministry of National Defense.

Dissemination and Penetration Methods

Most biological agents can be disseminated and transmitted in different ways, including:

- airborne dispersion (aerosol spray cans, airborne application, etc.);
- the contamination of food or water;
- the contamination of objects (blankets, clothes, etc.);

- the infection of animals that will subsequently transmit the agent to human beings either directly (through contact or ingestion) or indirectly (through fleas, mosquitoes, etc.);
- human-to-human contagion through air (sneezing, etc.) or fluid transfer (i.e. contact with mucous membranes, saliva, perspiration, blood, wounds, sexual intercourse, and so on).

Several entry pathways are thus possible when it comes to the body:

- through oral means (ingestion, etc.);
- through the respiratory tract (inhalation, etc.);
- using cutaneous/ocular routes (wounds, directly through tissues, and so on).

5. Pandemic

A small thing can generate a huge disaster.

— Tuscan polymath Leonardo da Vinci (1452–1519)

Biological warfare is a dirty business.

— US Secretary of War Henry L. Stimson, in a letter to President Franklin D. Roosevelt, 1942

Epidemics have shaped human civilisations and cultures and have resulted in countless deaths throughout history—far more so than all wars combined! Here is a non-exhaustive list:

DATE	NAME	CAUSE	DEATH TOLL
429–426 BC	Plague of Athens	A variant of typhoid?	75,000–100,000 deaths in Greece
AD 65–180	Antonine Plague	A kind of smallpox	30% of the Mediterranean basin
AD 541–542	Plague of Justinian	Bubonic plague	40% of the European, Middle Eastern and Mediterranean population
1330–1350	Chinese Plague?[1]	Plague?	30% of the Chinese population?

1 AN: Sussman G.D, *Was the Black Death in India and China?* Bulletin of the History of Medicine, 85, 2011.

DATE	NAME	CAUSE	DEATH TOLL
1346–1352	The Black Death	Plague	30%–70% of the European population
1545–1548	Cocoliztli	Haemorrhagic fever	80% of the native population of Central America
1576	Cocoliztli	Haemorrhagic fever	50% of the native population of Central America
1629–1631	Italian Plague	Plague	280,000 deaths in Italy
1663–1664	Plague of Amsterdam	Plague	24,000 deaths
1665–1666	Great Plague of London	Plague	100,000 deaths
1668	Plague of 1668	Plague	100,000 deaths in the Mediterranean and France
1679	Great Plague of Vienna	Plague	76,000 deaths
1738	Plague of 1738	Plague	More than 50,000 deaths in the Balkans
1770–1772	Russian plague epidemic	Plague	More than 50,000 deaths in Russia
1816–1826	First cholera pandemic	Cholera	More than 100,000 deaths in Europe and Asia
1829–1851	Asiatic cholera pandemic (also known as the second cholera pandemic)	Cholera	More than 100,000 deaths in Europe and Asia
1847–1848	Typhus epidemic	Typhus	Canada, 20,000 deaths
1852–1860	Russian cholera epidemic (or 'third cholera pandemic')	Cholera	More than a million deaths in Russia
1889–1890	Influenza	Influenza	More than a million deaths worldwide

DATE	NAME	CAUSE	DEATH TOLL
1899–1923	Cholera	Cholera	More than 800,000 deaths worldwide
1918–1920	Spanish flu	Influenza H1N1	30 to 75 million deaths worldwide
1957–1958	Asiatic flu	Influenza H2N2	2 million deaths in Asia
1968–1969	Hong-Kong flu	Influenza H3N2	1 million deaths in Asia
1980–?	AIDS	HIV	30 million deaths worldwide
2002–2003	SARS epidemic	Pneumonia	349 (official) deaths in China
2009	Swine flu	Influenza H1N1	14,000 deaths worldwide
2013–2015	Ebola outbreak in West Africa	Ebola	12,000 deaths in West Africa

Unlike seasonal influenza, which is considered random and grows rapidly and exponentially before declining a few weeks or months later, some epidemics may grow more slowly but maintain their presence for a long time, as in the case of AIDS. As regards emerging diseases such as Ebola or new strains of influenza, the spread can be very rapid and cause genuine social and economic devastation.

An epidemic often has a major effect on human societies. According to Herodotus, for instance, one of them decimated the Assyrian army besieging Jerusalem, the capital of the kingdom of Judah, in 701 BC. This disease, which is referred to in the biblical account as the 'Angel of Death', is said to have allowed the Hebrew religion, which was but a small cult among many others, to establish its credibility and to consolidate and sustain itself.

In addition to having a devastating effect on the morale of the population, a high number of deaths can cause a slowdown or complete paralysis of transport, trade, tourism, city supplies, investments, and thus significantly reduce the entire economy of a society. And yet, it is precisely the continuity of exchanges that is essential to allow an effective and rapid fight against the epidemic: laboratories, health systems, communication, transport and distribution are all dependent on the presence of qualified personnel. When fear prevents employees from going to work, or when they fall ill or even die, the whole system is brought to a halt.

The total direct or indirect cost to the economy of a nation can truly be considerable and weigh heavily upon the proper functioning of human activities. In addition, one notices that the societies characterised by the lowest living standards and the weakest economies are also the most impacted in the event of an epidemic. This is due to the lack of available resources (researchers and laboratories, doctors and nurses, security services) that would allow them to contain, isolate and fight the disease and treat the afflicted. Last but not least, such events can have major effects on both the culture and the beliefs of affected populations. History enables us to study some of these epidemics, which were well documented either by their contemporaries or by historians.

The Plague of Athens

We are not referring here to a particularly ill-mannered Greek girl, but an epidemic that struck the capital city of Attica and head of the Delian League at some point between 429 and 426 BC, as Spartan armies besieged it. (To this day, the culprit is still being debated:[2] typhus, smallpox or something else.) The event was studied and reported by the Athenian historian Thucydides,[3] who gave details of the epidemic's impact on the inhabitants' social and religious behaviour.

It seems that the risk of contracting the disease and being killed by it was so great that the population stopped concerning itself with the

2 AN: https://www.ancient.eu/article/939/the-plague-at-athens-430-427-bce/.

3 AN: Thucydides, *The History of the Peloponnesian War.*

law and the long-term effects of its own actions. Some poor people suddenly inherited wealth from affluent relatives and squandered their fortune over and over again. Care of the dead and the sick was quickly abandoned — indeed, since doctors, caregivers and gravediggers were among the first to be affected, they were not replaced, for no one wanted to take the risk of being contaminated by touching patients and corpses. Unfortunately, the latter probably contributed to adding further diseases to the plague epidemic by simply decomposing. Furthermore, many patients were abandoned and died due to a lack of water or food. The few people who found themselves immune (without being aware of this at the time), and who therefore continued to deal with the corpses, simply stacked them up in heaps to let them decompose or, wherever possible, piled them up to burn in pyres. This was strongly against the epoch's traditions and religious practices and the inhabitants of Athens felt abandoned by the gods. In addition, the disease affected indiscriminately both the less pious and the more devout. The temples, now filled with refugees from the countryside, turned into major infection sources, including the temple of Apollo, the god of medicine and disease. From this, the Athenians inferred that the gods had sided with Sparta.

In his comments, Thucydides criticises this reaction, labelling it a superstition. At the time, medicine had already correctly deduced that it was a contagion rooted in natural causes, particularly through the observation of birds and animals that had perished after consuming infected carcasses. Be that as it may, Athenian morale in the Peloponnesian War was thus affected on a long-term basis.

The Black Death

The Black Death, known as 'the plague of 1348', impacted the European continent, North Africa and a part of Asia from 1346 to 1352, causing the deaths of 30% to 70% of the population, depending on the region.

It had a very important effect on the culture, religion and economy of European peoples.

What follows is a contemporary testimony, namely that of Agnolo di Tura, an Italian merchant from Siena, who describes the horror of the situation afflicting his city:

> The mortality in Siena began in May. It was a cruel and horrible thing. … It seemed that almost everyone became stupefied seeing the pain. It is impossible for the human tongue to recount the awful truth. Indeed, one who did not see such horribleness can be called blessed. The victims died almost immediately. They would swell beneath the armpits and in the groin, and fall over while talking. Father abandoned child, wife husband, one brother another; for this illness seemed to strike through breath and sight. And so they died. None could be found to bury the dead for money or friendship. Members of a household brought their dead to a ditch as best they could, without priest, without divine offices. In many places in Siena great pits were dug and piled deep with the multitude of dead. And they died by the hundreds, both day and night, and all were thrown in those ditches and covered with earth. And as soon as those ditches were filled, more were dug. I, Agnolo di Tura, buried my five children with my own hands. And so many died that all believed it was the end of the world.[4]

People's ignorance of the causes, the impotence of the doctors, and the speed at which this plague spread awakened a nameless terror and triggered considerable social upheaval. Entire villages simply vanished, and families were decimated. Although the rising death toll was followed by an upsurge in marriages (celebrated at a younger age) and then births, the Europe of the 1400s was only half as populous as in the 1300s. This example attracts our attention to some very interesting social and economic effects:

- The first is flight, which often allowed the epidemic to spread as a result of the arrival of infected people. For a few years, the number

4 AN: George Deaux, *The Black Death, 1347*, New York, Weybright and Talley, 1969.

of wanderers and vagrants increased, favouring the development of banditry.

- Many people viewed the epidemic as a manifestation of divine wrath and sought to appease God using various means, including the organisation of pilgrimages. In 1350, more than a million pilgrims went to Rome, most of whom died on the way. Rites of collective penitence such as the processions of flagellants[5] thus surfaced.

- Social inequalities worsened in the face of the epidemic, as well-off people were able to organise their own escape or take shelter and keep away from large cities or contamination zones.

- Religious antagonisms were exacerbated, and it was particularly Jews who served as scapegoats: since they frequently traded from city to city, a connection was made between their presence or activities and the progression of the epidemic. They were accused of deliberately poisoning people, and pogroms took place in many cities and regions of Europe.

- Due to the death toll, the amount of available agricultural land increased while its price dropped, allowing surviving peasants to own more land and work less on seigniorial lands. A rise in employment costs was then added to the decrease in the number of available subjects. The need to repopulate the gradually deserted farms was implemented through a reduction in fees and services.

---

5 TN: Flagellants were people who practised an extreme form of self-inflicted mortification of the flesh by whipping it with certain instruments, a practice which became quite commonplace when the Black Death began to rage. Those people chose to mortify themselves in such a way because they were convinced that hurting themselves would drive God to have mercy on them and they would thus not contract the disease. There were others, however, who did this because they believed that mortification would somehow drain or remove infected blood from their bodies.

This dealt a severe blow to the rural seigniory and the feudal model, both of which had already weakened as a result of the increased social mobility of European populations, thus accelerating their decomposition.

In cities, the reduction in the number of workers triggered an increase in wages and thus in the prices of manufactured goods. Simultaneously, however, this also led to an intense period of innovation, in an effort to find suitable techniques and tools to increase the productivity of the workers.

- The transformation of much of the cultivated land into pasture land increased the amount of meat (and thus that of proteins) available to the population, improving its general health.

- Frustrated by the dubious explanations of both astrology and superstition, scientists of the era began to rediscover medical science and established medical schools rooted in observation, deduction and comprehension across the whole continent.

- Due to both their shared experiences of great suffering (stemming from the massive number of deaths in their family, circle of friends and close relationships) and the fact of witnessing the inability of the Church and clergy to offer an understandable explanation for that ordeal, survivors began to doubt the usefulness of a religious structure. Some heretical movements were thus born. Some felt that since God had abandoned them, they would, in turn, renounce him. Others, happy to still be alive, engaged in frenzied hedonism and debauchery, as described by Boccaccio[6] in his book

---

6 TN: Giovanni Boccaccio (16 June 1313–21 December 1375), was an Italian author, poet, and an important humanist of the Renaissance period. He wrote numerous works that earned him well-deserved fame, including the *Decameron* and *On Famous Women*. His writing style was highly imaginative and was characterised by the inclusion of realistic dialogues that contrasted with his contemporaries' writings, whose medieval style typically comprised highly formulaic models in terms of both characters and plot. Thanks to this fact, the *Decameron*

Decameron, which some artists have allegorically portrayed in their 'dance of death' (*Danse Macabre*).

- Priests, who, alongside doctors, had most contact with the sick, were actually more affected by the events than anyone else and their socio-professional class one of the most devastated. Many believers thus began to practice their faith outside the usual places of worship — whether at home or in private chapels.

- Another negative consequence was the diminishing of European genetic diversity,[7] causing a sort of genetic bottleneck.[8]

'Danse Macabre', illustration by Michael Wogelmut, taken fom the *Liber Chronicarum* by Hartmann Schedel, Nürnberg, 1493.

---

manages to evoke highly vivid images of people's lives, suffering, depravity and general behaviour during the days of the Black Death. It is a collection of novellas with topics ranging from the erotic to the tragic.

7 AN: Colin Barris, *Black Death Casts a Genetic Shadow over England*, New Scientist, August 2007. More information is available here: http://www.newscientist.com/article.ns?id=dn12393.

8 AN: https://en.wikipedia.org/wiki/Population_bottleneck.

In the end, European populations had no other choice but to develop new practices and push for scientific and innovational renewal. These measures, driven by people's need to survive, probably acted as the basis of the '*Renaissance*'. They undoubtedly also constituted the matrix of all that gave rise to the Reformation a century and a half later, as well as a certain form of proto-capitalism and free trade of both goods and ideas, the culmination of which was what would later be termed 'the Enlightenment' and the 'Industrial Revolution'. Paradoxically, despite suffering a very large number of deaths, Europe found itself strengthened, more innovative, more dynamic and richer. Our modern world stems largely from the effects of this great epidemic.

Spanish Flu

The 1918 flu, better known as the 'Spanish flu', raged from 1918 to 1920. It is said to have claimed up to seventy-five million lives worldwide, although the consensus is thirty million deaths.

It is considered the deadliest epidemic in human history over such a short period of time, even surpassing the estimated thirty-four-million death toll of the Black Death. Its nickname, 'Spanish flu', comes from the fact that only Spain, which was not involved in World War I, was able to freely publish information related to this epidemic in 1918. Even the King of Spain, Alfonso XIII, fell ill as a result of its spread. French newspapers thus spoke of 'Spanish flu' wreaking havoc '*in Spain*', without mentioning the numerous French cases, all of which were kept secret so as to prevent the German enemy from finding out that the army had been weakened.

Ever since the 1970s, various theories have been proposed imputing the Spanish flu to military experiments[9] or to possible side effects

9 AN: http://conscience-du-peuple.blogspot.ch/2011/11/la-grippe-espagnol-une-arme.html.

of the major vaccination campaigns[10] conducted at the time. However, according to genetic studies[11] involving the use of tissues taken from Inuit and Norwegian cadavers that had been preserved in frozen ground, the most likely hypothesis is that a particularly virulent strain of the H1N1[12] influenza virus is actually responsible. This strain is alleged to have emerged in China and was seemingly transmitted from ducks to pigs and then to people, or perhaps from birds to humans.

From Asia, this flu is believed to have spread to the United States, where the virus mutated to become fatal for 3% to 20% of all patients, compared to less than 0.1% in the case of the common seasonal flu. By spreading across Europe and then the world through the colonies, it subsequently turned into a pandemic.

Officially, it was responsible for about 210,000 deaths in France,[13] but the censorship of war is certain to have strongly minimised these figures. Nevertheless, its pandemic 'title' was fully justified, since one billion people (half of the world's population at the time) were afflicted by this flu during the winter of 1918–1919. Eventually, the disease claimed more lives than World War I, which ended during that same year of 1918, and some countries were still affected by it in 1919 and 1920. The only exception was Australia, which was only slightly impacted thanks to its introduction of strict quarantine procedures.

The spread of the virus was blistering fast: outbreaks of infection took place on both sides of the United States in a period of merely seven days, and in other countries and continents in less than three months. The incubation period was also very short, usually lasting two to three days and followed by three to five days of symptoms that

---

10 AN: http://www.alterinfo.net/La-grippe-espagnole-de-1918-est-due-aux-vaccins_a67138.html.

11 AN: Taubenberger JK, Reid AH, Lourens RM, Wang R, Jin G, Fanning TG, *The 1918 Flu Virus Is Resurrected. Characterization of the 1918 Influenza Virus Polymerase Genes*, Nature 437, October 2005.

12 AN: http://wwwnc.cdc.gov/eid/article/12/1/05-0979_article.

13 AN: http://www.1914-1918.be/grippe_espagnole.php.

included fevers, pneumonia, and weakened immune defences that sometimes led to life-threatening complications. The speed at which the virus spread was not only connected to the development of ship and rail transport (and, more particularly, that of troops), but also to people's overcrowding in trenches, factories and urban dwellings.

During the last year of the war, this particularly contagious epidemic had major economic, human and military consequences. A note from the French internal security police also points out that at the *Palais de Justice* in Paris, 'there is much more talk of the flu, and of the devastation it exerts, than of war and peace.' As for what was reported by doctors, the situation was dramatic:

> We counted the dead by the thousands. At Joigny Hospital, one person departs this world every hour. Lyons, a city that lacks hearses and coffins, is forced to transport the corpses in improvised shrouds, even carts, and bury them at night — I have seen this with my own eyes. Identical scenes are taking place in Paris, where, during the last week of October, 300 people died per day, and burials take place very late in the evening. In a single month, the Spanish flu has caused more damage to the capital than planes and guns have in four years of war.[14]

The low death estimate[15] for 1918 alone is 257,363 in Japan, 1,257,082 in Spain, 48,082 in the Netherlands, 390,000 in Italy, 225,330 in Germany, 100 000 in Hungary, 2,458 in Austria, 272,158 in the United Kingdom, 59,000 in Portugal, 23,277 in Switzerland, 34,374 in Sweden, 675,000 in the United States,[16] and between 13 and 18.5 million in India.[17] All

14 AN: Meyer (J.), Ducasse (A.), Perreux (G.), *Vie et mort des Français 1914–1918* [TN: The Lives and Deaths of the French, 1914–1918], Paris, Hachette, 1960, p. 357.

15 AN: Johnson and Mueller, *Updating the Accounts: Global Mortality of the 1918–1920 'Spanish' Influenza Pandemic* (2002), chart 4, page 113.

16 AN: http://www.flu.gov/.

17 AN: K. Davis, *The Population of India and Pakistan*, Princeton University Press, Princeton, 1951.

the doctors and people who experienced the events report that it was the most terrible disease the world has ever faced.

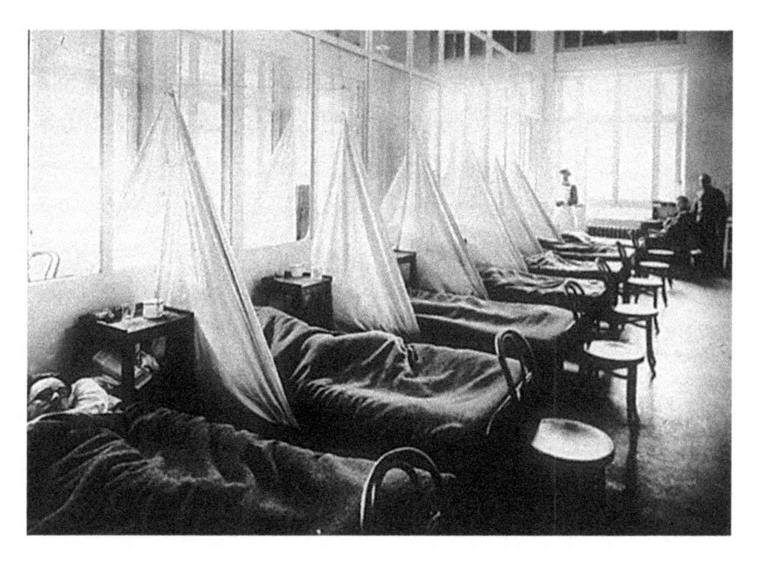

American victims of the Spanish flu, US Army field hospital 45, Aix-les-Bains, France, 1918.

The casualties were mostly young adults,[18] which was both surprising and incomprehensible at the time. Indeed, this part of the population is usually more resistant to disease. This development could thus be explained by the fact that this age group was more exposed to contagion as a result of its increased presence in the trenches or in factories, where the largest numbers of people gather and increased contact exacerbates the risk of contamination. In actual fact, it is the immune system of such young adults that reacted most vigorously to the new virus. The secretion of cytokines[19] was so intense and considerable

18 AN: http://wwwnc.cdc.gov/eid/article/12/1/05-0979-f2.

19 AN: Cytokines are substances synthesised by the cells of the immune system and crucial during the process of cellular signalling. Their remote action on other cells regulates the latter's activity and function. For more information, see https://fr.wikipedia.org/wiki/Cytokine#Cytokines_et_grippe.

that instead of regulating inflammation, it sometimes caused fatal organic failure.

The virulence of the virus, by contrast, remains a mystery. Could it be that this particular strain of H1N1 mutated just before 1918, turning it into a particularly effective killer, and another mutation followed just after 1920 reducing its dangerousness in its successive forms?[20] Were other factors involved? Were all deaths caused by the same virus strain? Did other diseases such as pneumonia also exploit the general weakening? Who or what was the virus' original host? There are many questions that remain unanswered. In addition to the colossal death count, the social and economic consequences were also considerable:

- The economy began to slow down because of absenteeism;
- In the United States, the claims dealt with by life insurance companies increased by an average of 745%;
- A large number of small and medium-sized enterprises went bankrupt or disappeared as a result of the deaths of their specialised or indispensable staff;
- Some historians[21] have demonstrated that the effect of influenza was proportionally greater on the populations of the Central Powers (Germany, Austria-Hungary...) than on those of the Entente Powers (France, the United Kingdom...), thus contributing to the latter's victory;
- The shortage of doctors, which was already felt because of the global conflict, had become so severe that, in many countries, the government asked companies to give their employees a day's paid leave so that they could help out in hospitals. Emergency and

---

20 AN: Reid AH, Taubenberger JK, Fanning TG, *Evidence of an Absence: The Genetic Origins of the 1918 Pandemic Influenza Virus*, Nat Rev Microbiol, 2004.

21 AN: Andrew Price-Smith, *Contagion and Chaos*, MIT Press, 2009.

makeshift hospitals were set up in schools and barracks. The health authorities of impacted countries decided to organise a centralised coordination effort comprising health care workers, medicines and medical supplies, and it was the International Red Cross that became indispensable with regard to this role of coordination and management;[22]

- The pandemic raised awareness of the international nature of epidemic and disease-related threats and the importance of establishing both mandatory hygiene standards and a control network to deal with them. In one of its clauses, the charter of the League of Nations (LN) thus stipulated the requirement to create an International Hygiene Committee, one that would later become the WHO;

- Most state ministries of health introduced, albeit too late, very strict laws to prevent contagion, including the obligation to wear gas masks in public, limiting the duration of burials to fifteen minutes, the issuing of health certificates for rail travel, etc. Government agencies thus grew considerably at the time, making the idea of centralised and expensive statal affairs acceptable, and therefore also the necessity for high taxes[23] to finance them. The Great War and its propaganda, the period's nationalism, and the flu epidemic eventually forced people to blindly accept, and for a long time at that, the authority of a centralised, strong, and expensive state that was characterised by an enormous and far-reaching governmental system often financed by debt and whose original

---

22 AN: Crosby A, *America's Forgotten Pandemic: The Influenza of 1918*, Cambridge University Press; Cambridge, 1989.

23 AN: These include the indefinite extension of the income tax, created 'temporarily' to finance the war.

matrix may be found in the Peace of Westphalia (1648),[24] two centuries earlier;

- The war and flu contributed to the prestige of scientists, especially in the medical sciences where new discoveries and developments saw the light. The theory of germs, surgery, vaccines, antiseptics and, ten years later, antibiotics all made great leaps at the time. States and peoples believed in both progress and the notion that these sciences could bring significant benefits to the population as well as competition between nations;

- The medical industry generated substantial profits and hired large numbers of new medical staff, the majority of whom were women;

- Trapped between the war and the influenza epidemic, European populations basically (and durably) surrendered many of their individual liberties to the state, agreeing to almost completely submit to it and subordinate their personal needs to those of the nation;

- From 1919 onwards, a type of collective amnesia allowed the epidemic to be 'forgotten'. The fact that this disease surfaced so quickly, killed so rapidly and disappeared as suddenly as the damage it had caused probably meant that its ravages were unconsciously associated with the collective woes of the Great War. It is the latter that 'bore the brunt of things', so to speak. In addition, propaganda and war censorship concealed the disease from public awareness during most of its spread, thus avoiding any and all media amplification and diminishing the impact on people's collective memory.

---

24 TN: The Peace of Westphalia (*Westfälischer Friede*) was a series of peace treaties signed between May 1648 and October 1648 in the Westphalian cities of Osnabrück and Münster. The signing was largely responsible for putting an end to the European wars of religion, including the Thirty Years' War.

The 2013–2015 Ebola Outbreak in West Africa

The Ebola outbreak in West Africa began in Guinea in December 2013, before spreading to Sierra Leone and Liberia. In 2014, Mali, Nigeria, Senegal were already affected as well, but to a lesser extent. For the first time since the Ebola virus was discovered, contamination occurred outside Africa, with cases in Spain[25] and the United States.[26] Furthermore, other countries such as Italy, Switzerland or France welcomed sick or already infected people without realising it.

Ebola is a virus that causes one of the most frightening diseases of all. It is transmitted from person to person through direct contact with blood, sweat, faeces, vomit, saliva, urine or mucus. A person afflicted with the Ebola virus produces a very large amount of these fluids, and can spread them uncontrollably through bleeding, vomiting, and blood-filled diarrhoea. Anyone who touches these fluids with their bare hands or gets some on their skin runs the risk of becoming infected.

Here is how journalist Richard Preston, who has studied the virus, describes its effects:[27]

> When a hot virus multiplies in a host, it can saturate the body with virus particles, from the brain to the skin. The military experts then say that the virus has undergone "extreme amplification". This is not something like the common cold. By the time an extreme amplification peaks out, an eyedropper of the victim's blood may contain a hundred million particles of virus. During this process, the body is partly transformed into virus particles. In

---

25 AN: Information can be found here — http://www.francetvinfo.fr/sante/maladie/ebola/une-infirmiere-espagnole-malade-d-ebola-premier-cas-de-contamination-en-europe_712711.html.

26 AN: Here is a link to the topic — http://www.lemonde.fr/planete/article/2014/10/15/ebola-l-etat-de-preparation-des-hopitaux-americains-en-question_4506424_3244.html.

27 AN: Richard Preston, *The Hot Zone — The Terrifying True Story of the Origins of the Ebola Virus*, 1994.

other words, the host is possessed by a life form that is attempting to convert the host into itself. … The seven mysterious proteins that, assembled together, make up the Ebola-virus particle, work as a relentless machine, a molecular shark, and they consume the body as the virus makes copies of itself. Small blood clots begin to appear in the bloodstream, and the blood thickens and slows, and the clots begin to stick to the walls of blood vessels. This is known as pavementing, because the clots fit together in a mosaic. The mosaic thickens and throws more clots, and the clots drift through the bloodstream into the small capillaries, where they get stuck. This shuts off the blood supply to various part of the body, causing dead spots to appear in the brain, liver, kidneys, lungs, intestines, testicles, breast tissue (of men as well as women), and all through the skin. The skin develops red spots, called petechiae, which are haemorrhages under the skin. Ebola attacks connective tissue with particular ferocity; it multiplies in collagen, the chief constituent protein of the tissue that holds the organs together. (The seven Ebola proteins somehow chew up the body's structural proteins.) In this way, collagen in the body turns to mush, and the underlayers of the skin die and liquefy. The skin bubbles up into a sea of tiny white blisters mixed with red spots known as a maculopapular rash. The rash has been likened to tapioca pudding. Spontaneous rips appear in the skin, and haemorrhagic blood pours from the rips. The red spots on the skin grow and spread and merge to become huge, spontaneous bruises, and the skin goes soft and pulpy, and can tear off if it is touched with any kind of pressure. Your mouth bleeds, and you bleed around your teeth, and you may have haemorrhages from the salivary glands — literally every opening in the body bleeds, no matter how small. The surface of the tongue turns brilliant red and then sloughs off, and is swallowed or spat out. It is said to be extraordinarily painful to lose the surface of one's tongue. The tongue's skin may be torn off during rushes of the black vomit. The back of the throat and the lining of the windpipe may also slough off, and the dead tissue slides down the windpipe into the lungs or is coughed up with sputum. Your heart bleeds into itself; the heart muscle softens and has haemorrhages into its chambers, and blood squeezes out of the heart muscle as the heart beats, and it floods the chest cavity. The brain becomes clogged with dead blood cells, a condition known as sludging of the brain. Ebola attacks the lining of the eyeball, and the eyeballs may fill up with blood: you may go blind. Droplets of blood stand out on the eyelids: you may weep blood. The blood runs from your eyes down your cheeks and refuses to coagulate. You may have a hemispherical stroke, in which one whole side of the body is paralysed, which is invariably fatal in a case of Ebola. Even while the body's internal organs are becoming plugged with coagulated blood, the blood that streams out of the body cannot clot; it resembles when being squeezed out of curds. The blood has been stripped of its clotting factors. If you put the runny Ebola blood in a test tube and look at it, you see that the blood is destroyed. Its red cells are broken and dead. The blood looks as if it has been buzzed in an electric blender. Ebola kills a great deal of tissue while the host is still alive. It triggers a creeping, spotty necrosis that spreads through all the internal organs. The liver bulges up and turns yellow, begins to liquefy, and then it cracks apart. The cracks run across the liver and deep inside it, and the liver completely dies and goes putrid. The kidneys become jammed with blood clots and dead cells, and cease functioning. As the kidneys fail, the blood becomes toxic with urine. The spleen turns into a single huge, hard blood clot the size of a baseball. The intestines may fill up completely with blood. The lining of the gut dies and sloughs off into the bowels and is defecated along with large amounts of blood. In men, the testicles bloat up and turn black-and-blue, the semen goes hot with Ebola, and the nipples may bleed. In women, the labia turn blue, livid, and protrusive, and there may be massive vaginal bleeding. The virus is a catastrophe for a pregnant woman: the child is aborted spontaneously and is usually infected with Ebola virus, born with red eyes and a bloody nose. Ebola destroys the brain more thoroughly than does Marburg, and Ebola victims often go into epileptic convulsions during the final stage. The convulsions are generalised grand mal seizures — the whole body twitches and shakes, the arms and legs thrash around, and the eyes, sometimes bloody, roll up into the head. The tremors and convulsions of the patient may smear or splatter blood around. Possibly this epileptic splashing of blood is one of Ebola's strategies for success — it makes the victim go into a flurry of seizures as he dies, spreading blood all over the place, thus giving the virus a chance to jump to a new host — a kind of transmission through smearing. Ebola (and Marburg) multiplies so rapidly and powerfully that the body's infected cells become crystal-like blocks of packed virus particles. These crystals are broods of virus getting ready to hatch from the cell. They are known as bricks. The bricks, or crystals, first appear near the centre of the cell and then migrate toward the surface. As a crystal reaches a cell wall, it disintegrates into hundreds of individual virus particles, and the bloodlings push through the cell wall like hair and float

> away in the bloodstream of the host. The hatched Ebola particles cling to cells everywhere in the body, and get inside them, and continue to multiply. It keeps on multiplying until areas of tissue all through the body are filled with crystalloids, which hatch, and more Ebola particles drift into the bloodstream, and the amplification continues inexorably until a droplet of host's blood can contain a hundred million individual virus particles. After death, the cadaver suddenly deteriorates: the internal organs, having been dead or partially dead for days, have already begun to dissolve, and a sort of shock-related meltdown occurs. The corpse's connective tissue, skin, and organs, already peppered with dead spots, heated by fever, and damaged by shock, begin to liquefy, and the fluids that leak from the cadaver are saturated with Ebola-virus particles.

Five types of Ebola virus are known, in addition to the Marburg virus, which is considered their close cousin. These different types of virus live inside their natural hosts, which are probably still unidentified species of bats, rats, reptiles or monkeys inhabiting the forests and savannas of equatorial Africa. When examined using an electron microscope, Ebola particles appear long and resemble a piece of wire or cooked spaghetti — this is one of the peculiarities characterising this family of viruses.

The first recorded infection occurred in July 1976 in Maridi, Southern Sudan. At the time, however, no one had heard of it because the disease regressed very quickly. Indeed, the medical staff of the bush hospital that had welcomed the patients was so panicked by the victims' physical condition that they fled, which broke the chain of contagion. Still, 151 people died. It was the so-called 'Sudan' Ebola strain, which results in death in 50% of all cases. A genuine killer.

In September 1976, the so-called 'Zaire' strain of Ebola was officially revealed to the world in the Yambuku mission hospital located near the Ebola River in northern Zaire. This hospital was run by Belgian nuns and, unfortunately, the doctors only used hot water to rinse the needles that had been stained with large amounts of blood, which is completely insufficient to kill such a virus. The epidemic began simultaneously in fifty-five villages across the region: it first killed the people who had received an injection and then their families, especially the women, who, in Africa, are tasked with cleaning the dead before burial. 280 people died as a result. For this strain of Ebola, the human death rate is 90% once infected. A real exterminator!

Other Ebola epidemics have erupted over the years, including the 1979 outbreak in Sudan (twenty-two deaths), the 1994 outbreak in Gabon (thirty-one deaths), the 1995 outbreak in Zaire (254 deaths), the 1996–1997 outbreak in Gabon (sixty-six deaths), in 2000–2001 in Uganda (224 deaths), in 2001–2002 in Gabon (96 deaths), the 2002–03 outbreak in the DRC, i.e. the former Zaire (157 deaths), the 2004 epidemic in Sudan (seven deaths), the 2007 epidemic in the DRC (187 deaths), the 2008–2009 epidemic in Uganda and the DRC (fifty-three dead), the 2012 outbreak in Uganda and the DRC (fifty-five dead), and last but not least the 2013–2015 outbreak in West Africa (nearly 12,000 deaths).[28] Anyone interested in these statistics will notice the increase in frequency and casualties characterising the epidemics. Could it be that Ebola is about to come out of its forests permanently and, thanks to our fast and modern transport, spread all over the world?

This was almost the case during the Ebola outbreak in West Africa. The latter, caused by the virus' Zaire strain, was much deadlier than the previous ones. According to the WHO, 'patient zero' would appear to have been Emile Ouamouno, a child that died in December 2013 in the small town of Guéckédou, in south-eastern Guinea.[29] It is likely that this child was contaminated as a result of eating infected bushmeat, possibly imported from Equatorial Africa. Then, the child is alleged to have infected his sister, his mother, his grandmother and a friend from Sierra Leone. All of them would perish a few weeks later,

28 AN: At the time when these words were written, the WHO had already stated that this number is but a low estimate, as can be seen on the following internet page: http://apps.who.int/ebola/sites/default/files/atoms/files//who_ebola_situation_report_16-09-2015.pdf.

29 AN: As confirmed here — http://www.france24.com/fr/20140409-propagation-ebola-afrique-ouest-inquiete-oms-sante/.

in January 2014. By the end of 2014, three inhabitants of Guéckédou had succumbed, followed by fifteen in January. Then, in February, nine new victims were reported, and two more in a nearby city. At the end of February, there were twenty-nine additional deaths in the region and five in the neighbouring area. All these cases were identified as cases of haemorrhagic fever but the link to Ebola was not formally established until the end of March 2014.[30]

The spread of the epidemic persisted, continuing to Liberia and affecting Conakry, the capital city of Guinea, which is inhabited by two million people. In April, Monrovia, the capital of Liberia, declared its first cases. In July, Freetown, the capital of Sierra Leone, also had its first cases of Ebola-related deaths. Everywhere, the number of afflicted people and deaths increased exponentially. Then came the first cases in Senegal and Nigeria, a country with more than 170 million inhabitants…

In light of this gradual spread and the alarming news stemming from the media, both *Médecins Sans Frontières*[31] and the WHO came to the aid of local health workers, as several countries began to take steps to contain the disease. The list of these measures is very well-documented on Wikipedia[32] and other informational websites:[33]

- On 24 March 2014, the French embassy in Liberia advises French citizens not to visit the affected areas in Guinea and northern Liberia, near the border between the two countries.[34]

30 AN: The related information is available at http://www.leparisien.fr/laparisienne/sante/guinee-la-mysterieuse-epidemie-est-une-fievre-ebola-22-03-2014-3697605.php.

31 TN: The usual English rendering is *Doctors Without Borders*.

32 AN: https://en.wikipedia.org/wiki/Ebola_virus_epidemic_in_West_Africa_timeline.

33 AN: http://www.bbc.co.uk/timelines/z9gkj6f.

34 AN: http://tempsreel.nouvelobs.com/monde/20140324.OBS0989/epidemie-de-fievre-ebola-en-guinee-le-point-sur-la-situation.html.

- On the 26 of March Mauritania closes all bridges and crossings across the Senegal River, which separates the country from Senegal. The latter then closes its own borders with Guinea on the 28 March.

- At the beginning of April, Morocco reinforces medical surveillance on all arrivals from West Africa.[35]

- On 2 August, an American missionary afflicted with the virus is taken from Liberia to Atlanta, USA, to undergo treatment. On 12 August, it is a Spanish priest who, in turn, is evacuated to Madrid. Other evacuations of this type take place during the crisis towards Germany, Norway, France, Italy, Switzerland and others lands. In August, Qatar decides to place a ban on the import of animals, food and meat from West Africa, while Sri Lanka rejects all travellers coming from the region. Algeria, the Ivory Coast and Kenya also implement increased surveillance measures. On 21 August, it is South Africa's turn to deny all access to its territory to any traveller arriving from the affected countries, as Chad announces the closure of its borders with Nigeria to prevent further spread.

- Still in August, Sierra Leone and Liberia deploy their army to contain the spread of the epidemic, as they attempt to stop the panic. Russia and China send substantial contingents of both doctors and equipment to assist the region's countries. On 18 August, Cameroon closes its 1,600-kilometre border with Nigeria.

- On 6 September, the government of Sierra Leone announces the 'home confinement' of its population from 19 to 21 September in order to fight the growing epidemic.[36] *Doctors Without Borders*

---

35 AN: Here is the link—http://tempsreel.nouvelobs.com/monde/20140401.OBS2147/fievre-ebola-controle-sanitaire-renforce-au-maroc.html.

36 AN: http://www.lefigaro.fr/flash-actu/2014/09/06/97001-20140906FILW-WW00093-ebola-en-sierra-leone-la-population-confinee-du-19-au-2109.php.

denounces the international aid for being 'slow, derisory and irresponsible'.[37] At the beginning of September, it is Cuba's turn to send a very large delegation of health experts. On 26 September, the UN announces the dispatch of equipment to West Africa (five helicopters, 470 all-terrain vehicles) so as to reach the most remote communities.

- On 14 October, prominent Western leaders label the Ebola virus 'the most serious health emergency in recent years'. On 15 October, the head of the UN mission tasked with coordinating the emergency response to the Ebola virus (UNMEER)[38] is pessimistic: 'The epidemic is moving faster than us and is winning the race. If we do not stop Ebola, we will have to face an unprecedented situation for which no plans are in place'. On that same day, US President Barack Obama declares himself alarmed by the spread of the Ebola virus and cancels most of his trips to focus on stopping the virus.

- On 16 October, the WHO announces that fifteen African countries located in the vicinity of the area most affected by the virus will benefit from increased assistance to prevent the spread of the epidemic. On 17 October, the East African Community announces the dispatch of more than 600 health professionals to the region, including forty-one doctors.

- On 17 November, Mali places 577 people under health surveillance to stop the spread of the epidemic. In early December 2014, the Geneva University Hospitals medical centre announces that the first thirty-four volunteers participating in a clinical trial of the Canadian experimental Ebola vaccine had tolerated the injection well.

37 AN: http://www.msf.org/article/ebola-failures-international-outbreak-response.

38 TN: The United Nations Mission for Ebola Emergency Response.

- On 29 January, 2015, the UN declares that the epidemic is 'slowing down but is not yet contained'.

- On 27 and 29 March, 2015, the entire Sierra Leonean population is confined to their homes in an effort to combat the epidemic.

- From March to May 2015, the number of new cases decreases before reaching the point of virtual cessation in June 2015, despite the presence of some sporadic cases.

- In the process, two countries managed the risk of contamination most effectively, namely Senegal and Nigeria:

- In March 2014, Senegal closed its borders with Guinea. Despite this measure, one case of Ebola was declared in a hospital in Dakar. The patient was isolated, treated and did not die. However, he was kept in quarantine for more than twenty-one days, while the staff caring for him and the seventy-four other people who could have contracted this first patient's illness were closely monitored. These strict measures were successful and the Senegalese authorities were congratulated by the WHO for their medical crisis management.

- On 20 July 2014, a case of Ebola was identified for the very first time in the city of Lagos, a metropolis of twenty-one million people, where an epidemic could have been devastating. The government immediately isolated the patient, who had willingly declared himself to the authorities, as well as all the people who had been in contact with him, including nurses. After twenty reported cases including eight deaths and a long confinement period, Nigeria was declared 'safe' on 20 October.

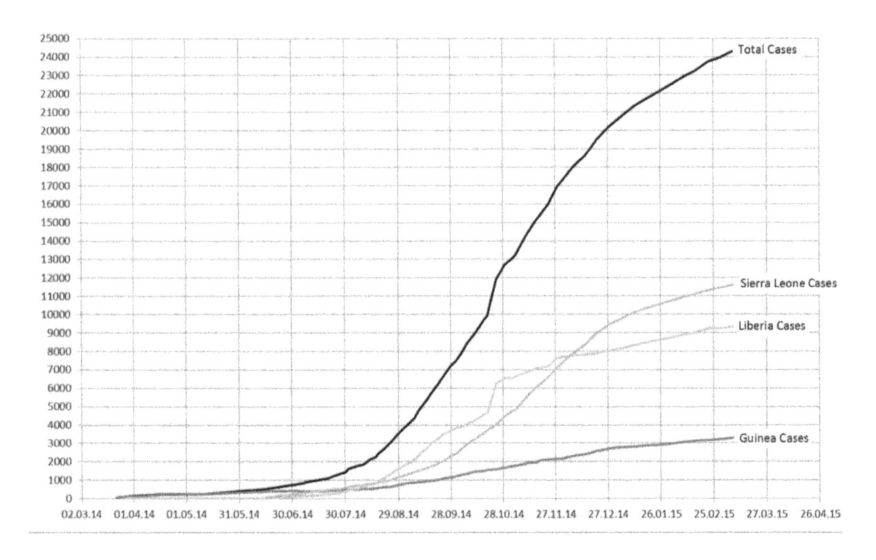

Figure 20. The Spread of Ebola in West Africa — Reported Cases by Countries, as of 1 February 2015 (2014–2015)

European countries initially managed this crisis in a riskier way. Indeed, no border closure or flight ban was required and only airport checks on flights arriving from high-risk countries[39] were carried out. These checks would certainly have identified a sick person, but not a person carrying the virus without exhibiting any symptoms. Moreover, they could not have identified a person who had stopped over in a third country. An outbreak threat was thus very real indeed, especially since patients might have been tempted to travel to Europe or the United States, hoping to find better quality care there (which would not have been the case, incidentally). As such measures are considered insufficient, it is but a stroke of luck that the epidemic did not spread through any European urban centres. And if Europe and Asia are said to have adopted, according to WHO officials, a risky

39 AN: http://www.lemonde.fr/planete/article/2014/10/16/ebola-la-france-met-en-place-des-controles-sanitaires-dans-les-aeroports_4506872_3244.html.

policy, the United States — where four cases were declared and one tragic death reported — was even slower to introduce such measures.[40]

It was also very dangerous for some countries to repatriate patients suffering from Ebola to be treated in their own hospitals. This was the case of the United Kingdom, Spain, Norway, the United States, France, Germany and Switzerland. These patients could have been treated in African hospitals, thus avoiding the risk of contaminating medical staff during their transport by specialised plane and their treatment in the hospitals of their destination country. Indeed, Spanish nurses were infected in Spain[41] while caring for one of the patients. With a virus such as Ebola, lugging highly infectious patients around the world seems to be an ill-considered sort of risk. Were these decisions taken in order to test, under real conditions, the teams and procedures established by the governments to counter this type of threat, in a display of utter contempt for the danger caused to the populations?

Serious protocol errors were identified[42] in American and European hospitals during the treatment of patients. Given the dangerousness of Ebola, these would never have been made by any staff working in Africa and accustomed to following precise and safe procedures.

In Switzerland, a Cuban doctor who had contracted Ebola in Sierra Leone in early November 2014 was transported to Geneva University Hospitals (HUG) for treatment in a level 4 care unit. According to Bertrand Levrat, the director of HUG, '200 people, teachers, doctors, nurses, and caregivers took turns caring for the patient day and night, for a period of 16 days. Every person who came in and out of the negative pressure room had a care plan and knew exactly what they had to do. They remained there for two and a half hours at most,

---

40 AN: http://www.ibtimes.com/ebola-spreads-europe-eu-doing-better-job-containing-ebola-us-1707148.

41 AN: https://en.wikipedia.org/wiki/Ebola_virus_disease_in_Spain.

42 AN: http://www.cbsnews.com/news/ebola-virus-response-in-dallas-had-mistakes-cdc-admits/.

defined as the maximal concentration time.'[43] The testimony of one of these caregivers, Professor Jerome Pugin, who acts as the head of the intensive care unit of the HUG, reveals the anxiety and stress inherent in dealing with such an illness: 'Because of our clothes, our actions were slower than usual at the time. Indeed, it is very hot under such a suit. One loses a lot of water this way and their ski goggles quickly become fogged. Possible improvements have, however, been identified. Treating an Ebola patient is a very stressful situation; but I was never frightened, and our preparations bore the desired results. The most complex process is that of undressing. It takes 30 minutes and, since one has been in contact with the patient, requires a lot of precision.'[44] The patient treated in Geneva survived, and no other case would be reported.

A vaccine known as VSV-EBOV was tested from September 2014 onwards, and clinical studies have been ongoing ever since. It is difficult to predict the effectiveness of a vaccine, let alone determine whether it will work against new strains of the virus.

What are the social and economic effects of this epidemic?

- Although no major restrictions on travel from either Europe or the United States have been issued, tourism in impacted countries — as well as in unaffected neighbouring countries, including Senegal and even Kenya and South Africa — has fallen sharply. In addition, business travel between African countries was greatly reduced as a result of the containment measures and the prohibition/restriction of certain flights.

- Trade, especially the export of agricultural products, also declined. This was due to two main reasons: the reluctance of farmers to visit markets and the highly pronounced absenteeism of farm workers, totalling 40% in the case of Liberia and Sierra Leone and 25 %

43 AN: *Le Matin Dimanche*, 14 December 2014.

44 AN: Ibid.

for Guinea. Economic sectors with strong ties to the export field, including diamond mining in Sierra Leone, found themselves affected, since many mines were located in the most heavily affected areas.

- The overall cost for Africa is estimated at one GNP growth point for the entire continent during the year of 2014.

- A high proportion of the medical staff[45] was infected because of the lack of personal protective equipment (masks and gloves); their misuse; the largely insufficient number of doctors; and the fact that they were overworked, which causes greater fatigue and therefore makes it easier to make mistakes.

- In some areas, official communication may have been misunderstood by the population or interpreted as an extension of postcolonial discourse, for instance when referring to the consumption of bushmeat as a source of contamination. Similarly, authorities or NGOs may have been perceived as abiding by a 'domineering sort of discourse' that 'stigmatises certain communities, whose members are victims of contempt or cultural prejudice which official messages only serve to reinforce'. This discourse urged the local populations to move away from the forest, although the latter is a resource, particularly on a medicinal level. Local communities were also aware of the fact that hospitals lack nurses and doctors, which is why some refused to send their family members there. In addition, many doctors were stigmatised by the population as possible vectors of the disease.

---

45 AN: According to the health services of Sierra Leone (11 August 2014), nearly 10 percent of the country's Ebola victims — i.e. thirty-two people — were nurses. In total, more than 240 health care workers were infected with the Ebola virus, of which at least 120 died in West Africa, according to figures released by the WHO on 26 August 2014.

- Many African countries coordinated their aid efforts and reinforced the idea of intra- and supra-statal collaboration and solidarity in the region, both in the field of health and that of economy.
- In some of the most affected areas, local social structures found themselves disrupted, with a high number of orphans and villages drained of their inhabitants.
- In the aftermath of this epidemic, the economies of impacted countries are now slowly recovering, especially in rural areas. However, this decline in activity has led to food shortages and increased unemployment in these countries, all of whose economies and social structures were already affected by the wars that had raged there during the 1990s and 2000s.
- Culturally, it is still too early to determine whether the presence of the virus will call into question traditional practices related to the preparation and burial of the dead and ensure that it is carried out by professionals and not family members. The importance of educating rural communities on the right procedures to follow for rapid access to health care was also highlighted.
- Lastly, it should be noted that there are rumours circulating both in relation to the origin of the Ebola virus and concerning its recent spread in West Africa. The most persistent ones claim that Ebola was allegedly created in American research laboratories during experiments conducted in Africa or that it is a strain that mutated spontaneously and thus became infectious for humans and large primates. Unfortunately, we cannot be certain that we will ever find out the truth about this. With regard to the 2014 Ebola outbreak, here are some elements that may seem disturbing — let everyone form their own opinion of things:

- The Ebola virus that struck West Africa is, at first glance, a mutant close to the Zaire strain (yet located at a distance of 3,500 kilometres);

- The United States has three level 4 research laboratories: one in Liberia, one in Sierra Leone and one in Guinea. These centres provide maximum security to work on the world's most dangerous biological agents, such as Ebola itself;

- At the time of the events, Guinea had not yet signed the International Biological Weapons Convention,[46] and although Liberia was a signatory, it had not ratified it yet. These two countries were therefore not very interested in the research carried out in the laboratories established on their own soil. The Convention would only be ratified by Guinea and Liberia on 4 November 2016;

- The laboratory[47] located in Sierra Leone (and subsidised by both the *Bill and Melinda Gates Foundation* and the *Soros Foundation*) was closed by the authorities. It would seem that patients who went there for a vaccine came out with Ebola...

- The United States sent 3,000 soldiers rather than doctors and health service members. Even if the stated purpose was one of logistical support, there is still doubt surrounding other missions: the protection of US interests in the country and that of certain laboratories (and thus the samples and data that they contained), etc.

---

46 AN: More information on the subject can be found by simply following this link: https://www.unog.ch/80256EE600585943/(httpPages)/7BE6CBBEA0477B52C12571860035FD5C?OpenDocument.

47 AN: https://birdflu666.wordpress.com/2014/07/08/us-bioweapons-lab-in-sierra-leone-at-the-epicentre-of-ebola-outbreak/#more-6323.

- The biologists responsible for various programmes, including Glenn Thomas and Humarr Khan, both of whom conducted research on the subject, died in the prime of lives…[48]

It is not yet possible to ascertain what the long-term impact of this epidemic will be on West African culture and economy. Today, the following question could indeed arise: was this Ebola wave a mere warning, or a rehearsal for the next pandemic? Viruses are lurking, waiting for the right moment to strike, as health organisations prepare for the challenge. Who will be the winner when the bell announcing the next round sounds?

48 AN: http://www.wikistrike.com/2014/08/des-experts-d-un-laboratoire-d-armes-biologiques-qui-fabriquait-une-nouvelle-souche-du-virus-ebola-tues-dans-le-vol-mh17.html.

6. Scenarios

> To be a good soldier you must love the army. But to be a good officer you must be willing to order the death of the thing you love. That is … a very hard thing to do. No other profession requires it. That is one reason why there are so very few good officers. Although there are many good men.
>
> — Robert E. Lee, Army Commander of the Confederate States (1807–1870)

Unknown Virus

Fiction

Dan finished putting on his white suit, using gestures that reflected long-term habit. He entered the airlock, waited a few seconds for the door to close behind him and put his swipe card on the console that controlled access to the laboratory.

A characteristic sucking sound was heard as the glass wall moved aside. Briskly, the man then entered the secure room.

Suddenly, he stopped short: Patricia was already there!

—'Hello Patricia,' he said in a voice tinged with annoyance. 'You're early today. Everything o.k.?'

—'Hi Dan. Yeah, don't worry. I was just looking forward to seeing the results.'

—'All the same, you could have waited!'

The young woman turned around.

—'And you? What are you doing here, an hour before everyone else? Did you want to warn me?'

—'Well, it was me, after all, who conducted all those tests. So, it's normal for me to find out the results first.'

—'You know very well that without Monsanto's money, the lab would not even exist. So put a cork in it, will you! Otherwise you'll be heading for a lot of trouble.'

—Patricia! Have you actually forgotten that you lack the ability to conduct these studies? You need me.'

—'Biologists—we can find them anywhere these days. Plus, the truth is you're wrong: I no longer need you!'

—'What? What do you mean?'

—'The tests are positive! I have all I came here for.'

Suddenly, Dan took a step towards the glove box.[1]

—'Let me see', he said, pushing his colleague gently but firmly aside.

The man put his eyes on the microscope lens and, with quick gestures, refined the focus.

—'Whoa! The parasitic fungus has been stopped dead. In fact, it even *looks* dead.'

—'Oh, I can confirm it—it actually is.'

—'But how is this even possible? In so little time? This would mean that our agent's virulence has proved extraordinary.'

---

1 TN: A glove box (sometimes also glovebox) is a sealed container intended to allow one to manipulate objects in a separate atmosphere.

—'Which is why it's a major discovery. Just imagine: in less than three days, we can rid any culture of a fungus that, until recently, resisted all our antifungals.'

The biologist nodded silently and approached another glove box.

—'And what do our immunological tests say?'

—'Nothing conclusive for the moment. Our bacterium does not belong to any known group.'

—'And the analysis of surface antigens?'

—'The wall shows the presence of lipids that are typical of the archaea domain. The proportions are also consistent with this.'

—'And the DNA study?'

—'The first phase will be done in an hour.'

Looking at the results sheets, Dan scratched his chin. After a few minutes, he began to talk to himself, as if the young woman no longer existed.

—'We need to run a deeper genome analysis; and look for any trace of protein synthesis. A selective media culture could also be interesting… And to me, observation under a scanning electron microscope seems…'

—'Stop it!' Patricia exclaimed suddenly. 'You can do all of that later. I have no time to waste now. I have a plane in less than three hours. And I need the sample immediately.'

—'Excuse me?'

—'Monsanto is holding its biennial conference tomorrow in New York. I had my boss on the phone earlier, he wants me to present the results.'

—'"The results?" Seems a little premature; and as for the agent itself, that's impossible. You will never get transport permits in time.'

—'Who's talking about permits?'

—'Have you lost your mind? I won't let you do that. It would be plain and simple recklessness! We do not know anything about this bacterium. You cannot take a sample and fly there!'

—'Look at me. Do I look like a fungus?'

— 'What's your point?'

— 'This archaeon is a killer of plant-parasitic fungi. I, on the other hand, am a human being, and there's no risk for me at all!'

— 'But nobody knows anything about it! You're extrapolating results without any...'

Suddenly, Patricia lifted her palm towards her colleague, signalling him to stop talking. With an annoyed gesture, she then took her mobile phone out of her pocket and quickly dialled a telephone number.

— 'I thought you were going to react this way. Don't move, I'll give you my boss on the phone,' she said, handing him the device.

For the next few minutes, Dan nodded, unable to formulate a single complete sentence.

Patricia saw the man's features harden, his face becoming impervious as the conversation progressed.

— 'So?' she asked as soon as her colleague had hung up.

'I have two options,' Dan said nervously, with a half-blank look in his eyes. 'I either persist and, in this case, lose my job with immediate effect, or give you the sample and keep my post with a one-year salary bonus thrown in and a lab budget twice as big.'

— 'Oh, I know all about that, but what interests me now is your decision.'

— 'I don't know. It violates the most basic safety regulations, as well as my principles.'

— 'Think about your wife; and your children, who attend the best school in Helsinki. You have a nice house, and you and your family just got back from Honolulu... Do you want to lose all that because of one miserable sample?'

— 'You're not getting it: we don't really know what it contains. It spent 70,000 years in the ice and we only reactivated it eight days ago.'

Patricia walked ahead until she stood in front of her colleague and put her face a few inches from his.

—'Listen to me, Dan. I don't care about your feelings or principles. I don't want to miss my plane, so I'll just take this sample, with or without your consent.'

—'I…'

'Shut up,' she said, interrupting him curtly. 'I suggest you go to the rest room. When you come back, I won't be here anymore. This way, you won't be responsible for anything and your conscience will be clear.'

—'Um…'

—'There's no point arguing about it. Go on, then! Go have a coffee!'

Dan went through the airlock again, as if in a trance. He did not like any of it at all.

Slowly, he walked to the cafeteria and ordered a cappuccino as soon he got there; he then drank it slowly, still lost in his thoughts.

Realising that the place was about to fill up, he suddenly rose to his feet. He was in no mood to talk to anyone.

He headed back to the lab, hoping to run into Patricia on the way. Why? He had no idea. Unconsciously perhaps, he wanted to try and change her mind again.

This was not to be, however, because he did not cross paths with the young woman. When he got there, the place was already empty. On the other hand, the Honolulu postcard he had sent the team on his last vacation was now glued to the airlock door.

Dan forced a smile. Maybe Patricia was right after all.

He equipped himself again, and then, for the second time that day, went into the lab.

Soon, he noticed that the sample that had been kept in the cold room was no longer there. A transport box had also disappeared.

There was nothing he could do about it anymore. What's more, Edward, his assistant, was bound to arrive soon. He thus decided to get to work and keep it all under wraps. He still had some things to do—indeed, his priority was now to carry out all the tests he considered appropriate on the bacterium. Forgetting his worries, he set himself to work…

By the time night had fallen, Dan had only taken a short break to go to the cafeteria and have some pizza. His assistant had left an hour earlier and he was alone again.

The fatigue and, of course, the stress of the entire situation were giving him headaches. He let out a long sigh and moved away from the electron microscope. The test could wait a bit. Right now, what he needed was a brief respite. He would take the opportunity to eat.

He followed the usual procedure to exit the lab and, once more, headed for the cafeteria. Upon entering, he waved to a group of people chatting loudly by the counter, grabbed his tray and ordered some tagliatelle alla carbonara.

As he was about to pay, Dan could not help noticing a discussion whose tone was gradually escalating.

'But I'm telling you that Karl was rushed to the hospital that night,' the tallest of the three men said.

'It's been two days he hasn't been feeling well,' said another. 'Yesterday, he told me he had terrible headaches and his body was painfully stiff. And I totally believe him — he was as white as a sheet and his forehead was covered with sweat.'

— 'I personally think it's all very suspicious,' the third man added. 'You don't get taken to the emergency room for having a fever.'

— 'I agree. I'd bet you my whole wage he caught some crap at one of our sites.'

— 'What makes you say that?'

— 'You remember what happened about ten days ago? When we did that special drilling job for the lab?'

— 'Sure, and?'

— 'Well, Karl took off his gloves, put his hands on the ice core said: "It's been 70,000 years since someone last touched you, isn't that right, my pretty?" I think he even ate a piece of it…'

Dan felt his body stiffen. Was it the same drilling job from which the sample had come?

His blood froze. He grabbed his cell phone and called his friend Erwan, who worked as the company's doctor.

After the usual greetings, he went straight to the heart of the matter.

—'You know about Karl?'

—'Yes,' answered Erwan. 'I am actually writing the report.'

—'And? What's your conclusion?'

—'He died of a brutal collapse of all his vital functions caused by a biological agent of unknown origin.'

—'What?! You mean he's dead?'

—'I thought that was the point of your phone call.'

—'Wait a minute, do you have any idea what kind of germ is responsible for this? He was surely infectious! It is imperative that we alert people before it's too late!'

—'I'm afraid it already is. I have three people here with the same symptoms…'

—'Exactly, so we must initiate the containment measures. You know! Apply emergency procedures and…'

—'I'm one of those three patients,' said Erwan, interrupting him.

Dan was speechless; it took him a few seconds to pull himself together.

—'My God! That's terrible!'

—'That's what I've been telling myself,' whispered the doctor faintly.

—'Do you have any leads? Antibiogram results?'

—'An antibiogram? But we haven't found any bacteria!'

—'No bacteria? What's responsible for all this, then?'

—'It seems rather viral. But hey, you're the biologist. It's up to you to tell *me*!'

Dan felt his heart pounding in his chest. He mumbled a few words of encouragement to his friend, hung up, and ran towards the lab, leaving his tray where it was.

Without wasting a single second, he equipped himself, passed the airlock and checked the latest results. After a few minutes, he went to the electron microscope and looked at the preparation.

Nothing abnormal.

The medium only contained the fungus-killing archaeon.

Perhaps Karl's death had nothing to do with the core sample in the end.

In his gut, however, Dan felt he had missed something.

He bent towards the microscope again… and suddenly saw it!

Hidden in the very heart of the bacterium lurked a tiny form, an assemblage of genetic material clearly foreign to the host.

Still, Dan remained cautious. He was not sure he had found the culprit. The structure was reminiscent of a virus, but it was way too big to be one.

He thus decided to examine the sample that had killed the parasitic fungus more thoroughly.

This time around, the message was clear. The electron microscope allowed him to see very large quantities of the above-mentioned structure. In addition to this, it was now visible everywhere, both in the archaea themselves and in the environment that surrounded them…

For long minutes, the biologist remained pensive.

Little by little, an idea began to take shape in his mind.

What if this element was a virus, but one that belonged to an unknown genus and differed from all that exists today? An entity with new and destructive capabilities…

And what if the archaea's awakening had somehow reactivated this sleeping agent?

Dan began to sweat. He had played the sorcerer's apprentice and the result did not take long to manifest…

If his analysis were confirmed, the whole planet would be in great danger. Considering its speed of action and contagiousness, the biologist realised that he may have awoken a monster more dangerous than Ebola itself.

The start of an epidemic within the company, or even in all of Helsinki, seemed likely. Moreover, a young woman was on her way to one of the most populous cities on the American continent, carrying a biological agent dating back to another age and capable of causing the largest pandemic the world has ever known.

Dan grabbed his phone and, for a few seconds that seemed to go on forever, thought hard.

In the end, he called his wife and gave her specific instructions: she was to load the car with the stuff mentioned in the list he had left in his office, take the children, not talk to anyone and drive without stopping until she reached their cottage a hundred kilometres to the north. Once there, she would have to focus on acquiring a maximum of non-perishable foodstuffs and amassing fuel reserves…

The Facts

The story above is, of course, a work of fiction. Nevertheless, recent events have shown that such a development is not impossible, and there are several points that support this.

On the one hand, accidents and terrorist attacks involving lethal biological agents are far more common than they seem.

On the other hand, laboratories that conduct research into microorganisms abound these days. Since most countries have signed the Biological Weapons Convention, they are obviously not allowed to develop such arsenals, but because there are no control mechanisms in place, unscrupulous governments could carry out some work in this direction, all under the guise of studying these pathogens in order to synthesise remedies. In addition, some scientists have set out to 'reconstruct' agents that caused the greatest calamities in the past. In 2011, a certain team thus reconstituted the plague bacillus, which, in the Middle Ages, was responsible for the death of one European person out of three. These biologists took pieces from the skeletons of people that had died of this disease nearly 700 years ago. Using modern methods, they were able to assemble 99% of the *Yersinia pestis* genome. Furthermore, it is not uncommon to see the financial interests of some multinationals override factors such as consumer safety or health.

Last but not least, in 2014, scientists brought back to life a giant virus that had been trapped in the ice for approximately 30,000 years. This completely new biological agent is the largest known virus to date. Fortunately, it only infects amoebae and remains harmless to both humans and animals. However, and as we have already seen, current lifeforms are said to only represent a small percentage of what the Earth has borne in the course of its existence... This allows for a considerable number of potential threats. If viruses can survive 30,000 years in the cold, who knows what scientists will awaken next? Following the same logic, what would happen if one of these biological entities resurfaced as a result of melting ice caps or oil drilling? If such an old agent returned from the past, the human immune system would not know how to deal with it. Would the result be the same as for those American Indians that were exposed to smallpox?

Analysis

Viruses are capable of infecting just about any living entity on Earth. Nevertheless, they do have a certain specificity: for example, an agent characterised by an affinity with plants will normally be harmless to animals, and vice versa. This peculiarity stems from the fact that these biological entities must be able to identify cell receptors in order to penetrate them. Otherwise, they remain outside and are unable to infect the host. This selectivity is quite accurate and, in most cases, the viruses can only trigger diseases in a specific type of species.

Nevertheless, some of them can still cross this species barrier as a result of some mutations or re-combinations, as seen with bird flu, which can impact birds, chickens, humans and even pigs…

When it comes to the fictional account related above, it seems unlikely that the giant virus infecting the bacterium could ever be dangerous to humans. However, since such entities are of an unknown type, nothing is impossible.

The first symptoms described in the case of patient number one correspond to a flu state and are quite representative of a viral infection. Similarly, the incubation time and fatal outcome in a period of eight days are both credible, albeit characteristic of a virus of extreme pathogenicity and virulence, as is the case with Ebola or smallpox.

The Flu

Fiction

Stephanie was driving recklessly. She accelerated hard as soon as she could, trying every time to overtake as many cars as possible by changing lanes unexpectedly.

On the radio, the news reports all repeated the same thing: that year, the flu epidemic had been different. Severe symptoms appeared out of the blue and the virus resulted in a much higher mortality rate than usual. In addition, it seemed to favour young people as targets. To make matters worse, it could, it seemed, be transmitted to humans by birds. Realising how many pigeons there were in Paris sent shivers down one's spine…

Suddenly, Stephanie swore as she got to a crossroads. Annoyed, she banged the steering wheel hard with her fist: the police had blocked the road she had been planning to take.

She clenched her teeth and began to sigh loudly. She knew Paris well; she had been living there for more than fifteen years and knew how to avoid the junction by going through the side streets of Saint Michel. Still, this inconvenience exasperated her.

The call she had received from the school twenty minutes earlier had terrified her. The headmistress had asked parents to pick up their children when one of the pupils had been taken to the infirmary with a temperature of 40 °C, complaining of a headache and body aches.

Still unable to believe it, Stephanie closed her eyes for a moment and took a deep breath through the nose. How could it have degenerated so quickly? The previous weekend, everything had been fine, but then, three days earlier, the first cases of flu had surfaced in the city. And that morning, the authorities had announced that it was preferable to minimise the use of public transport, limit one's travelling and avoid public gatherings.

The young woman suddenly emerged from her thoughts. The line of cars had just come to a stop. It seemed that some driver was trying to somehow park his old Mercedes into too narrow a parking space.

Stephanie felt her heart rate increase and her forehead sweat. She could not afford to wait any longer, knowing that her little girl was in a school where she could catch a deadly sort of flu from one moment to the next. Imagining the worst, she began to honk the car horn and was quickly imitated by those around her.

After a few minutes that seemed like an eternity, the traffic finally resumed. Relieved, she once again started driving as fast as she could and nearly knocked over a cyclist crossing the last junction. Hands trembling, she looked at the street in front of her, which led to the school. It was too crowded with vehicles parked in all directions: it seemed impossible to get through.

Hesitating for a few seconds, Stephanie turned to her right and, after a dozen meters, stopped on the lane reserved for buses. She locked her Clio and, having cast a quick glance at both sides of the road, ran across.

The closer the young woman got to the school, the more she felt the tension intensify and trigger the onset of a migraine. Around her, many moms were rushing in and out of the school. Nobody was paying attention to anyone else — or, to be more accurate, everyone was actually *avoiding* everyone else…

Her heart pounding in her chest, Stephanie briskly entered the premises. One single, hastily created sign indicated the way to where classes had been assembled.

Without actually realising it, she ran to where the second-graders were. After about a hundred meters, she reached the gym that was supposed to be sheltering her little girl. Running her eyes across the crowd, she shouted Caroline's name in a loud voice.

No answer.

With anguish gripping her throat, she called out again.

Suddenly, a shout echoed through the air and a blonde head broke through the crowd, rushing into his arms.

— 'Mommy, what's going on? I'm scared.'

— 'Nothing at all, love,' replied her mother, casting worried glances all around her. 'Come on, we're going home.'

— 'But why is everyone gathered here, waiting and not doing anything?'

— 'There are some little boys or girls that have caught the flu… So, to prevent the whole class from getting ill, the director has asked the parents to pick their kids up.'

Having scarcely finished her sentence, Stephanie grabbed Caroline's hand and led her away, making sure they didn't touch anyone.

Once outside the gym, the two alternated walking with running until they reached the Clio. The young woman gave a roar of rage when she noticed that her rear-view mirror had been torn off. Perceiving the growing disorder spreading across the avenue, however, she decided not to linger and return home without further ado.

The traffic was now chaotic. Drivers no longer respected traffic signs, their faces betraying a growing state of fear.

Stephanie thought it better to go back by taking side roads. In the space of twenty minutes, she had arrived at the bottom of her block of flats. She quickly exited the vehicle and urged her daughter to do the same, before locking it.

As she was preparing to push the door open and enter the lobby, a disturbing thought crept into her mind: her food supply was running very low.

For a few seconds, she remained indecisive. Finally, she grabbed Caroline by the hand and strode to the small grocery located at a 100-metre distance from there. As she was about to enter the store, the door opened to reveal an old woman coughing and sputtering continuously. Reacting reflexively, Stephanie pushed her daughter behind her as she stood paralysed before this elderly person. Without really being aware of it, she allowed her pass while staring at her in fear, using her foot to stop the door from closing. Frozen in this position, she desperately tried to determine what her next move would be.

Realising suddenly that onlookers were staring at her with a puzzled look on their faces, she decided to go into the shop. Advancing carefully, she was relieved to find that no other customers were present. Not wasting any time, she made sure she only took food from the bottom shelves—the ones she thought were least likely to have been touched. In this manner, she picked up some tins, pasta and rice. A minimal amount of food, enough to last four or five days.

She forced herself to smile as she arrived at the counter.

—'Hello, René, how are you today?'

—'I'm fine myself, though I must admit that people are particularly tense right now. And I've already been robbed twice this morning. At this rate, I think I'll be out of business by tonight.'

—'Yeah… It's this thing about a deadly flu spreading.'

—'Oh, speaking of that, I still have a bottle of disinfectant if you're interested…'

Stephanie felt her heart leap in her chest. She'd found a solution to her problems! Well, for the time being, at least.

—'What a great idea! You wouldn't happen to have two of those by any chance, would you?'

—'The last two, actually. Sure, it's fine by me; that'll be 57 euro. For the lot.'

—'Here you go, keep the change. Have a good day,' said the young woman, heading for the exit.

No sooner did she find herself outside than she hurriedly opened one of the bottles and proceeded to smear both her own hands and her daughter's with the antiseptic product. She also cleaned the handles of her plastic bags, and then rushed towards her block of flats. Once she had climbed all three floors, she entered the three-room flat and locked the door behind herself.

At long last, she felt safe! Her daughter Caroline was there with her now. And, taking into account the supplies they already had in their flat, they had enough food to last more than a week. This gave her time to anticipate things…

Suddenly, cries of distress rose from the street, echoing in her ears. Stephanie rushed to the window and realised that one of her neighbours had been attacked. Judging by the bag now lying at her feet, she seemed to have been robbed of her shopping items…

A wave of anguish rose through her chest. It all seemed incomprehensible to her; after all, her neighbourhood was considered safe and problem-free. How could someone be assaulted and robbed in the street, and in broad daylight at that?

The young woman then began pondering her own physical safety. She knew that when people were scared, they sometimes reacted very violently. Furthermore, if chaos were to set in on a long-term basis, groups of unscrupulous individuals would not hesitate to rob ordinary citizens out of sheer greed or so as to get their hands on food and water…

Suddenly, her mobile phone began to ring, startling her. It was Clément, her husband.

She picked up the phone and started talking to him. Slowly, the tone of her voice rose as distress came over her.

Once the phone call had ended, she sat down on a stool and leaned against the wall, weeping softly—Clément couldn't come home. His flight had been cancelled until further notice, and he was stuck in Algiers…

How long would she be alone with her daughter? How many days could she remain hidden without having to go out to find food? What would happen if criminals entered her home?

With a sudden shaking of the head, Stephanie tried to chase her dark thoughts away before slowly rising to her feet. As she passed in front of Caroline's room and saw her flicking through her favourite book, she flashed her a thin smile and headed for the flat's entrance. She found the pepper spray she had bought when Clément first started going on his trips and gripped it tightly in her hand.

God, how little protection that offered…

Worried, she went back to the window. Many questions were racing through her mind, simple and basic questions which she could not answer: would she keep the light on at night? If the apartment remained dark, it could eventually attract burglars. In the opposite scenario, criminals would know that there were people and therefore money and resources there… And what would she do in the event of a power cut? What would happen to the bit of food she still had in her fridge? And what if the crisis lasted a long time? She would have to go out sooner or later…

The Facts

The account above is obviously another tale of fiction. And yet, in 2009, the situation almost degenerated: a new virus, of the influenza

A kind (and H_1N_1 subtype) came close to causing global panic. This agent had, indeed, a specific characteristic that frightened the scientific community: in addition to its affinity with the human species, it also possessed aspects that derived from bird flu and swine flu viruses. In theory, the disease could have spread and been transmitted within and among the following three groups: human beings, birds and pigs.

This flu outbreak quickly contaminated the entire planet and, on 11 June 2009, the WHO declared it to be the first pandemic of the twenty-first century.

Although of considerable magnitude, this type A (H_1N_1) influenza spreads like any other:

- through the air — when a person coughs, sneezes or splutters, they allow a large number of virus particles to become airborne and perhaps be inhaled by people nearby;
- through contact with infected people — when kissing, shaking hands, and so on;
- through indirect contact — when touching objects that have been contaminated by ill people, such as doorknobs, escalator railings, banknotes etc.

Fortunately, the mortality rate characterising this biological agent was not as high as had been expected. Although it is difficult to make accurate estimates, it seems to total 0.2 fatal cases for each 1,000 infected people, which is lower than the common seasonal flu (with 1 fatal casualty for each 1,000). Nevertheless, a study conducted in 2012 by the Center for Disease Control and Prevention (CDC) puts the figure in excess of this and estimates that around 280,000 individuals succumbed to this virus on a worldwide scale during the first twelve months of the pandemic (which makes its mortality rate equal to that of seasonal flu). The major difference is that in the case of influenza type A (H_1N_1), young populations have been much more impacted, whereas with seasonal influenza, the main victims are among the ranks of the over-sixty-fives.

This threat has, however, brought about a general sort of awareness. It has forced many governments to take exceptional measures, including the purchase of antiviral stock (Tamiflu, etc.) or the launching of mass vaccination campaigns. These decisions, all of which were made in a context of utmost urgency, were the subject of great controversy.

Nevertheless, it should be noted that this first pandemic, whose impact was similar to that of electroshock therapy, had at least one positive effect: in the years that followed, governments learned from this episode and proceeded to adapt or create prevention and control plans so as to improve their future readiness and reactions.

In the case of France, a document with free access is available for download.[2]

Analysis

As far as that tale of fiction is concerned, the focus is on the psychological aspect and not on the clinical symptoms. Indeed, the story describes a variety of reactions that a citizen could have at the beginning of such a crisis. It is thus possible to note:

- **A strong emotion, namely fear.**

  Faced with the unknown or with something outside their control, many individuals could be gripped by such a feeling. Their reactions then become more instinctive and, generally speaking, more violent, even irrational at times. Group phenomena,[3] or the mere

---

2 AN: Those interested in this document can download it by accessing the following webpage: http://www.risques.gouv.fr/sites/default/files/upload/sdgsn-plan_pandemiegrippale_octobre_2011.pdf.

3 TN: A group phenomenon, or 'groupthink', is a notion developed in 1972 by research psychologist Irving Janis, who defined it as 'a deterioration of mental efficiency, reality testing, and moral judgment that results from in-group pressures'.

sight of ill (or supposedly ill) people, only serve to amplify and stir up such fears.

- **Behaviour: isolation.**
  Such behaviour stems in part from the previously described emotion. Unprepared people tend to keep away from others and shut themselves away, either at home or in a shelter. They will only venture outside out of sheer necessity or when they believe that the situation has improved.

The following point is also interesting to note: the vast majority of people with families will try to protect their own first. This is especially true of parents with regard to their children. As described in the previous scenario, it would thus not be surprising to see pupils/students gathered in a school, even if this does seem contrary to logic in the case of a contagious disease that is transmitted through contact or the air. Indeed, since teachers themselves are parents like any others, some of them will certainly have taken to the road to collect their own offspring, leaving insufficient staff on the school premises to ensure its normal functioning.

Similarly, although institutions do function properly under normal circumstances (as tranquillity prevails and access to both food and services is easy), the situation can sometimes change drastically. In the event of a major crisis, for instance, public transport could be discontinued, banks closed, and supermarkets and service stations left without supplies or perhaps looted. If the situation persists, the theft, assault and murder rates tend to increase considerably. Most of the time, the perpetrators are isolated individuals or groups taking advantage of the prevailing chaos either to get their hands on other people's property — in order to survive or satisfy their greed — or to take control of a certain zone.

INTERVIEW WITH PROFESSOR FRANÇOIS CHAPPUIS, HEAD OF THE DEPARTMENT OF TROPICAL AND HUMANITARIAN MEDICINE, GENEVA UNIVERSITY HOSPITALS

—Throughout history, attempts to eradicate diseases have, at times, been successful, as was the case with smallpox. However, with the emergence of new diseases and the increased risks of globalisation, how can one prepare for pandemics?

F.C.: Here at the HUG, we monitor the evolution of the world's infectious agents on a daily basis. This is a normal task in a hospital department like ours, since our duty is not only to give advice to travellers visiting countries or areas where these outbreaks tend to occur and possibly provide them with potential prevention means, but also to remain very attentive in relation to those who return from such areas. We are always fully alert. If one has to deal with a potentially contagious disease of the Ebola type (which, although not very contagious, is still extremely serious), the bird flu type or the MERS-CoV type, we collaborate with the infection control and prevention services that issue written recommendations to potentially exposed facilities, i.e. to adult and paediatric emergencies, ambulatory care services, etc. An information note is thus given out to these facilities, stating what needs to be done, the type of protection measures to be taken by the personnel and the type of investigation that must be launched if a person returning from such a region exhibits a given symptom or symptoms. For example, if one suspects that a person arriving from the Middle East has contracted MERS-CoV, the person should be placed in respiratory isolation, respiratory samples taken and molecular testing carried out to confirm the presence of the virus. In the case of the Ebola outbreak, all patients returning from affected areas and suffering from fever and other typical symptoms were routinely screened, but there were no confirmed cases except for that Cuban doctor, who was repatriated from Africa in a highly organised and safe manner.

And yet, many of those who returned from Guinea, Sierra Leone or Liberia were either taken to the emergency room or placed into our care.

We had to be ready all around the clock, twenty-four hours a day, so that we could respond very quickly and place the suspected patient under observation, put on the appropriate protective equipment to examine him, and determine his status. We did so rigorously, even though the risk was minimal, until any and all possible contagion diagnosis had been eliminated.

These diseases are known to us, and we can thus prepare ourselves to fight against them.

It is not like dengue, which is a disease that originates from Southeast Asia and is neither contagious nor transmissible as yet in Switzerland, because mosquitoes are not present here in sufficient numbers. It therefore does not require us to take any special precautions. Here in Switzerland, I mean.

A more delicate situation would involve, for example, the bird flu that has been emerging sporadically for approximately ten years now. If someone came back from their stay in Egypt exhibiting fever and respiratory symptoms, there is no guarantee that their primary care physician would think of bird flu and ask the patient whether he/she had had any contact with birds or poultry.

—When a new disease emerges, how does one implement such measures in cooperation with other health-related services (ambulances, doctors) or security-based ones such as the police, firefighters, customs, etc.?

F.C.: I have already described the information and information transfer procedures that a hospital is to conduct. Outside this spectrum, it is the responsibility of the cantonal doctor to inform private hospitals, emergency services, ambulance companies, etc. of the measures to be taken. During the Ebola crisis, recommendations were issued at the federal level[4] before being transmitted at a cantonal level and, from there, to the various health actors across the territory as well as to the NGOs involved in the field, such as Médecins Sans Frontières (MSF). This is done to ensure that if any members of their staff, who are all more at risk of coming into contact with the disease, find themselves afflicted, they are effectively identified and treated.

MSF, with whom we cooperate very closely to guarantee their staff's medical care and whose members are sometimes exposed to contagious tropical diseases, asked us very early on during the 2014 Ebola epidemic to have doses of the ZMapp5 experimental treatment at the ready. We were actually the first hospital in Europe to have this medicine.

—From a medical point of view, was it not risky to transport an Ebola patient by plane from an African country to Europe or the United States?

F.C.: Compared to measles or the flu, Ebola is not very contagious. Last year's extensive Ebola outbreak was due to a huge delay in responding to the epidemic. Despite the warning signals given by MSF in April 2014 and even before that, the international authorities failed to react quickly enough. They no doubt thought that MSF's experience in dealing with and containing such an epidemic would suffice, which was not the case, however. This insufficiency, combined with the health deficits inherent in these West African countries, which are only just emerging from long periods of conflict and economic crisis, rendered this crisis particularly acute.

4 AN: Switzerland is a confederation of cantons governed by a federal state that exercises some prerogatives and centralises certain information. In the healthcare domain, the Federal Office of Public Health (FOPH; http://www.bag.admin.ch/org/?lang=en) is responsible for public health issues and the implementation of healthcare policies in cooperation with the cantons.

5 AN: https://en.wikipedia.org/wiki/ZMapp.

On the other hand, some reports have praised the excellence and professionalism of Senegalese and Nigerian health services, which managed to contain the epidemic. It is, therefore, possible to both deal with and contain such outbreaks.

In Europe and North America, thanks to people's protective capabilities, the thoroughness of the hospital systems, the use of adequate infection control methods and effective risk management, the threat was extremely minimal. Repatriating patients is therefore not a great risk and increases the patient's chances of survival. This is due to two reasons: first of all, the general healthcare ability to maintain vital functions is more efficient in a Western hospital, and secondly, access to experimental treatments could improve a patient's survival chances, although this has admittedly not been formally demonstrated yet. But as long as one only welcomes patients gradually…

— Is it because there isn't enough room?

F.C.: Yes, because there is not much room, and because providing that specific patient with the necessary treatment used up a lot of resources. It would have been difficult to offer another patient that same level of care. In truth, it would have been difficult to even welcome a second one shortly afterwards. Indeed, we had to allocate staff from other departments and strengthen the intensive care team so as to not compromise the other patients' healthcare level. And this happened in spite of the high level of preparation and training attained since the 2000s, in the aftermath of the Lassa fever alerts.[6]

— What, according to your scientific understanding, would be the worst possible scenario?

F.C.: A bird-flu type of disease that would become highly contagious as a result of involving airborne transmission from person to person. This kind of virus would have a pandemic potential

6 AN: http://www.who.int/mediacentre/factsheets/fs179/fr/.

similar to that of influenza and be associated with a very high virulence level bordering on the 50% mortality rate characterising bird flu. If such a disease mutated to become contagious through the air, while losing none of its virulence (indeed, there is no rule stating that a mutation necessarily retains its virulence, as there are cases when a virus' virulence is reduced by its mutation), we could be faced with a disastrous and global pandemic. Although the same could be said for Ebola or MERS-CoV, this is less likely from a virological point of view.

A return of diseases such as smallpox would be unpleasant, since it is no longer customary to carry out this very specific vaccination, but vaccine doses are still available to the general population, and such an outbreak could therefore be contained within a short time.

—In the event of a pandemic, what measures could the authorities take to limit contagion?

F.C.: In general, communication and informational transparency are paramount. Rumours travel fast and can trigger inappropriate reactions. Other than that, everything depends on the infectious agent and its transmission method, because the implemented measures are different: Ebola, for instance, is not the same thing as a respiratory virus. All front-line health facilities must be trained and equipped very quickly. The quarantine measures that were recommended back in the day are no longer acceptable and, in practice, difficult to abide by. We have since moved on to training and information strategies that ensure the ability of hospital structures to care for their patients. Mentally, it is very important for the population to know that the hospital system will take care of the afflicted.

If, however, the number of patients were to exceed certain thresholds of care capacity within the hospital system, the latter would have to adapt. If there were hundreds of serious cases in a canton like Geneva, we would be clearly overwhelmed, but we would adapt by establishing tent camps, using private hospital facilities, offering home care etc., all in coordination with our army and civil protection department, both of which have anti-disaster plans.

—What can citizens do to prepare themselves and thus limit the risks threatening their own person and others?

F.C.: Let us be comforting and reassuring: people must have confidence in their own health system. Although you will surely tell me that such an attitude is normal coming from a doctor!

The decisions that have been taken by the FOPH or the cantonal authorities, and that I myself have witnessed, seem reasonable and adequate to me, so do not give in to panic. As for any specific recommendations allowing people to protect themselves or their families, they depend on the type of virus and the transmission method. Self-quarantine measures are known to be ineffective, but aerial transmission can be avoided, for example, through simple measures such as wearing respiratory protection masks. In this very instance, since it is uncertain whether the authorities will have enough stocks for the entire population, it might be useful for people to have some masks of their own. Personally, I have absolutely nothing like that at home, neither for myself nor for my children.

—Is there anything one can do to strengthen their immune system?

F.C.: I would advise everyone to stay up to date with regard to vaccines—I'm thinking especially of measles, as fatal cases are still encountered nowadays.

Other than that, I suggest being positive about one's life, doing some sports, eating healthily, having proper personal hygiene habits, and so on. This increases your chances of surviving an illness compared to when you are in poor health.

— Are these emerging diseases likely to be more prevalent in a world of growing populations where tropical/equatorial deforestation and the destruction of micro-ecosystems are putting humans in contact with organisms such as viruses which had, until recently, remained isolated?

F.C.: The overexploitation of resources exposes us both increasingly and more frequently to new infectious agents that are mostly of a viral nature, sometimes bacterial and, in a rarer number of cases, parasitic. As a result of global urbanisation and the development of passenger transport, which also concerns these regions, the risk of dissemination is bound to increase both regionally and internationally. And one must also bear in mind the effects of climatic variations and changes such as lower or higher rainfall, or higher or lower temperatures, which can sometimes allow an agent to spread over a larger area or reach greater altitudes — although the very opposite can sometimes apply as well! It may even be the case that a change of environment would destroy the ecosystem that enables the disease to thrive, thus destroying the disease itself. Urbanisation, for example, significantly reduces the spread of malaria. Aedes mosquitoes, by contrast, which act as chikungunya[7] or dengue[8] vectors, are very fond of it! It is a very complex issue indeed.

There are two types of disease emergence. The first is the emergence of a known agent that has been transported to a new region where it can thrive, as has been the case with chikungunya or West Nile virus in America.[9] Such a development is very difficult to counter, because once Aedes mosquitoes have established their presence, the struggle is a very complex one and prevention has never proved effective. Then there is the case of an infectious agent which is present in a wild animal reservoir and which, as

7 AN: www.who.int/mediacentre/factsheets/fs327/fr.

8 AN: www.who.int/mediacentre/factsheets/fs117/fr/.

9 AN: www.who.int/mediacentre/factsheets/fs354/fr/.

a result of changes in the ecological environment, surfaces at the human level. Here, the solution is not only medical but also multisectoral and global in nature, as suggested by the UN's seventeen *Sustainable Development Goals.*[10] If we do not embrace such a global and multisectoral approach, if we lack the necessary strong and international political will, we will continue to be faced with an increasing number of epidemics and emerging diseases over the next years and decades.

— On 1 February 2016, the WHO declared a 'public health emergency of international concern' with regard to the Zika virus. If there are words which, as part of the very same sentence, can arouse a great deal of anxiety, it's 'virus', 'emergency' and 'international'! What's your view of this 'alert'? Is it all a mere media psychosis or a new emerging disease?

F.C.: Only the future will tell whether it is a media psychosis or not, depending on the scientific answer to two fundamental questions: 1. What is the proportion of infected pregnant women with a congenital infection (and what are the clinical consequences?) 2. What proportion of the microcephaly rate reported in Brazil is actually due to Zika? Although originally discovered in 1947, Zika has, from an epidemic perspective, been an emerging disease since the 2007 Yap Islands outbreak, yet it is especially emerging within our own consciousness…

10 AN: http://www.un.org/sustainabledevelopment/fr/.

CHEMICAL RISKS

I do not understand this squeamishness about the use of gas. I am strongly in favour of using gas against uncivilised tribes. The objections of the India Office to the use of gas against natives are unreasonable. The moral effect should be so good that the loss of life should be reduced to a minimum. Gas is a more merciful weapon than [the] high explosive shell, and compels an enemy to accept a decision with less loss of life than any other agency of war. Why is it not fair for a British artilleryman to fire a shell which makes the said native sneeze?

— British statesman WINSTON CHURCHILL (1874–1965), in a letter sent to the Royal Air Force in 1919

[There can be no] doubt that chemical terrorism is emerging not as an abstract threat, but a grave reality of our time

— Russian Foreign Minister SERGEY LAVROV, 1 March 2016

1. A History of Chemical Warfare Agents

I will not attack them with chemicals just one day, but I will continue to attack them with chemicals for fifteen days.

— Iraqi Defence Minister Ali Hassan al-Majid, aka 'Chemical Ali' (1941–2010), in a reference to the attack against the Kurdish village of Halabja (5,000 fatal casualties)

Ancient Times

It may seem an exaggeration to state that man has always used chemical weapons. And yet, history is rife with numerous examples of such use extending from antiquity to the present.

One can, for example, go back to 428 BC, during the Peloponnesian War, when Athenian leader Demosthenes burned pitch[1] with added sulphur to produce toxic fumes. Carried by the wind, these fumes poisoned his Spartan enemies, who were entrenched in the city of Sphacteria. In 187 BC, the Romans generated a suffocating cloud by using their cavalry to upheave caustic ash during the siege of Ambracia.

1 AN: A sticky and flammable material made from vegetable resins or tars.

Later, in AD 256, the Roman city of Dura-Europos was, in turn, subjected to a chemical attack. Recent excavations have shown that the attackers used toxic gases (probably sulphur) to asphyxiate legionnaires working in the tunnels to counter the sappers' efforts. At a time when anything inexplicable was synonymous with divine intervention, the combatants' mysterious death and the absence of visible wounds was certainly the source of the wildest possible rumours…

Although there are probably many other examples of chemical weapon use in antiquity, it is quite possible that only a few have ever been reported. Indeed, this kind of attack was already considered immoral at that time, especially among Greek intellectuals and thinkers.

A few centuries later, one witnessed the emergence of specific incendiary materials that can be classified as chemical weapons. Greek fire, for instance, which was invented in AD 673, is a highly flammable substance that burns even in water and generates toxic fumes. It allowed the Byzantine Empire to repel Turkish attacks for more than five hundred years, until the Turks put their hands on it and used it for their own benefit.

From the fourteenth century onwards, however, Greek fire was gradually replaced by black powder and firearms. It fell into disuse and its formula was lost for several hundred years. Nowadays, historians and chemists propose different compositions that vary according to the sources being studied. The following mixture is an example: *pitch, naphtha, sulphur, saltpetre, oil and quicklime*. In some cases, iron filings, sulphurised arsenic or ammonium salts may also have been added.

Following the discovery of black powder (a mixture of sulphur, saltpetre and charcoal), new chemical weapons were invented, including arsenic bombs, which were used by those who defended Belgrade against the Turks in 1456.

Similarly, during the Thirty Years' War (and especially between 1635 and 1648), the 'stinking pot', which consisted of faeces, turpentine, sulphur and a plant characterised by a strong stench of rotten eggs (*Asa foetida*), was used in abundance.

From the eighteenth century onwards, technical breakthroughs enabled the creation of more sophisticated devices. Other explosive devices with toxic properties also surfaced in many military treatises. These bombs or grenades, containing for example mercury, lead, antimony, or even biological poisons (poisonous plants, venoms, etc.), would, nonetheless, only have an anecdotal use. One thus had to wait until the twentieth century and the advent of modern chemistry to witness major advancements in these unconventional weapons…

World War I

What is certain is that the mass use of chemical weapons began during World War I: on 22 April 1915, ten kilometres to the north of the city of Ypres (Belgium), German troops placed 6,000 steel cylinders containing nearly 170 tonnes of chlorine in front of the Allies. Towards the end of the afternoon, the opening of all these containers caused the release of the volatile compound in a matter of five to ten minutes. Heavier than air, the gas thus formed a toxic cloud which, carried forth by the wind, spread along the allied lines, reaching high concentrations within the trenches and creeping into every nook and cranny. The French and Canadian troops defending the area found themselves unprotected in the face of this terrible threat: the vapours turned into hydrochloric acid upon contact with the victims' eyes, lungs and respiratory tracts. This marked the onset of intense suffering for the defenders, incapacitating 15,000 soldiers and resulting in 5,000 deaths among them.[2] A month later, on 31 May, an attack involving chlorine-phosgene contained in 12,000 bottles resulted in the deaths of more than 1,000 soldiers on the Russian front. What then followed was a

2 AN: And yet, this considerable death toll may vary according to the sources. In addition to this, it is quite possible that the actual figure was rigged for reasons of propaganda.

chemical weapons race, as none of the sides hesitated to use this type of arsenal. Here are some examples:

- During the month of March 1916, in Verdun, the French army fired phosgene shells (a gas that is even deadlier than chlorine);
- On 22 June 1916, still in Verdun, it was the Germans' turn to resort to phosphene-loaded shells;
- In July 1916, during the Somme offensive, the French made use of shells containing hydrocyanic acid…

With chemical attacks following one another in quick succession, the two sides came up with increasingly effective means of protection. Initially comprising mere pieces of cotton or wet sponges, and even rags impregnated with urine, the equipment evolved and, very soon, the first masks appeared. Beginning in 1915, the latter were gradually put into service, offering effective protection against the gases used in combat (chlorine, phosgene, hydrocyanic acid, and so on).

The arms race, however, was not yet over. On the night of 12 to 13 July 1917, the German artillery at Ypres resorted to a new type of chemical agent. The Germans used shells loaded with 'mustard gas' or yperite, whose very name is based on that of the above-mentioned city. The compound was much more persistent than those that had hitherto been used. It resulted in both a toxic atmosphere and contamination of the general area. What is more, yperite could pass through clothes and equipment and cause burns and severe lung and skin damage a few hours after contact. Although death could be brought about through as little as a few grammes, its main function lay in putting entire troops out of action.

Germany would produce 12,000 tonnes of this compound, and by the end of the war, all the nations involved in the conflict had their own stock. Ultimately, between 1915 and 1918, nearly 83,000 soldiers died as a result of the damage caused by chemical weapons: 56,000

Russians, 9,000 Germans, 8,000 Frenchmen, 8,100 Britons and 1,500 Americans. To these figures, one must also add approximately one million combatants who were wounded by these unconventional weapons.

The Modern Era

The Great War of 1914–1918 was etched into a lot of people's minds because of the massive use of chemical weapons. The very scale of ammunition produced and used was considerable; which is why, even today, a farmer or casual walker sometimes still comes across a shell dating back to this period and loaded with yperite.

Despite the devastating effects of these unconventional weapons, their use did not cease in the years that followed. It is thus likely that, after 1920, the Spaniards used yperite sporadically in Morocco. It also seems plausible that, during that same period, chemical weapons were used in the Russian Civil War, by all sides. In 1936, Mussolini-controlled Italy made use of mustard gas[3] (especially by means of aerial spraying), which led to the deaths of several thousands of Ethiopian fighters. Later, between 1937 and 1943, it was the turn of the Japanese, who used ammunition loaded with yperite, lewisite or phosgene against the Chinese. It was, furthermore, during these years that the first neurotoxic agents, i.e. the most powerful toxic chemical substances, were invented: tabun in 1936 and sarin in 1938.

It is interesting to note that although most of the warring parties had large stocks of chemical weapons[4] (including nerve agents) during World War II, they did not make use of them — with the exception of the Japanese in China[5] and, according to a theory presented

3 AN: Another name for yperite.

4 AN: The Germans, for instance, produced and stored up to 78,000 tonnes of chemical munitions. In the case of the Americans, the amount was 146,000 tonnes.

5 AN: Particularly through Unit 731, which carried out numerous experiments on prisoners between 1932 and 1945 and dropped many bombs on the Chinese

by Russian biological weapons researcher Kanatjan Alibekov,[6] the Soviets, who allegedly resorted to *Francisella tularensis* (a category A biological agent) against the German forces advancing on Stalingrad in 1942. In view of the quantities involved, it is likely that the mere threat of their use had a sufficiently significant deterrent effect.

After 1945, existing stocks were not destroyed, and some, such as those of Germany, were moved to Russia. Likewise, many Third Reich scientists were offered employment in the United States (and others in the USSR). In addition to this, the beginning of the Cold War between the Americans and the Soviets in the 1950s led to an insane arms race on a chemical, biological and nuclear level. Since then, several cases involving the use of chemical weapons have been reported:

- During the Yemeni War of 1963–1967, Egyptian forces used both phosgene and yperite;
- During the Vietnam War (1963–1970), the Americans used about 50,000 tonnes of herbicides and defoliants such as the 'infamously famous' Agent Purple and Agent Orange;
- During the Iran-Iraq war of 1980–1988, Saddam Hussein's regime conducted chemical attacks against enemy forces. Attacks on Kurdish civilian populations were also reported, including the one in Halabja, where the use of several compounds (apparently a mixture of tabun, sarin, and yperite) resulted in the deaths of at least 5,000 people.

More recently, chemical events of the 'terrorist' type have surfaced:

---

territory, with a death toll of 300,000 to 480,000 victims. https://fr.wikipedia.org/wiki/Unité_731.

6 AN: Kanatjan Alibekov (Ken Alibek), *La guerre des germes* [TN: *Biohazard: The Chilling True Story of the Largest Covert Biological Weapons Program in the World — Told from Inside by the Man Who Ran It*], Presses de la Cité, 2000.

- The *Aum Shinrikyo* sect perpetrated two sarin gas attacks. The first, which took place in Matsumoto (in June 1994), claimed the lives of seven people and injured 600. The second, conducted in the Tokyo subway in March 1995, resulted in the deaths of twelve people and the poisoning of about 5,000 inhabitants and intervening individuals;

- In 2006 and 2007, suicide bombings involving the use of vehicles were carried out in Iraq (mainly in Baghdad and Ramadi) through a combination of conventional explosives and chlorine containers. They did not apparently result in any fatal casualties, but wounded many people and aroused fear within the population;

- Attacks involving the use of chlorine and sarin have also been committed and been repeating sporadically in Syria since 2013. Given the nature of the conflict and the political pressures, it is still difficult to determine their origin with certainty (the rebels? the regime? an external entity?). It seems more prudent to avoid hasty conclusions; indeed, we all remember the claims regarding Iraq's alleged weapons of mass destruction, which led to the country's invasion and occupation at the hands of Western forces…[7] On 12 February 2016, Paul Brennan, the director of the CIA, stated[8] that the Islamic State/Daesh possessed chemical weapons (chlorine, sarin, mustard gas and corrosive products such as

7 AN: On 17 October 2005, Barry Lando, who previously worked as a journalist for the American channel CBS, told the French daily *Le Monde* that 'Iraq's chemical weapons were supplied mainly by French, Belgian and German companies whose engineers and chemists knew exactly what Saddam was preparing to do, and the United States had previously provided Saddam with satellite imagery to attack Iranian troops with chemical weapons.' http://www.agoravox.fr/tribune-libre/article/chemical-wars-history-pages-140994.

8 AN: http://www.dailymail.co.uk/news/article-3443721/ISIS-used-chemical-weapons-make-mustard-gas-chlorine-warns-CIA-director.html.

vinyltrichlorosilane)[9] stolen from Syrian stocks and had purportedly used them in battles waged in both Syria and Iraq, possibly also smuggling some into Europe.

Industrial Toxic Agents

As we have just seen, many chemical attacks have been perpetrated in the past. Due to the increasing industrialisation of our current society, however, ***new threats*** are now emerging. The average person thus faces an almost permanent menace. This can be embodied by the daily handling of ordinary chemicals (such as certain cleaning products), but also, on an entirely different scale, by major accidents or serious malicious actions involving ***industrial toxic agents***.

In view of the actual quantity and diversity of chemical compounds that can be encountered in a developed country (as part of its industry, hospitals, laboratories, nuclear power stations, transport of hazardous materials, etc.), it is clear that this risk is truly omnipresent. Moreover, such a risk is comparable to that generated by the chemical weapons used in the past (toxic vapours in antiquity, shells loaded with lethal gases, napalm bombs, Agent Orange, etc.). It can entail effects that are:

- mechanical in nature (due to the explosion of pressure containers or mixtures of incompatible products, etc.);
- thermal in nature (and are due to the ignition of a plume of steam, as well as to fires and arsons);
- toxic and/or corrosive (resulting from inhalation, poisoning, contact, etc.).

9 AN: http://fr.sputniknews.com/international/20160225/1022064724/daech-armes-chimiques-corrosives.html.

This type of risk, known as 'technological', is not necessarily limited to just one of these effects. In some cases, it can accumulate several of them, whether simultaneously or successively. An explosion, for example, which usually generates a blast effect triggered by overpressure, as well as splatter or splinter injuries, can also lead to burns,[10] cause a fire and result in poisoning through dangerous fumes…

Unfortunately, history proves that major technological accidents are not as rare as one would like to believe. Indeed, they have already caused the deaths of thousands of people. This is all the more remarkable as these terrible incidents can occur without there actually being any particular threat and at a most inopportune time — when people are asleep or in a gathering, for instance. Here are some examples:

- **In July 1976,** in Seveso (Italy), two kilos[11] of dioxin (2,3,7,8-tetrachlorodibenzo-para-dioxin)[12] escaped from the Icmesa chemical plant and spread through the atmosphere. No direct deaths were reported and about 200 people were poisoned, but the impact on the fauna, flora and the environment was enormous (with more than 3,000 domestic animal deaths, 70,000 heads of cattle slaughtered, 250,000 m^3 of contaminated soil, 15,000 evacuees, etc.);

- **In July 1978,** in Los Alfaques (Spain), the explosion of a propylene-containing pressure tank left 216 fatal casualties and numerous wounded victims in its wake, some of whom were badly burned;

- **In November 1979,** in Mississauga, Canada, a train transporting dangerous goods derailed. Several cars containing propane exploded and others spilled their cargo (caustic soda) on the ground; one of them became the source of a chlorine leak (a deadly gas

10 AN: Some mechanical explosions, for instance, do not cause any burns.

11 AN: The figure varies according to the source.

12 AN: https://fr.wikipedia.org/wiki/2,3,7,8-Tétrachlorodibenzo-p-dioxine.

that is heavier than air). The leak would last more than four days and lead to the evacuation of 240,000 local residents;

- **In December 1984,** in Bhopal (India), the accidental leakage of thirty-five to forty tonnes of toxic gas (methyl isocyanate) in the middle of the night caused 3,800 deaths in the hours that followed, 6,500 to 10,000 after a period of ten years[13] and 25,000 after two and a half decades. 350,000 people were poisoned to some degree, many of whom were left disabled for life;

- **In May 1988,** the explosion of 4,000 to 4,500 tonnes of ammonium perchlorate in Henderson (USA) caused a 3.5 magnitude earthquake (on the Richter scale) and left two fatal casualties and nearly 400 wounded victims in its wake;

- **In March 1989,** in Jonova (Lithuania), a tank containing ammonia exploded as a result of a rise in pressure. The generated vapours ignited and the fire reached the fertiliser plant's storage sites (comprising 55,000 tonnes of fertiliser). Toxic fumes were produced for a period of three days and contaminated an area of 400 km^2. The accident resulted in seven fatal casualties, 57 wounded victims and the displacement of 32,000 people;

- **In March 1992,** in Dakar (Senegal), the explosion of a truck loaded with twenty-two tonnes of liquid ammonia killed around 150 people and poisoned more than 1,000;

- **In June 1996,** a train carrying vinyl chloride derailed in Schönebeck, Germany. One of the cars exploded and the fire spread to other cargoes. Black and toxic fumes (hydrochloric acid, phosgene, dioxins, etc.) were generated and liquid leaks permeated the soil, as the residents were confined to their homes

13 AN: These figures take into account the victims who died from the side effects and diseases that resulted from their exposure to the gas.

for several days. Sixty-five people were injured and a number of above-normal chromosomal aberrations and genetic mutations were observed in the following years;

- **In September 2001,** an explosion of 300 to 400 tonnes of ammonium nitrate at the AZF plant in Toulouse (France) caused the deaths of thirty-one people and injured 2,500 others. In addition, the fumes were toxic and contained either nitrous vapours (resulting from the combustion of ammonium nitrate) or ammonia (due to the breached tanks containing this compound). Fortunately, the toxic cloud rose above the city and had very little impact on the population;
- **In February 2004,** a train transporting sulphur (seventeen wagons), petrol (six wagons), fertiliser (seven wagons) and cotton wool (ten wagons) spilled its load in Nishapur (Iran). The subsequent fire led to the load's explosion, causing the deaths of 289 people and destroying two nearby villages;
- **In March 2004,** the collision of two trains in Ryongchŏn, North Korea triggered the explosion of a petrol wagon and two containing ammonium nitrates: the result was 161 fatal casualties, 1,300 wounded people, and the destruction of 2,000 apartments and 129 public buildings (= 40% of the town);
- **In April 2004,** a chlorine leak in a chemical plant in Chongqing (China) lasted several days. Nine people died as a result, and 150,000 were evacuated;
- **In March 2005,** a tanker accident involving thirty-five tonnes of chlorine led to the deaths of twenty-seven people, the poisoning of 400 others and the evacuation of 3,000 families in the Jiangsu region (China);

- **In May 2008,** a major earthquake caused the deaths of approximately 80,000 people in China. Two chemical plants collapsed and eighty tonnes of ammonia leaked out, resulting in the demise of around 600 residents of the city of Shifang;

- **In October 2010,** a caustic waste reservoir containing red mud (which is both toxic and caustic) at an aluminium production plant in Kolontár, Hungary was breached: as a result, 10 people died and 286 suffered 'chemical' burns. Additionally, 300 homes were destroyed;

- **In September 2012,** the accidental opening of a valve in a cosmetics factory in Gumi (South Korea) caused a spill of hydrofluoric acid (a very toxic and corrosive compound). The resulting vapours were responsible for five deaths, the hospitalisation of eighteen people, as well as nausea, rashes and pulmonary pain in more than 4,000 residents. An area stretching across a twenty-kilometre radius was left polluted and had to be decontaminated;

- **In April 2013,** a fire of undetermined origin erupted north of the city of Waco (USA). It caused an explosion at a nitrogen fertiliser plant, killing fifteen people and injuring more than 200;

- **In August 2013,** a leaking ammonia pipeline in Matias Romero (Mexico) led to the evacuation of 1,500 people, leaving nine deaths and forty poisonings in its wake;

- **In August 2015,** the explosion of chemical stocks in the port of Tianjin (China) resulted in the deaths of approximately 200 people, in addition to injuring more than 700. Furthermore, the presence of 700 tonnes of sodium cyanide (among other substances) presented a major potential environmental threat. Firefighters would require several days to extinguish the various fires.

The few examples listed above only represent a small sample of such occurrences. In addition, they do not include the innumerable accidents caused by the petrochemical industry (i.e. leaks and explosions of gas or oil, etc.).

The effects of these last, moreover, although perhaps less 'spectacular' in some cases, remain very pernicious indeed, as witnessed in Côte d'Ivoire in 2006: the *Probo Koala*, a Panamanian-flagged oil tanker owned by a Greek corporation (Trafigura), manned by a Russian crew and rented by a Swiss-Dutch company, discharged around 600 tonnes of toxic waste following a series of 'cleaning operations'. The mixture (comprising oil, hydrogen sulphide, phenol, caustic soda…) was dumped in truckloads across various parts of Abidjan where, in addition to the nauseating odours, it causes major pollution affecting the health of the inhabitants. For a full week, the population suffered from various ailments without understanding their source. According to official figures, seventeen people ended up dying and a total of 40,000 to 75,000 were poisoned.

Despite not acknowledging the facts, the Trafigura corporation was found guilty by the international court of the Netherlands. In 2007, however, the lawsuits were dropped in exchange for a compensation of 185 million euros, of which only a quarter was to be paid to the victims (with the rest pocketed by the government and various entities!).

In the next parts of this book, the risks pertaining to explosions and fire/arson will not be developed, thus allowing us to concentrate on the toxic aspect.

2. The Chemical Threat

I consider nature a vast chemical laboratory in which all kinds of compositions and decompositions are formed.

— French chemist ANTOINE-LAURENT DE LAVOISIER (1743–1794)

History proves that the threat is both real and constantly present. Nevertheless, not everything is negative in this regard. Indeed, international treaties such as the Chemical Weapons Convention[1] have put a noticeable brake on the chemical arms race and have progressively banned the use of unconventional weapons, even in times of war. To date, of the 197 recognised nations (including Palestine and the Holy See), only four have failed to sign these agreements (namely Angola, North Korea, Egypt and South Sudan); two others (Israel and Burma) have signed them but not ratified them. In addition, the laws and regulations pertaining to commercial chemical products and industrial sites (Seveso, the transport of hazardous materials, REACH,[2] etc.) are constantly evolving and tend to learn from past mistakes.

1 AN: Signed in Paris in 1993 and ratified by the United States and Russia in 1997, this convention prohibits the use, development, manufacture, acquisition, stockpiling and transfer of chemical weapons, and requires the dismantling of production plants.

2 AN: http://www.developpement-durable.gouv.fr/REACH,30375.html.

Nowadays, furthermore, the risk of being targeted by toxic chemical warfare agents such as sarin or yperite is probably quite limited, especially in Western countries. Of course, in a land plagued by armed conflict or located in the vicinity of nations waging war upon each other, the likelihood is bound to increase. The main threat of such usage, however, remains rooted in terrorist action. With regard to industrial toxic agents, moreover, accidents remain possible. Although standards have been raised and many devices are now installed to restrict potential consequences, there is no such thing as zero risk. In addition, there is always a chance that the population will indeed be impacted in the event of a malicious or terroristic act… Let us start by trying to understand how a toxic chemical agent can affect an individual.

Means of Penetration

Generally speaking, the impact of a toxic agent, and more specifically that of a chemical, depends on many factors including the received dose, the product's dangerousness, the target's health, etc. Nevertheless, for the compound to take effect, it must, above all, be in contact with the body or enter it. **Four main entry pathways** are thus recognised when it comes to toxic chemicals:

1. **The respiratory route** (through inhalation)
   This is the most common entry pathway.
   Taking on different shapes (gas, steam, aerosol, some fumes), toxic agents can cause damage to the entire respiratory system. On the other hand, dust particles larger than one micrometre fail to penetrate one's pulmonary alveoli.

2. **The oral route** (through ingestion)
   Ingestion phenomena are commonly connected to domestic accidents in which the subject absorbs the toxic agent through the mouth (e.g. the drinking of bleach). One must not forget, however,

those cases where the person's contaminated hands, for instance, can also lead to oral poisoning (putting your hands in your mouth, eating a sandwich, etc.).

3. **The cutaneous route**

Two phenomena can come into play here:

=> Transcutaneous permeation: depending on their actual state (solid, liquid, gaseous) and their physico-chemical properties, some toxic agents have the ability to pass through healthy skin and penetrate deeper into the body. This ability obviously varies depending on the thickness of the *stratum corneum* (cornea layer), the number of hair follicles, etc.

=> Penetration through skin wounds (cracking, cuts, accidental pricking, and so on). Burns also weaken the barrier embodied by the dermis.

4. **The ocular route**

Spatters and/or vapours can cause severe localised lesions (involving temporary or permanent loss of vision). Similarly, the penetration of a toxic agent into the body through the eye can lead to general effects (which are, however, normally more moderate than in the case of other entry pathways).

Note: The ocular route is sometimes connected to the cutaneous one.

Types of Poisoning

Depending on the received dose and the substance involved, **different types of poisoning** may occur:

=> Acute or superacute (usually involving high doses)
This is the case with short-term exposure (less than twenty-four hours) where absorption of the substance leads to rapid signs of poisoning (and possibly the person's death).

=> Subacute (usually low doses)
This is the case with repeated exposure over several days or months, resulting in specific symptoms.

=> Chronic (typically involving very low doses)
Toxic exposure is repeated for a long time (during an individual's entire life or a part thereof). The effects may surface after several years, even if the subject is no longer in contact with the toxic agent.

Main Effects

Disorders caused by toxic chemicals can range from a simple tingling sensation in the eyes or throat to the appearance of irreversible effects that can lead to death. Three major categories of damage are generally recognised:

1. **Damage to the respiratory system**
   The mechanisms involved are highly variable, but can all lead to the same result — **death by asphyxiation.**

   - **Anoxic gases**
     These gases are not, strictly speaking, toxic, but they can cause anoxia through a substitution effect of the oxygen present in the air. Examples include nitrogen, carbon dioxide, hydrogen, etc.

   - **Narcotic gases**
     These can cause a narcosis (sleep) effect evolving into a deep coma and respiratory distress. Examples: varnish, ether, trichlorethylene, etc.

   - **Cellular toxic gases**
     Being highly toxic, these can alter the functioning of cells in certain target tissues and organs and lead to their asphyxia by impacting the transport of oxygen (e.g. carbon monoxide, arsenic, etc.) or its use (e.g. hydrogen cyanide, hydrogen sulphide, and so on).

- **Choking/pulmonary agents**

Corrosive or caustic in most cases, these toxic agents result in tracheal and bronchial irritation and may cause damage to lung tissue (oedema).

Examples include chlorine, phosgene, methyl isocyanate, ammonia, and others.

2. Damage to the nervous system

The main culprits behind this type of disorder are organophosphorus insecticides and nerve agents that are classified as chemical warfare agents (and are actually insecticide derivatives). They act through the inhibition of tissue cholinesterases. The **severity** of the disorders and the **speed of their onset** depend on the absorbed dose and the entry pathway. Effects may surface a few seconds to several hours after contact with the agent.

MINOR EFFECTS	MAJOR EFFECTS
▪ Vision disorders ▪ Headaches ▪ Nausea ▪ Vomiting	▪ Hypersecretions (salivary, bronchial, etc.) ▪ Involuntary urination ▪ Tremors and convulsions ▪ Loss of consciousness ▪ Paralysis of the respiratory system ▪ DEATH

3. General damage

This type of disorder can affect the skin, the eyes, the respiratory tract, the digestive tract, the bone marrow, the central nervous systems, etc. Some toxic agents, including vesicants (yperite, lewisite, and so on), will cause this kind of affliction through contact in liquid form (as they can pass through many materials such as wood, leather, latex, etc. before touching the skin) or in vaporous form. The symptoms vary depending on the toxic agent's means of penetration and the absorbed dose. The effects may sometimes be delayed and only surface twenty-four hours later (mainly in the case of yperites).

IMPACTED ORGANS OR TISSUES	MINOR EFFECTS	MAJOR EFFECTS
Eyes	Tingling, redness, conjunctivitis...	Photophobia, oedema, blindness...
Skin	Burns, erythema...	Vesication, blisters, necrosis...
Respiratory Routes	Inflammation of the mucous membranes, coughing, burns...	Pulmonary oedema, alveolar haemorrhaging...

The toxic agent will subsequently trigger effects of a gastrointestinal (nausea, vomiting...), haematopoietic (a decrease in leukocytes...), and neurological nature (miosis,[3] tremors...), etc.

Lists and Classification of Chemical Toxic Agents

The number of existing chemical compounds is so great that many classifications/lists have emerged. It is for instance possible to categorise the different agents according to their chemical characteristics, their effects (vesicant, choking/pulmonary, etc.), their origin (military, civil), their availability, and so on.

Effect-Related Classification

When classified according to their effects, toxic substances fall into two categories:

3 TN: Miosis/myosis is defined as 'excessive constriction [i.e. shrinking] of the pupil.'

- **Lethal toxicants**
  - **Vesicants** are aggressive compounds that cause chemical burns, swelling, blisters and other vesicles on people's skin, mucous membranes and eyes.
  - **Choking/pulmonary agents** are toxic and corrosive gases that affect the respiratory tract, sometimes leading to pulmonary oedema and death by asphyxiation.
  - **General toxicants** are compounds that can lead to death by asphyxiation using various mechanisms such as the poisoning of cells that thus become unable to use oxygen (cellular toxic agents) or the destruction of red blood cells (haemotoxic agents).
  - **Nerve agents** are extremely toxic compounds which, as suggested by their name, impact the nervous system. Very small quantities are enough to trigger a person's death. These agents include:
    - **G series nerve agents** (tabun [GA], sarin [GB], cyclosarin [GF], and soman [GD]) which, at room temperature, are in a liquid state and emit much denser vapours than air.
    - **V series nerve agents** (VX or A4), which are oily compounds characterised by low volatility. They primarily penetrate the body through the skin. Vapour-related danger is low in such cases, except in the event of extreme heat or when one breathes in their immediate vicinity.

The following chart gives us some examples of lethal toxicants:

Lethal Toxicants

VESICANTS	CHOKING AGENTS	GENERAL TOXICANTS
Sulphur yperite (HD)	Chlorine (Cl_2)	Hydrogen cyanide (AC)
Nitrogen mustard/yperite (HN)	Ammonia (NH_3)	Cyanogen chloride (CK)

VESICANTS	CHOKING AGENTS	GENERAL TOXICANTS
Lewisite (L)	Phosgene (PG)	Arsine (SA)
	Diphosgene (DP)	Phosphine (*PH_3*)

Note: The letters in brackets represent either the NATO code or the compound's chemical formula (in italics).

- **Non-lethal toxicants**
  - **Neutralising agents** are substances that cause a temporary and reversible **physical disability** (irritation of the skin, eyes, and respiratory tract; vomiting, etc.).
  - **Incapacitating agents** are substances that cause a temporary and reversible **mental disability** (behavioural disorders, deliriousness, hallucinations, and so on).

Non-Lethal Toxicants

NEUTRALISING AGENTS
CS (tear gas)
Phosgene Oxime (skin irritant)
Adamsite (vomiting agent)

List of Industrial Toxic Agents of Operational Importance

Of the more than 4,000 most common toxic industrial chemicals (TICs), twenty-one were defined by a trinational group (Canada, United States and United Kingdom) as operational-level toxic, based on:

- the probability of encountering the product in a theatre of operations;

- vapour pressure (vapour hazard);
- the toxicity of the substance.

List of the Twenty-One Toxicants of Operational Importance

1. Chlorine	11. Hydrogen sulphide
2. Formaldehyde	12. Sulphuric acid
3. Phosgene	13. Ammonia
4. Arsine	14. Sulphur dioxide
5. Boron trichloride	15. Ethylene oxide
6. Boron trifluoride	16. Hydrofluoric acid
7. Carbon disulfide	17. Phosphorus trichloride
8. Diborane (B_2H_6)	18. Nitric acid
9. Fluorite	19. Hydrogen bromide
10. Hydrogen cyanide	20. Hydrochloric acid
	21. Tungsten Hexafluoride (WF_6)

Classification according to the Chemical Weapons Convention of Paris (13 January 1993), which entered into force on 29 April 1997

Chemical Compounds and Toxic Agents (Chart 1)

- List of chemicals and their precursors with no civilian industrial application
- Synthesis authorised only for medical or pharmaceutical research purposes and for human protection studies
- Annual production limited to **one tonne per country**
- Export prohibited to countries that have not ratified the CWC

SUBDIVISION A	= Chemical warfare agents These products are characterised by their high toxicity or incapacitating power (nerve agents, vesicants)
SUBDIVISION B	= Precursors having reached the final technological manufacturing stage

Chemicals Compounds (Chart 2)

- Toxic agents and precursors used in limited quantities in the civil sector
- Products that can be used as aggressive chemical warfare agents
- Industrial facilities subject to declaration if their annual production exceeds a certain number of tonnes

SUBDIVISION A AND A*	Chemical compounds of no commercial interest that can be used for military purposes.
SUBDIVISION B	= Precursors of Chart 1 and Subdivision A of Chart 2.

Chemicals Compounds (Chart 3)

- Products that are widely used by the chemical industry, but some of which could be used for military purposes
- Facilities producing more than thirty tonnes per year of a substance listed in chart 3

SUBDIVISION A	= Former chemical warfare agents now manufactured in large industrial quantities for purposes not prohibited by the CWC
SUBDIVISION B	= Precursors of the product synthesis in Chart 1 or Section B of Chart 2.

3. Conclusion

> Nothing is born or dies; on the contrary, everything is assembled out of existing things and then dissolved.
>
> — Greek philosopher ANAXAGORAS (500–428 BC)

The chemical threat is certainly the most commonplace of all. Every day, millions of people use corrosive or potentially toxic cleaning products. Every year, many children[1] and even adults fall victim to such products. Nevertheless, when we approach this threat from the point of view of a major accident (the Seveso site, the transport of hazardous materials) or that of terrorist action involving industrial or warfare-related toxic substances, the perspective becomes quite different: the consequences can indeed be dramatic, whether in terms of the number of victims or the possible impact on the environment.

It is interesting to note that, unlike biological[2] and radiological[3] threats, a chemical incident is likely to cause symptoms quickly. For example, an infectious agent will take several days to a few weeks to bring about the disease; exposure to or contamination by radiological elements will result in observable yet delayed effects. By contrast,

1 AN: Poisoning resulting from the ingestion of medicines, household products or cosmetics is the second most common cause of accidents among children in France (second only to physical trauma, but preceding burns).

2 AN: Except for toxins.

3 AN: Unless the dose in question is a massive one that could have such effects.

exposure to an irritating gas or corrosive liquid will lead to an immediate reaction on the part of the subject.

This characteristic, along with the fact that a chemical substance is mostly visible (unlike a biological agent or radiation), will *generally* allow a potential victim to realise that an abnormal phenomenon is taking place. One can thus become aware of a chemical accident through:

- **Observation:**
    - crowd movements and panic;
    - people displaying certain symptoms, the presence of dead animals;
    - leaks of liquids, white or colourful vapours.

- **Smell:**
  Most compounds have a characteristic odour. Some products, however, are odourless, while others are so toxic that the moment someone smells them, they have already received a dose that is likely to be fatal…

- **Certain sensations:**
  Tingling; itching; and irritation of the skin, lungs and eyes are all warning signs which the body takes into account instinctively.

  Such elements are only rarely encountered (or perhaps are encountered in a delayed manner) during incidents of a biological or radiological nature. The advantage is that in most cases involving a chemical incident, people *do* have the opportunity to react — in the hope that it isn't already too late and that the choices they make are adequate! The section entitled 'CBRN Incidents: Reactions and Protection, which one can find further on, will provide you with the necessary details on the topic.

!!! WARNING !!!

Although some gases such as carbon monoxide are both invisible and odourless, their lethality is very swift indeed. Others such as hydrogen sulphide are so toxic that a single puff can lead to coma and then to death. Last but not least, some products (including yperite) only trigger the first symptoms after dozens of minutes, with a potentially fatal result.

AN INTERVIEW WITH SERGE WALTER, A CHEMICAL SAFETY PROFESSOR AT THE MULHOUSE NATIONAL SCHOOL OF CHEMISTRY AND THE TEACHING MANAGER IN CHARGE OF THE CGE'S[4] SPECIALISED MASTERS[5] IN CBRN-E RISK AND THREAT MANAGEMENT.[6]

— What do you think are the chemical risks affecting the population? Would a second Bhopal disaster be possible today? And what about a Tianjin scenario in France?

S.W.: In the case of Bhopal, when one takes into consideration the installation's state of maintenance, the accident was due to a series of foreseeable malfunctions. The toxic gases could easily have been destroyed upon emission by means of flaring combustion, which was the last barrier to their actual emission following the runaway of a hydrolysis-based degradation reaction in an aqueous environment impacting a reaction medium that contained methyl

4 TN: The *Conférence des Grandes Écoles* (CGE), literally 'Conference of Higher Education Schools', is a French national institution founded in 1973 and representing all the engineering institutions accredited by the *Commission des Titres d'Ingénieur* (CTI) to deliver the French engineering degree, as well as all business schools that can grant a Master's degree.

5 TN: The *Mastère Spécialisé*, rendered in English as 'Specialised Master' (or, alternatively, 'Advanced Master'), is a French post-graduate specialisation degree created by the *Conférence des Grandes Écoles* in 1986.

6 AN: Professor Walter has developed these topics in two books: *À l'Aube de la Lumière* [TN: At the Dawn of Light] and *Le Développement Durable… autrement* [TN: Sustainable Development — from a Different Perspective].

isocyanate. Due to a lack of technical maintenance, however, the gas flare[7] was not able to kindle the emitted toxic gas ignition system. Several safety devices had been provided to ensure that potentially toxic gases did not trigger a situation where they would have to be discharged by the flare stack. And yet, even in the case of such a situation, their combustion through flaring would have overcome their toxicity. Unfortunately, the operators' ignorance (leading to their misunderstanding the events in progress during the preliminary phase that actually allowed the accident to take place), the lack of maintenance characterising the entire facilities (which had virtually been abandoned, yet continued to be exploited), and the negligence displayed by the management in relation to the installation's security requirements prevented all these devices from functioning correctly, although they were largely redundant. Had a single one of them been operational, it would have sufficed to avoid the catastrophe.

Secondly, the meteorological conditions which caused the vapours released by the flare to stagnate in a very densely populated area (this happened in India, after all) represented an extremely aggravating factor in this dramatic situation.

In the case of Tianjin, even though it is difficult to accurately determine the origin of the disaster and the course of events that led to its exacerbation, it is clear that three major aggravating factors allowed the events to reach extremely serious proportions. These included the first responders' lack of training, their ignorance of the nature of the products stored at the site where the initial fire had broken out and where the authorities had sent them to intervene, and especially the sheer enormity of the port's cargo storage facilities, which lacked any and all of the usual, adequately

7 TN: A gas flare or flare stack is a gas combustion device used in industrial plants, where its role is to burn off any flammable gas released by pressure relief valves during unplanned over-pressuring of plant equipment. During full-scale or partial plant start-ups and shutdowns, flare stacks are also often used in connection to the planned combustion of gases over relatively short periods.

sized protective barriers whose purpose is to prevent a disaster from spreading by means of a domino effect.

Each of these events is characterised by its own location factors—from geographical and temporal to cultural and political. And these factors are themselves connected to certain societal, technical and economic aspects.

These location factors are completely independent and highly significant in the genesis of such disasters, as they create conditions which generate multivariable situations whose evolution depends on the behaviour of chaotic systems.

It is important to understand what sort of consequences this can have on the genesis of major accidents: with a maximum of three independent variables acting on a system and sufficient to describe its behaviour, the system can be oriented and its evolution remains predictable with a high level of certainty. But the appearance of an additional variable (a fourth one) generates a loss of orientability within the descriptive space of the behaviour in question: the space initially located on the right or the left becomes a space comprising two right sub-spaces and two left sub-spaces that allow the coexistence of two opposing yet equally true statements, whose truthfulness depends on the observer's position, with the transition taking place when the observer traverses the plane defined by the observed variable and another systemic command variable.

This may seem abstract to anyone who has not worked with such complex systems. To admit that we can have two opposite answers to the same question and that they can be both correct is no mean feat. However, this constitutes a fact that is so true in complex spaces that it applies with absolute rigour to the calculation of spacecraft ballistic trajectories, for which we can only make predictions within the framework of a model involving three bodies. When we have to take into account more than three bodies (such as the Earth, the Moon, the Sun and Jupiter, but also Mars and Saturn, and even other celestial bodies) on a flight to Jupiter, for example, we always consider the three most important ones for the trajectory in question. Once this trajectory has been determined, it is recalculated by integrating into this initial calculation the perturbations induced by the bodies of lesser importance. If one is lucky, the iterated calculations converge and the trajectory proves predictable. Otherwise, in case of divergence, the trajectory becomes chaotic and unforeseeable.

This could come across as being very pessimistic in a context where one strives to predict serious accident events through predictions which, in most cases, relate to systems with more than three command variables. Fortunately, following the pessimistic understanding that stemmed from the discovery of chaotic systems in the mid-1980s, it was proven in the late 1990s that if we had a sufficiently early ability to detect a deviation from the ideal trajectory for reasons pertaining to the presence of chaotic systems, a trajectory correction as minimal as its precocious implementation would enable us to bring the system back to its ideal trajectory in such a way that it encountered sufficiently small deviations on a more or less periodic basis, thus allowing us to consider that the pursuit of that trajectory had been corrected in an almost ideal fashion. (It is also in implementation of this process that a motor vehicle is driven, which is a good example of a general case in which we know how to do things before actually understanding why we have to operate in this way. Another example is that of aviation; we were able to fly objects heavier than air at a time when physics still maintained—wrongly of course!—that it was physically impossible to do so. And yet, all physicists had surely noticed that birds could indeed fly.)

However, if the trajectory correction is not made early enough, the means necessary to enable this correction can be rendered so prohibitive that it becomes impossible to carry out the required trajectory correction within the framework of available means.

The conclusion resulting from these remarks is, therefore, that if one remains vigilant and remedies possible deviations sufficiently early, one can control the overall system with enough precision to avoid major accidents. But should one allow an excessively significant deviation to set in, there is always a point beyond which a major disaster can no longer be avoided. Although this phenomenon had already been demonstrated by N. Semenov in 1928,[8] it was widely disregarded for a long time. It was not until 1976 that a tragic accident, which claimed the lives of sixteen people three days before Christmas in a chemical industry located in Basel, Switzerland, ultimately freed it from oblivion, reducing the number of chemical accidents to less than 3% of their earlier frequency in less than ten years.

As for the likelihood of a major chemical accident occurring in France and having a significant impact on the population, it is relatively low *at present*, though not inexistent.

The Toulouse accident, which occurred on 21 September 2001 and whose causes I will not discuss here, is an example of a large-scale accident in France. It claimed relatively few lives (around thirty), because of a fairly common phenomenon associated with 'thermal' accidents that result from combustion, fire or explosion: a large part of the products released by such phenomena come in the form of hot gases, which, due to their high temperature, quickly rise above residential areas and thus only pose a small risk to neighbouring populations located at ground level.

The most serious accident of this type took place in 1921 at the BASF chemical plant in Ludwigshafen, Germany. Due to an explosion involving about 3,000 tonnes of ammonium nitrate, it left more than 500 casualties in its wake. Just like in Toulouse, where 300 tonnes (i.e. ten times less) were responsible for the explosion, only 10% of the product actually blew up, with the decomposition

8 AN: SEMENOV, N.N., *Zur Theorie des Verbrennungs prozesses*, Zeitschrift für Physik, 48, 571–82, 1928.

reaction of ammonium nitrate stopping spontaneously as soon as the product loses its confinement following the dispersion that results from the explosion of the first 10% of the substance.

On 6 August 1947, the explosion that shook the cargo ship *Ocean Liberty* in the harbour of Brest[9] (with 3,200 tonnes of ammonium nitrate on board) claimed the lives of thirty-three people and left more than 1,000 others heavily injured.

The most serious accidents of this type are very rare. Their severity depends on the presence of inhabited areas close to the disaster's epicentre, but also on the prevailing conditions (the wind, watercourses, food distribution networks, communication circuits such as streets and underground transport tunnels, etc.), which could direct dangerous substances to densely populated areas.

We therefore cannot completely exclude the occurrence of a major accident of chemical origin in France, considering that, for techno-economic reasons, some industries do have to store large quantities of flammable materials (liquid fuels, particularly those of a liquefied or gaseous nature, but also organic gases and liquids or flammable or explosive minerals); toxic or even deadly substances (ammonia, chlorine, phosgene…); or explosive ones (notably ammonium nitrate, which, as part of its various applications, must be available in large loads totalling many tonnes, especially in the agricultural field) in order to be able to permanently respond to industrial demand, in connection to which these products act as the very basis of sustained activity.

Another aspect to take into account is the transit of hazardous materials transported by railway through densely populated areas (the outskirts of cities, the Rhone Valley, etc.). Even if derailments are indeed rare, the sheer amounts transported by rail do make such situations dangerous and can render any intervention

9 AN: This is a reference to the explosion which occurred on board the *Ocean Liberty* cargo ship in 1947—https://fr.wikipedia.org/wiki/Explosion_de_l%27Ocean_Liberty.

operations on the part of rescue forces particularly difficult. Inland waterway transport can also present major hazards because of the quantities likely to be carried by a single barge or narrow-boat, whose collision potential cannot be completely dismissed. The danger stemming from such situations concerns both river environments and approaches to rivers, especially those located downwind of the source of an accident or malicious act.

To draw a clear conclusion, I would say that we are in no way immune to a major chemical accident in France, regardless of whether it is linked to a breakdown in a large chemical enterprise or the result of an accident involving the transportation of dangerous goods. Road transport uses smaller unit quantities than rail transport and even more so than inland waterway transport. A river accident would also have severe environmental consequences downstream of the accident site. As for road transport, the flaring-up of a TDG lorry (i.e. one used for the Transport of Dangerous Goods) could quickly degenerate into a major problem on a road-side parking facility where transport vehicles are often parked too close to each other to prevent a fire-transfer-based domino effect involving one of their immediate neighbours from taking place. It is, however, unlikely that conditions similar to those of Bhopal could currently be met in France, since many regulations have been adopted to eliminate the presence of safety-related negligence in classified installations, i.e. the kind of negligence responsible for the gravity of the Bhopal or Tianjin accident.

The word 'currently' is extremely important: in a world where financing is synonymous with reasoning, manoeuvring space is rather narrow, and under the pretext of 'economising' in the broad sense of the word, there is a tendency to gradually replace rigorous knowledge and intelligence with accounting economics, skill-levelling from below, and regulation-based security. These regulations are rooted in *ad hoc* criteria and past feedback, with one-off objectives in mind. Security is the result of anticipating the behaviour of global systems while taking into account the entire system as part of a *long-term* vision of the future.

— What mechanisms have been provided by the world's states, cities and industry to prevent and respond to a CBRN type of incident?

S.W.: The preventive measures put in place are extremely different from one region of the world to another. This is undoubtedly the reason why the risks are probably lower in European countries, and especially those of western Europe, compared to other regions of the world.

This is not only due to technical development factors, but also to geographical issues pertaining to the country's size, the density of its population, and the territorial distribution of the latter's density.

Since population density is very high in Western Europe, the same principle is applied to the density of rescue systems (firefighters, police or gendarmerie, doctors and medical devices, road or air access routes, etc.). This structure allows dense networks of evenly distributed backup means, *thus concentrating their resources in an extremely short amount of time* in the event of a major incident. If we take the Haut-Rhin department in France, for example, first-aid emergency services can reach any location in that department in less than half an hour. In addition to this, any specific departmental technical reinforcements can be sent to the sites in question in less than an hour, even those with the most restricted means of access. Thanks to national coordination services such as the COGIC (i.e. the Operational Centre for Interministerial Crisis Management) and many others that remain on constant alert twenty-four hours a day, regional and national resources can be mobilised within a very short period of time ranging, generally speaking, from one to three hours.

In addition to these measures, the periodic inspections carried out by DREAL (a French supervisory and control body) in various facilities help to draw the attention of companies to any safety points requiring improvement, to advise them and, if need be, to force them to remedy certain security defects in the event of a serious lack.

Despite the high-performance technical means available in countries such as the United States of America or Canada, the per capita surface area to be covered by rescue services is immense, and access distances are, as a result, often considerable (almost ten times greater than in Western Europe). This is also true of the countries of the former Soviet Union, whose technological environment and technical approach are of the same level but characterised by a different conception of things: the American approach is to resort to highly specialised technology that is often more fragile than that of Eastern countries, while the latter favour the use of proven and reliable technologies that can be depended on in the face of unforeseen consequences yet remain inexpensive to operate and maintain.

As for countries such as India or China, their major issue lies in the density of their population as well as in the dispersion of their technological capabilities: in both countries, one finds extremely educated and cultivated populations in some cities or regions, with others characterised by great monetary or cultural poverty resulting in extremely reduced means of coping with unforeseen technological crises, which are all the more likely to occur in such areas since, account taken of the very limited technological culture typifying such populations, the implementation processes often involve staff whose members lack sufficient training to be able to cope with the dangers of their profession — a situation which gradually leads to a downward slide that paves the way either for the gradual deterioration of the safety barriers that have been put in place to avoid technological disasters (as was the case in Bhopal) or for the inadequate training of the first-responder rescue teams that deal with technological and chemical risks, as witnessed in Tianjin, where sodium stocks were sprayed with water, displaying ignorance of the danger posed by the release of deadly hydrogen cyanide resulting from the contact of water and (fire-generated) carbon dioxide with particularly alkaline cyanides. As for African countries, their rescue means are all simply too limited to cope with such technological risks and, considering the vastness of the territory to be covered and the sheer size of certain mining, gas or oil operations, their rescue services would only be able to protect limited areas anyway.

Compared to this, Europe is therefore still a privileged region of our planet.

However, there is a major danger threatening these countries — current European governments, regardless of their nature, are more or less devoid of scientific and technical advisers that at once have the necessary transversal knowledge and are the only ones with the ability to make reasoned decisions in the face of multivariable systems.

Worse still, widespread ignorance of what conditions are required for value to be measurable has gradually allowed a fundamental mistake to invade the minds of all nation leaders: indeed, no matter where you are, development is assessed in monetary terms. But a measurement can only measure one single (one-dimensional) quantity or value. And yet currency alone has at least three such values: the available money supply (balance sheet); the cash flow, which is a quantity derived from the previous one; and the relevance of the action governed by this flow.

And all our authorities thus find themselves in a state of extremely serious confusion between the notion of money supply and that of wealth, which are independent quantities. Money is worthless — it is but a means of exchange, just as it would be foolish in chemistry to confuse the catalyst (= the currency that enables trading to take place) with the product formed in a reactor (= societal goods), which represents the genuine and useful wealth that one seeks (or rather *should seek*) to produce.

Across Europe, it's all a matter of balancing budgets. What a mistake! It is the budget that must be adapted to suit the necessary actions; under no circumstances should actions be restricted to fit into one's budgetary limits. Money is a virtual value created by man, and due to our having adopted it as a control variable in our system by ascribing to it a real property instead of an effective virtual one, we are allowing the scientific ignorance of economists to exponentially destroy the wealth that humanity has taken decades, centuries and even millennia to create. As part of this monetary (and in no way economic) blindness, one ends up sacking people in order to artificially inflate the financial balance sheet of a given year, for the sole purpose of increasing the value of the shares embodied by the company. This virtual wealth offered to shareholders diverts the company entirely from its original purpose, which is to produce useful goods that contribute to the general functioning of the society in question. As for the company itself, it thus deprives itself of some members who all have their own know-how, a know-how that represents its real wealth, i.e. its unique technical approach to a complex problem.

Let us give a few examples. I have had the opportunity to look at the different approaches adopted by Philips and Siemens with regard to the birth of colour television. These two companies had two fundamentally different concepts when dealing with the issues of brightness and contrast. Philips offered two controls, one of which regulated overall image brightness (by adjusting the mid-grey colour between pitch black and full white) and the other allowed users to set the desired contrast (i.e. the difference between total black and total white around medium grey). In contrast to this solution, Siemens chose to replace these two controls with two others which, although completely different, enabled users to obtain equivalent settings, with one allowing them to adjust the black colour and the other the white — overall brightness could be set by moving these controls in the same direction, and contrast by moving them in opposite directions. Over time, vacuum tubes were replaced by transistors and then by integrated circuits, but the technology of the devices manufactured by these two companies remained unchanged during the three to four decades of CRT television: in each of the two companies, a specific know-how had established its presence with regard to the specificity of the philosophy behind the design of the systems they were developing, and the philosophy was thus passed down from generation to generation, with know-how transmission taking place from old employee to new 'apprentice'. Things were no different when it came to the very high voltage (VHV, 24,000 V) required for cathode ray tubes, which Philips adjusted by using, in the case of images that lacked brightness, a regulated short-circuit tetrode to burn off any excess current that had not been consumed by the cathodic tube (generating, in passing, X-rays on the tetrode). By contrast, Siemens chose to regulate the low-voltage supply of the VHV generator transformer, thus avoiding the production of X-rays on the tetrode, albeit at the cost of a regulatory system that was both harder to design and to run once time-worn.

These differences are quite comparable to those between a conventional aircraft, whose stabiliser provides directional control and stability through vertical action upon the empennage and longitudinal control and stability through horizontal action (similar to the above-mentioned levels of black and white), and a *Fouga Magister*, formerly used as a Patrouille de France[10] aircraft and providing both controls through a v-shaped tail (similar to the brightness and contrast control mentioned above).

10 TN: The *Patrouille Acrobatique de France* (French Acrobatic Patrol), also known as the *Patrouille de France or PAF*, is the precision aerobatic demonstration unit of the French Air Force.

These examples reflect the durability of a technological philosophy within a given company and the importance it can have when it comes to mastering the advantages and disadvantages of the products derived from it. Nothing is worse than the current tendency to make everything uniform and get rid of anything that is only useful in specific cases, cases considered too few to be financially profitable.

The decision criterion allowing one to determine whether something must be done or not should not be a financial criterion, but one of societal utility. We must not make our decisions according to their monetary profitability, but on the basis of their societal necessity. And it is because we have forgotten that money was originally created by people with a specific purpose in mind, namely to facilitate the exchange of goods, that it has become our master, a very bad master that is granted permission to make decisions in our stead. By elevating it to the rank of sole measurement criterion for a multidimensional system, one has forgotten that this alone makes it unmeasurable, and the whole world now watches helplessly as self-proclaimed and greedy economists, eager for material gain and profoundly ignorant of the dynamics of complex systems, plunge us ever deeper into programmed decadence and an unprecedented economic crisis of which we have only seen the tip of the iceberg.

And so tomorrow, under the impact of the budgetary savings which every government is trying to make in all areas (including education, training, technology, rescue and intervention resources, hospital systems and medical environments), accidents resembling those that occurred in Bhopal and Tianjin are likely to pose a serious threat to our Western countries in the short to medium term, no matter how improbable they remain at present.

— Is our educational system suited to dealing with the threat or risks?

S.W.: This question is a very broad one and requires us to consider many aspects. First of all, one must specify the nature of the threat(s) and/or risk(s). Secondly, there is the issue of defining the educational system. In it, we must also distinguish the notion of knowledge from that of education. For they represent orthogonal quantities (in the mathematical sense of the term) insofar as one can easily attain a very high level in one without having the other one at all. In other words, improving one of them does not presuppose an improvement of the other. And what is more than regrettable is that what is termed 'National Education' in France is but an institution whose sole objective is ultimately the teaching of knowledge. This state of affairs has gradually established itself in our country through a semantic blur between the notions of education and teaching. Whereas teachers of the early twentieth century still had a very important role to play as educators, this role was gradually taken away from them and transferred to parents. Nowadays, however, these parents are often not up to the task themselves, since, for historical reasons, they themselves were no longer given any proper education in the aftermath of 1968.

To ensure that I am understood correctly in the next lines, it is important for me to first redefine these two notions in their semantic context.

Education has an essentially societal role: it has an important part to play in the behavioural development of a child. By bestowing upon the child an awareness of belonging to groups (an individual in a brotherhood, family, school class, commune, region, nation, ethnicity, humanity, etc.), its role is to develop within him/her notions of abilities (ranging from the more basic ones to the more elaborate: perception, action, reflection, interpretation, contextualisation, conceptualisation, etc.); notions of limits (physical ones relating to one's body, affect, and mind, but also those of a societal, regulatory and legal nature); notions of duty (physical, relational and intellectual contribution to the group to which one belongs); notions of freedom, whose limits are imposed by duty; and notions of integration both into a local environment (with its rules and customs ensuring a harmonious functioning within their scope of spatiotemporal application) and, on a much larger scale, into a single global system shared by all, thus having to take into account the management of our planet's resources which, although admittedly significant, remain limited and renewable at speeds specific to each resource.

Education, therefore, corresponds to a primordial aspect of the education system whose aim is to develop in the individual a sense of community and an understanding of his/her role within it. And this process begins in childhood. The transmission of education should essentially be the responsibility of parents who, by passing it down from generation to generation, inculcate in their children the subliminal bases of what their own culture will be, ultimately defining their societal behaviour. Education thus plays a role in children's development and behavioural control as they grow up in their own society, which elevates them from childish frailty to their role of responsible adults, ultimately ensuring mankind's development and perpetuation on the planet, with all the richness of its cultural diversity, which can be expressed through everyone's respect for everyone else.

And it is in this framework that an important limiting factor surfaces, namely that of respect for life at all levels. Nowadays, however, the notion has been implanted in our minds that acquired knowledge is superior to the education which our ancestors once possessed. Through the development of ignorance and its concealment under the mask of science and technicality (both of which we unsuccessfully strive to pass off as all-powerful when, in actual fact, they are no more than a type of scientism limited to the superficial appearances of scientific knowledge that is beyond partial and shallow), we have gradually dethroned education in favour of a materialistic culture that reduces the vain measuring of the unmeasurable to the commercial dimension of money and profit.

In parallel to this, the intention was to pass on and impose a message implying that freedom is a complete and limitless universal right, one that can be incremented through a material sort of literacy generating monetary wealth and therefore a power of material possession and constraint over others by depriving them of financial affluence. Alas, these simulacrums of truths, which creep into people's minds through the current media's constant rehashing of stock market values, the economic power of large groups, the public debt, growth measured by monetary indicators, and the soundness of policy decisions based solely on financial profitability, are completely indemonstrable because they are fundamentally wrong. They do not include any data on the measurability of quantities, nor on the dynamics of complex systems. By ignoring fundamental things and relying on errors, such axioms can only plunge us into regression and decadence.

And any attempt to offer a scientific demonstration in this respect would be only be regarded as proof of its author's ignorance. Indeed, as a result of the simultaneous presence of four (or more) independent factors, a complex system loses its right or left orientability and every assertion thus becomes, in fact, simultaneously true and false both as a function of the subspace in which the observer is operating (a tetra-rectangular tetrahedron generates two sub-spaces comprised of right trihedra and two sub-spaces consisting of left trihedra, none of which is equivalent to another, and with each leading to different conclusions that are all true within the trihedron in question, yet different or false in the others).

These notions are very difficult for our minds to process, as the latter are usually faced with a three-dimensional representation of our physical space, which, thanks to its three-dimensional and thus orientable character, leads to mutually exclusive certainties between the existence of correct conclusions (generally sought by science) and that of false conclusions (acknowledged as such by those very same scientists).

However, life is full of examples where opposing phenomena are non-contradictory but rather only valid in a context that requires clarification and definition, in the absence of which we would arrive at senseless conclusions (in the etymological sense of the term, where referring to sense in a non-orientable universe is, strictly speaking, nonsense, the kind of nonsense that many of our decision-makers cannot perceive as such, as a result of their own ignorance). Thus, when moving in a line (i.e. a geodesic) on Earth (Riemann's universe), we actually rotate in a circle as we travel round the planet, yet believe ourselves to have moved in a straight line. And a New Zealander does not walk on his head, of course, even though his body is oriented in a different direction to ours on a planetary level.

Secondly, teaching plays a fundamentally different part: its role is not that of regulating behaviour within a given society, but of providing an individual with knowledge that will enable him to achieve his goals. While education is essentially the responsibility of parents, teaching is the responsibility of teaching staff in the broadest sense of the term, ranging from primary school, even kindergarten, through secondary school all the way to university and research, in addition to contact with 'knowledgeable' and experienced professionals that pass on their knowledge and skills to younger people who, through such contact, acquire the ability to take things over. This learning needs to occur at all levels and is extremely important insofar as the school system —which is often mistakenly referred to as our 'educational system' or 'National Education', when its role is not actually to educate but to teach, since the objectives behind these two actions remain fundamentally different — must bestow not only knowledge upon people but especially know-how.

Knowledge links facts in simple subspaces (defined by a maximum of three simultaneous variables), with any other variables only acting as small-scale variables compared to the main control variables (of which there must never be more than three if one does not wish to pave the way for chaotic behaviour in a system that one would like to control). In these simple systems, mathematical proof is possible, and the true and the false are indeed opposites, which allows us to establish certainties and demonstrate things. In more complex systems characterised by a number of variables which, albeit limited, is in excess of three, a new logic surfaces, one where we can no longer be certain that the correct counters the false and demonstrations can no longer serve as a corollary of this duality; instead, what we have are aspects that may be concurrent, orthogonal or opposite. Any and all demonstrations become impossible. However, one can still pursue a controlled trajectory, provided that one constantly and closely monitors its deviation from the desired path. With the maximal tolerable deviation making it possible to reduce the observed drift in a manner coinciding with the desired trajectory (which is how one drives a car on a road, for example), we can, depending on the correction means available, control the chaotic aspect of multivariable behaviour — provided, of course, that the trajectory is known at all points. Beyond these limits, which can all be 'compensated for', we lose control of the system, whose characteristics then reflect those of a deterministic chaotic evolution. Last but not least, when including a very large number of variables, a system becomes deterministic again and therefore predictable thanks to the general mutual compensation of the components' random individual drifts. Such systems are described by sciences such as thermodynamics, fluid flow, and so on, which offer general behavioural certainty amidst individual behaviours that remain completely random.

In complex systems, in no way can knowledge lead to certainties; it can only allow understanding of local aspects, of which all non-essential variables must be established to leave a maximum of three that grant the system a predictable character. While making these choices, observers often forget, whether intentionally or unconsciously, the essential elements that enable them to correctly define a system with all its internal contradictions, because any other hypothesis would hinder them in their conclusions, which would then develop various internal paradoxes. The certainty at which they can arrive through simplified systems is at the source of conflicts with other observers who, by resorting to similar yet still disparate simplifications, draw different conclusions and thus acquire their own certainty of being right (which is out of the question in the common paradigm of exclusive dualistic logic, where the correct and the incorrect are opposites). Their contention is, of course, correct from their local perspective, but completely absurd in a holistic description of the observed system, in which all the described aspects are simultaneously correct, even in their apparent contradictions, yet only in limited areas that the observer must determine beforehand in order to acquire the ability to use his knowledge correctly, from his own point of view.

This awareness is absolutely essential when one wants to ascertain whether the educational system is adapted to threats and risks or not. For both are aspects of an extremely complex system that only has local and very *partial* approaches (in both meanings of the word, perhaps).

From a strictly holistic point of view, around six people die on our planet every second, which, in absolute terms, represents a continuously reloaded machine gun firing six shots per second round the clock, with each bullet hitting its target with fatal consequences. This translates to 6 × 24 × 3,600 fatal casualties, or approximately 520,000 deaths a day. Viewed from a different angle, this would ensure a constant population level, provided that the birth rate was strictly equal to that of deaths. Unfortunately (for lack of a better word!), since there are more babies (a very affectionate term that I have chosen to use here intentionally, so as to reinforce the paradox) born every day than there are deaths across the planet, the world population is, in fact, constantly increasing, which is a major threat to our limited world... One would thus be right to say that those 520,000 daily deaths are a real blessing; and that the terrorists who kill two hundred people contribute to mankind's sustainable development, just like all those road accidents which, on a daily basis, claim the lives of thirteen people in France (counting only those victims who die within twenty-four hours of the accident).

In short, a terrorist attack, however serious it may be, ends fewer lives than an airliner accident; and in this context, one can indeed wonder whether it would not be preferable to invest all the energy we spend on the prevention of terrorism in the prevention of aviation accidents. After all, a plane crash involving an Airbus 380 can kill nearly a thousand people simultaneously, depending on the occupancy rate.

This example highlights the manner in which the analysis of complex systems can actually lead to heresies if we do not extract entire subsets of data from their general context. This is due to the unmeasurable nature of multidimensional values, to the absence of a measurement unit for all quantities of a subjective nature, and to the lack of an order theory[11] connecting the descriptive quantities used, many of which are orthogonal to each other in the mathematical sense of the word, meaning that that they are characterised by the absence of a variation effect that one of them could have on the measurable magnitude of another.

---

11 TN: Order theory is a branch of mathematics that uses binary relations to study our intuitive notion of order.

The point of this entire preamble is to make you understand the relativity of my answer: our education system is currently very far from taking into account the sheer magnitude of the risks.

It is quite appropriate to speak of education, because if it were provided properly (beginning in early childhood), the problematic question we are asking ourselves would not even exist. I was fortunate enough to have parents that set clear limits for me to abide by, and a deserved spanking never traumatised me. I am now a researcher at university, and my courses are tools in the hands of firefighters, police officers, gendarmes, emergency doctors, industrialists, and soldiers who, in a display of admirable courage, risk their own lives for the safety of our civilian society as they fight against completely uneducated people who no longer have any faith, laws, nor limits. Through lack of education and false teachings that are assembled under the illusion of divine truth, leaders whose sole purpose is to amass power and possessions of all kinds enslave them without their knowledge, combining this inculcation with enough knowledge for these people to become extremely dangerous for the civilian population.

We have not yet reached a stage where surface-to-air missiles are being fired against civilian aircraft, but this will soon change. As for nuclear weapons, the truth is that if we remain idle, these lawless people are bound to target a symbolic city someday. It is not a matter of establishing *whether or not* this will happen, but of determining *when and where it shall*. Terrorists are scattered across the globe in such a way that they now form an elusive hydra that will be extremely difficult, if not impossible, to combat militarily and effectively; it is as if septicaemia had spread its harmful genes throughout the body. Perhaps it is not too late, as gangrene has not yet set in everywhere, but its progression is fast, very fast, *too* fast, and no one knows where the point of no return actually lies.

The problem of the current struggle is that it fights the branching leaves instead of tackling the problem at its root. Our children's parents, who were themselves born to parents that experienced the May 1968 events, have, alas, in many cases, suffered the deleterious consequences generated by the maxim, coined in those days, which states that 'one is forbidden to forbid'. These parents, whose false understanding of freedom drives them to imagine that freedom is synonymous with the ability to do anything to anyone, had better read Montaigne's definition of liberty: 'To be free is to be able to do anything with oneself'. I was born in 1953, my brother in 1945, and the last real president I have experienced was Charles de Gaulle, who, after the war, set an example as to what a head of state should be like: someone who has plans for a nation, not only in the run-up to the next elections, but for a period of fifty years or even a century. This was the case during the peace process with Germany, in which Konrad Adenauer, who was just as great a leader as de Gaulle himself, managed to seal an impossible agreement with him: that of making peace in the aftermath of the atrocities committed during the last world war, thus putting an end, once and for all, to the permanent conflicts in which the populations on both sides of the Rhine had hitherto clashed. He thus launched our nuclear programme, our space programme, our road and rail communications construction programme and many other social and societal developments that we still enjoy today and which continue to bestow upon our small French homeland an aura that still shines (a bit?) upon the whole world.

He also set limits to our freedom in order to guarantee our complete liberty: although many things were indeed prohibited so as to ensure the proper functioning of our society, everything was possible and we ventured far, very far indeed, in our cultivation of effort and our respect for the freedom of others. Freedom is the result of accomplished duty and not the other way around. Within the illusion of freedom where parents no longer have the right to punish their children to show them the error of their ways, and where everything seems to be allowed, we find ourselves entirely confined to a straitjacket of regulations that is so constraining that it completely stifles all those whose actions could still free our country from the foul marshes into which it now sinks and which rob it of its very soul. Even a cat strikes her kitten when the latter bites her nipple a little too hard. As for a child under seven, most of his/her cognitive abilities are found in his/her physical body. Following a transgression, a spanking does not have to be hard for the message to be clear to the child. A long theoretical explanation of what should or should not be done would not sway him/her in the slightest, because a child's intellect-based cognition can only develop when he/she has acquired a sufficiently advanced semantic arsenal to establish links between abstract (intellectual) notions and his/her physical cognition faculties, which relate to the physical senses that he/she has had since birth.

We have thus given rise to generations that speak of rights without understanding that rights are solely the consequence of accomplished duties. And as a result of this, respect for others has fallen into oblivion, and our heads of state and associate states are now required to lavish students, who no longer understand the meaning of work, with such enticements as a baccalaureate which, truncated to an 80% rate by the Gaussian curve[12] representing it, is awarded in 80% of all cases, which, ipso facto, discourages capable students from working hard, since they will be able to pass this exam without making the slightest effort... And having attended courses at our School of Chemistry, which, thanks to the reputation of quality that its chemical engineers had back when I myself was accepted, was once the focus of so much envy (I myself am

12 TN: In probability theory, the normal (or Gaussian/Gauss/Laplace–Gauss) distribution is a very common continuous probability distribution represented by a typical bell-shaped curve. The term 'bell curve' stems from the bell shape created when a line is plotted using the data points for an item that fulfils the conditions of normal distribution. In a bell curve, the centre comprises the highest number of a given value and is thus also the highest point on the line's arc.

both happy and proud to belong to their ranks), what they possess upon graduating is a level of knowledge that would not even have allowed us to be admitted in the first place. I now find myself unable to do my job in this country, which was to ensure that our outgoing engineers contribute to our society's advancement: indeed, the capable ones are bogged down in the mediocrity of the rest, and our young technological elites are subjected to insulting attitudes that our sports champions would never have to face. Before being allowed to participate in competitions, athletes undergo a strict selection process and are given intensive training in teams that only comprise the best. In our schools, however, under the false pretext of an illusion of equality, we have been robbed of our right to resort to streamed classes. What a mistake—the good ones are no longer pushed to the very limits of their abilities, while the weaker ones, who could make excellent craftsmen, are lost in excessively theoretical studies which, instead of enhancing their talents and manual abilities and bestowing upon them the positive self-image that they deserve, only manage to undermine the flair and manual skills which would often enable them to work as painters, carpenters, musicians, tailors, and bakers, as well as in other trades where they could excel without theoretical knowledge, thus sharing this great wealth of theirs with the rest of our country. Instead of this, they are stuck on school benches for increasingly longer periods of time, often lasting up to twenty-four years, i.e. more than a quarter of their entire lives (according to the most optimistic lifespan forecasts).

It is thus hardly surprising that they choose to seek different ideals by turning to people who promise them a readily accessible paradise if they agree to put on an explosive belt or help destroy our society—a society which, in terms of a future outlook, fails to offer them anything more than the nothingness of a world in which money (the very same money that they will never have anyway) is the only driving force motivating people in their decisions, with the more or less clear purpose of wielding an illusory power that it still succeeds in generating.

Based on all these facts, our education system is far from being able to respond to those threats or risks. At best, palliative measures are being taken at the level of crisis-related respondent training. But this only helps us fight against the consequences and not the root causes of such threats and risks, both of which stem from ignorance. When will we understand that, just like top athletes, our children must also undergo a selection process according to their own skills and passions, one that would turn them into something to be proud of when they look at themselves in a mirror? When will we understand that they must become adults who will be able to say to those around them: 'Here's what I have done; here's how, in my own field, I contribute to the functioning of the society that sustains me; I have given it all my very best and no one can claim to be better than me in this domain'?

When we have set ourselves the goal of making each of our fellow citizens proud, all our problems will be fade away, regardless of whether they are of a societal, racial, religious or any other nature. I grew up in a multiracial and multi-faith environment in a small town to the north of Mulhouse, in Alsace, where people of all social, national, ethnic, cultural and faith-based backgrounds co-existed in a state of complete respect for one another, working in the then thriving textile and potash mining industry sectors. Yet all of them spoke French, which they learned quickly upon settling here, and many went as far as to speak the local Alsatian dialect, while simultaneously abiding by habits that allowed them to keep their roots, albeit only within their own group of origin. From a general social perspective, however, they settled there in a display of respect for local customs, a fact that enabled their complete and constraint-free integration into the old native community that had welcomed them with kindness and benefited from the multicultural wealth they brought. Our current societal problems stem from an excess of freedom which actually destroys liberty, as well as from a lack of constraints that only gives rise to the decrepitude caused by our society's growing weakness, which is, in turn, rooted in an absence of constraint that acts as a seed of weakness. An explosive substance or mechanism requires confinement in order to generate destructive power, and this physics principle is well worth pondering here.

So, to conclude this chapter, I can say that our education system is, in my opinion, completely unsuited to dealing with those threats and risks, not only because it is unable to alleviate them, but because it actually engenders them at their very source, which is embodied by basic education and teaching.

Only certain palliatives are suitable for emergency treatment, but if we do not remedy the situation at its very core, at the source itself, these palliatives, however powerful they may be (armies, elite units, technological means, the shaping of high-quality adults, etc.), will not, in the long run, suffice to contain the tidal wave which our global governments are preparing for us everywhere as a result of their weakness and/or corruption. Offering people of all origins a real future, one that they can be proud of having shaped themselves by the sweat of their brow, is the only way to resolve the inextricable problems burdening our current world, including the issue of immigration.

— Are populations sufficiently prepared?

S.W.: This point will be much shorter to respond to than the previous one; for it follows that if the previous point is resolved, this one automatically becomes obsolete.

For the time being, however, since the above-mentioned aspects have not yet been resolved at all, two contradictory aspects must be kept in mind: preparing populations is implicitly synonymous with admitting that they run a risk. And yet, for psychological reasons, many people are unable to exercise self-control once they realise that what they are facing is a significant risk or threat. The panic that such tension arouses in these people may, in some cases, render them completely unable to abide by logical and consistent behaviour, which is even more the case when it comes to a group that experiences severe stress in the face of imminent danger.

The current position of our public protection and response services is one of focusing the preparedness effort on the role that these very services play and on reassuring the masses with regard to the effectiveness of the government's response capabilities, rather than assigning responsibility to those masses for any part of the intervention, including any steps pertaining to individual protection initiatives.

In this regard, Switzerland entrusts more responsibilities to its citizens than France does. It seems to me that both choices are justifiable, and we do not have enough perspective to determine whether, in a generic case, one of the two solutions would prove superior to the other; only the future can give us the necessary feedback. Crowd management is a very delicate matter and involves the use of chaotic systems. Virtually anything—the actions of a single group leader, for instance—can drive a crowd to embrace uncontrollable behaviour or, alternatively, reasoned civic behaviour.

From my point of view, I would say that that we cannot do things much better than what is currently being done in the context of our present society.

4. Scenarios

And now? We bury the dead and treat the living.

— Sebastião José de Carvalho e Melo, 1st Marquis of Pombal, after the 1755 Lisbon earthquake

Toxic Gas

Fiction

A broad crescent moon shone high up in the sky, illuminating the flat and still countryside with its pale glow.

Greg was driving his *Mégane* hurriedly on the road to the city. He had been working at the PVC plant for five years, and he knew the way by heart. It was three o'clock in the morning and he had just finished work. He had but one desire — to get home as soon as possible.

Stopping at the first intersection, he tapped his fingers on the steering wheel as he waited for the light to turn green. Although in a hurry, he remained cautious. There were still a few cars moving here and there, mainly driven by party-goers returning from the city's nightclubs. The city itself was certainly not a large metropolis, but the presence of its 50,000 inhabitants turned it into the main attraction for those residing in the surrounding countryside.

Having resumed his journey, Greg carefully crossed the bridge that spanned a lazy river — the fog patches slowly drifting around him could indeed be hiding a vehicle or a cyclist. It would be a terrible waste to have an accident or knock someone over just to gain a few seconds.

With the bridge now behind him, he left the main artery and entered the street that gently climbed towards his neighbourhood. After two minutes, he arrived at his destination and parked his car in front of the small brick house he had bought two years earlier.

Entering the house without making any noise, he went straight into the kitchen, opened the fridge and poured himself some fruit juice. Instinctively, he grabbed his tablet computer and settled comfortably in his chair. He knew his wife would not be happy about it, but it was the weekend, and he deserved a moment's relaxation.

After spending a few minutes surfing the news sites, he started playing his favourite game.

A whole hour passed quickly, without him even realising it. Increasingly overwhelmed by fatigue, he finally decided to go to bed.

Putting his glass in the dishwasher, he was surprised to notice an unpleasant and pungent smell. His mind half-asleep, it took him some time to realise that he knew what it was. It struck him like a whiplash: it was chlorine!

Spontaneously, he placed his nose above the sink, then opened the cupboard below.

Nothing! No mixture of cleaning products or anything like that…

Not wasting any time, he set out to look for the source of this irritating odour. It only took him about ten seconds to understand that it was coming from the outside.

How was that even possible?

Suddenly, he felt shivers run up his spine. Could there have been a leak at the PVC plant?

Determined to ascertain what was going on, he opened the door and took a step towards the outside. Immediately, the corrosive gas affected his breathing. Turning back, he closed the door behind him and spent a few moments coughing and trying to catch his breath.

Immediately, he thought of his wife and two little boys, who were all sleeping upstairs.

What could he do to get them out of this deadly trap?

His senses heightened, he climbed the stairs and headed towards the rooms. Having reached his wife, he woke her up without wasting a single second.

— 'Sorry to wake you at this time, honey, but we have a big problem on our hands!'

— 'Huh? What're you talking about?' asked Lysa, still half asleep.

— 'The air outside is full of chlorine. There must be a leak at the PVC factory.'

— 'So what? Is it serious?'

As he was preparing to answer, Greg had a sudden coughing fit. He managed to light the bedside lamp, then turned back to his wife.

— 'Is it serious? You bet it is! It's a lethal gas! Tens of thousands of people died during World War I, when it was used as a chemical weapon! On top of that, it…'

— 'Ok, ok!' exclaimed Lysa, cutting him off. 'I get it, I get it! What do we do now?'

— 'I don't know yet…'

— 'Well, you'd better think of something! I'm going to wake the children,' she said, getting up.

Frozen to the spot, Greg watched her go. His mind seemed to be racing, yet he was unable to make a decision, and his body remained paralysed.

Suddenly, the city's sirens resounded. He stiffened at first, then left the room as well. Taking a series of firm steps, he headed to the room below.

The smell of chlorine was now a little more pungent than before.

Instinctively, he pulled his T-shirt up until it covered his mouth and dashed off towards the window. From there, he'd be able to see a large part of the city.

Immediately, he spotted his car, which was parked a mere few feet from the entrance. If he held his breath, he could surely reach it. Then, if they took the back road, it would take less than ten minutes for them to leave the city and reach the highway. It was worth a shot!

Suddenly, he noticed lights coming on in most houses. The city was awaking in spite of itself, with some having been roused by the sirens, others by the effects of the acid gas. In addition, the dancing headlights that began to fill the streets proved that the inhabitants were fleeing in their cars.

As he was about to abandon his observation post, he saw his neighbour leave her house with her little dog in her arms. Having reached the door to her garage, the woman, who was in her sixties, was seized by a strong coughing fit and dropped her pet. She began to follow the Chihuahua, but the latter ran away at full speed. Distraught, the woman finally decided to return to her car. The pain she felt in her lungs and eyes undoubtedly reminded her that the situation was very urgent indeed.

Still coughing, she barely managed to get to her vehicle and somehow settled into it. Reversing the car, she tore a rear-view mirror off as she bumped into the garage wall and damaged her bumper by ramming it straight into the lamp post on the other side of the street.

Greg was flabbergasted. The poor woman was simply unable to drive. Her stress, pain or irritated eyes were certainly at the root of this incredible behaviour.

Suddenly, he caught himself screaming 'No!'

His neighbour had just started heading towards the river. Without consciously realising it, he had noticed the thicker green mist hovering over that part of the city.

It was like an electroshock.

In an instant, he remembered the courses he had attended at the factory, which had dealt with the dangers posed by different gases. He remembered that chlorine was heavier than air and that inhaling it could cause an agonising death, even days after exposure.

He then realised that, even if she did manage to leave the city, his poor neighbour would not survive.

Suddenly, he felt his chest tighten. Once again, he thought of his own family.

To flee his home, he would have to cross half of the city, and the already numerous headlights proved that the traffic had intensified. It would take time to get out of there. Too much time! Not to mention the likely accidents that could block the traffic completely…

A sudden noise behind him tore him out of his thoughts: Lysa was coming down the stairs with their two boys.

Everything was now becoming clearer in Greg's mind. Without wasting a single second, he rushed towards them.

— 'Go back upstairs,' he shouted.

— 'Why?' asked Lysa, worried. 'We have to leave. It's too dangerous to stay.'

— 'No! It's hitting the road that could kill us all.'

The woman froze for a moment, her eyes fixed on her husband.

— 'Go back to the upstairs room, I tell you,' Greg insisted loudly. 'We need to put some adhesive tape wherever the outside air can seep in.'

Lysa felt disconcerted. She did not understand why the family had to stay there, instead of running away from the mortal danger threatening them. She was convinced that shutting themselves in would be fatal to them all.

— 'We can't stay here!' she shouted.

— 'It's all going to be alright,' the man replied gently but firmly as he pushed his wife and two little boys, taking them upstairs.

Lysa complied without offering any resistance. Yet deep down, she felt she was ready to crack and run towards the car.

Greg must have sensed her panic.

— 'It's okay, I tell you. We'll put the children in my office. The room is large and lacks any contact with the outside. Then we'll put adhesive tape all around the windows and doors. Alright?'

— 'Alright then,' said Lysa, who seemed to be recovering bit by bit.

— 'Come on, turn on a cartoon for them while I get things started. Look at them, they're scared.'

It only took Lysa a few moments to get *Ice Age* playing on the tablet computer. Soon, she'd joined her husband and set out to help him isolate the house from the outside.

Five minutes later, Greg felt relieved.

— 'Looks good enough to me,' he said. 'Go back and stay with the children. And whatever you do, keep the door closed and try to seal it well.'

— 'And what about you?'

— 'I'm going to plug the vents in the bathroom. Won't take long at all.'

No sooner had Lysa entered the office than her little boys threw themselves into her arms.

'Mummy, I'm scared,' said the older one, echoed by the younger.

— 'It's nothing, my darlings. Just imagine that we're playing a game together. The whole family will stay in this room, and the first one to leave loses. Sound o.k.?'

— 'Is daddy playing too?'

— 'Yes, he is, and we'll start as soon as he comes back. In the meantime, just watch the cartoon…'

Lysa put some adhesive tape all around the door and sat down by the desk.

It was only then that she realised that her throat was burning up. 'At least the air seems breathable in this room', she comforted herself.

Minutes passed, empty and endless, with no sign of Greg to break the silence.

What was her husband doing?

Anxiety began to get the better of her. He should have been back already.

Powerless, she stood up and started to pace the office, regularly casting anxious glances at the door.

It had already been ten minutes.

Soon, the anxiety became unbearable. She felt truly distressed. What would she do if he didn't come back? Perhaps he'd fallen and she should rush to his aid. But what if it was more serious than that? What if the gas had overwhelmed him and invaded the house? Leaving the room would allow chlorine inside and be fatal to all three of them.

With her fingers still moving about nervously, Lysa began to mentally pray. She couldn't go back on the decision she had made… At least she and her children were safe in the office — but how long for?

Suddenly, Lysa froze. She thought she'd heard a noise.

No, nothing! Apart from the cartoon on the tablet, all one could hear was a heavy silence.

Finally, after two or three minutes, she decided it was time to take a look down the hall. As she was about to open the door a bit to stick out her head and shout Greg's name, she heard footsteps coming from the stairwell.

When her husband had entered the office, she let out a long sigh of relief. As soon as the door closed, she threw herself into his arms.

— 'My god, you reek of chlorine,' she said, edging away.

— 'Yes, its concentration is increasing on the ground floor.'

— 'What were you doing all this time? I was worried sick.'

— 'I sealed the rooms downstairs and brought back a couple of things for us to drink and eat. There's no telling how long we will have to stay isolated. And I also took the radio.'

The young woman grabbed the baguette, fruit and three bottles of water and led the way for her husband to sit.

— 'You look exhausted.'

— 'Yes, I am,' he whispered between coughing fits. 'I have trouble breathing.'

— 'Your eyes are red and tearing up!'

— 'It's nothing, it'll pass,' he replied, wincing in pain.

— 'Why that look on your face, then?' she retorted, her voice filled with anxiety.

— 'It's my throat. It's burning. My lungs too...'

— 'Ok, stop talking then. Here, have a bit of water.'

Lysa looked at Greg, unconsciously trying to assess the damage. He looked exhausted, as if he had run a marathon.

Stroking his neck, she remained silent for a few minutes. However, her mind was flooded with questions, and she couldn't restrain herself for too long.

— 'What are we going to do now?'

'Wait,' said Greg hoarsely.

— 'How long?'

— 'Don't know. That's why I brought the radio.'

— 'Which channel shall we listen to?'

— 'No idea. Luckily, we have an internet connection here; it should not be too difficult to get some information.'

No sooner had he said that than he turned on his computer and waited for it to be ready for use.

He did some quick research and, after five minutes, came across a government website containing several articles on technological risks. All the information was there: the alarms emitted by the sirens, the intervention plans and, most importantly, the behaviour to adopt in case of an accident...

He leaned back in his seat and let out a sigh of relief. He had made a good decision — staying confined seemed to be the right choice.

Suddenly, a coughing fit made him wince in pain. He really felt as if he had got burnt from the inside.

Remembering a packet of throat lozenges he had seen in a drawer, he took it out and swallowed one in the hope that it would soothe him, at least for a while.

He then picked up the radio and looked for a suitable station. When he had finally found one, the commentator was talking about the PVC plant.

'... the explosion of a chlorine tank. Although the cause of this accident has not yet been determined, a preliminary investigation suggests a short circuit and a subsequent outbreak of fire. Furthermore, the authorities have announced that the situation is now under control. Priority is obviously being given to intervention and the fire brigade's Chemical Intervention Unit is already on the premises. Residents are urged to stay at home and confine themselves as best as they can. Stay tuned, we will be informing you of any developments regarding the situation...'

Greg turned slowly and looked at his wife and two boys. Despite the growing pain in his chest, he somehow found the strength to smile. His family were going to live. As for his own survival chances, he didn't have a clue. The coming days would be decisive.

— 'Come, take my seat,' he softly said to Lysa. 'Write an email to your parents — and mine, too.'

— 'Don't worry, I'll take care of it,' she said, gazing at her husband as he settled into the armchair. 'You look terrible — you know that, right?'

— 'Yeah, but it's nothing,' came the lie. 'As soon as the authorities give us the green light to come out of our confinement, we...'

Another coughing fit cut his sentence short.

'We'll go to your parents,' he said. 'They'll be able to babysit the kids while I go to hospital.'

— 'Honey, but they live more than an hour away from here.'

— 'Maybe so, but at least the hospital will still be in operation there. Also, if we take a few secondary roads, we won't have to cross the entire city. I can't even begin to imagine the chaos reigning there right now...'

He paused for about ten seconds to catch his breath, then continued in a low, hoarse voice.

— 'Not to mention the sight of all the damaged vehicles and corpses on the way, which could traumatise the children…'

Lysa looked at her husband again. She had a bad feeling.

Deep inside, she was afraid that Greg's poisoning was much worse than it seemed.

With tears in her eyes, she turned towards the computer. As soon as she had sent her emails, she would look for information on chlorine and the nearest hospitals …

The Facts

Accidents that lead to toxic gas leaks are relatively common in industry. Although they do not, generally speaking, reach the kind of magnitude described above, it is not uncommon for surrounding populations to be confined to their homes for hours or even days in response. The outbreak of a fire in a chemical storage facility can also result in similar consequences.

The fictional events related above are, in fact, similar to one of the most dramatic cases in all of human history—that of Bhopal, where tens of thousands lost their lives. Here is a short summary:

BHOPAL, 1984

On the night of 2 to 3 December 1984, an incident occurred in India at a pesticide manufacturing plant owned by Union Carbide Corporation (UCC), on the outskirts of the city of Bhopal.

Around 10.30 pm, following a series of washing operations, nearly 3000 litres of water found their way into the tank containing more than 40,000 litres of methyl isocyanate. What then ensued was a gradual and constant pressure rise in the tank itself. At the time, many security devices and systems were out of order in an attempt to save money. Soon enough, low emissions of toxic fumes began to impact the staff, whose members chose not to pay any attention to the incident, however.

By approximately midnight, pressure in the tank had reached the maximal authorised limit… and continued to climb. Half an hour later, it had already doubled. Shortly afterwards, the safety valve exploded, causing a significant leakage of methyl isocyanate.

Due to instructions not to wake the appropriate officials during the night, much less to advise any authorities of the incident, the alert was raised too late—at about 1.00 a.m. in the factory, and 3.00 a.m. in Bhopal.

In the meantime, for a period lasting approximately two hours, thirty to forty tonnes of highly toxic gas had been released into the atmosphere. Emissions heavier than air (with a density of 1.42) were spewed out and carried towards the city of Bhopal, whose outskirts lay at a distance of merely five kilometres.

The city's population (about 800,000 people) awoke to the first symptoms: tingling and a burning sensation in their eyes and respiratory tracts… Soon, chaos and panic had taken over the city. The inhabitants did not understand what was happening. Distraught, they ran through the streets, which allowed further amounts of deadly gas to seep into their lungs. Having surrendered to fear, they tried to escape both death and the invisible enemy that was decimating their family members one after another and mowing down their friends and strangers in the dozens… The weakest people fell within minutes… The gas in question triggers blindness, which added to the collective hysteria… The streets were strewn with children's corpses…

By early morning, about 3,800 lifeless bodies covered the city's surface. And yet this death toll was far from being final. Indeed, within three days, the number of fatal casualties had risen to 6,500. Nearly ten years later, it reached 10,000, and then 25,000 after approximately twenty-five years.

Unfortunately, the tragedy did not end there. The descendants of the victims of the Bhopal disaster are still subject to terrible evils: genetic mutations and infertility issues have become a recurrent phenomenon, and the rate of birth-related malformation is seven times higher than in the rest of India. Furthermore, infant mortality has increased by 300% since the accident…

As if this were not enough, the area is now contaminated. Toxic compounds continue to seep into the soil and groundwater; and tens of thousands still drink this water. In some places, mercury levels six million times higher than the norm have been detected. To date, we have yet to witness any willingness whatsoever to take responsibility for the disaster or bear the financial burden resulting from a depollution project.

Analysis

The previous example is an extreme-case scenario. The chlorine quantities required to produce lethal concentrations for a city of 50,000 inhabitants would have to be considerable. Thanks to the redundancy of the numerous security systems, an accident with such severe consequences is implausible in a Western European context. Nevertheless, the possibility of terrorist action aimed at generating chaos and claiming a large number of victims remains realistic.

The main purpose of the scenario was to show that, in most cases, confinement is indeed the best response. Although flight is a reflex that springs to everyone's mind, it comes across as unreasonable under the circumstances described in the fictional account above. The amount of gas one would inhale when fleeing would certainly be much higher than the one absorbed by any residents that would choose to stay at home and isolate themselves. It is thus likely that Lysa's worries are unreasonable. Greg will, of course, feel some pain in his respiratory tract and probably require medical care, but his life is in no danger whatsoever.

There is, however, another element that weighs in favour of confinement. Indeed, the chemical nature of the incident exacerbates the fears of those involved in it. Perhaps it is our human unconscious, which pictures the effects of the gases used in World War I, or the mere fact that the danger is not really visible. Whatever the case, one experiences maximal panic. This may lead to traffic accidents that could block possible escape routes and result in irrational and often violent behaviour on the part of the individuals present there. When a family find themselves in the midst of such chaos, the resulting danger could be greater than the threat they are trying to evade…

Nevertheless, things are not always so simple and many factors can influence one's confinement. First of all, the buildings available in European or North American countries are of a different standard compared to those in developing countries (as was the case of India in 1984). It is much easier to seal off homes or flats that are well-isolated to begin with than to do the same with a hovel which, more often than not, possesses numerous openings other than its own windows. The climate is also a determining factor, since any dwellings that are subject to harsh weather conditions are bound to be better isolated from the outside.

The period during which a chemical incident occurs is also crucial. In the summer, people tend to sleep with their windows open. During a colder season, it is rather the opposite that applies. In the daytime, the wind is usually stronger than at night and thus enables the toxic cloud to spread more quickly. If one takes this thought a little further, a temperature inversion, for example, can actually constitute a 'ceiling' that will prevent the gas from rising higher into the atmosphere.

Last but not least, the characteristics of the substance itself are also important. If the gas is lighter than air, it will easily dissipate. On the other hand, the heavier it is, the more it will tend to fill gaps. Under such circumstances, it is obviously better to shut yourself in in an upstairs room than in the cellar.

To conclude, it should be noted that, in most so-called 'developed' countries, the warning systems in place are generally effective (including the use of sirens, of police or firefighter vehicles equipped with loudspeakers, etc.). In addition to this, materials comprising adequate instructions and the appropriate actions to be taken are distributed to the population residing in risk areas.

Chemical Warfare

Fiction

Colonel Cialdini checked his watch carefully. The chosen time was drawing near.

Slowly turning around, he looked at the dark landscape barely lit by the first quarter moon still high in the sky. On that day, there was neither wind nor fog, a rare occurrence in Lombardy.

Despite the CBRN suit he was wearing, he felt that the air was actually cool. He must have remained motionless for too long. Raising his head, he took a deep breath, causing his filter cartridge to produce a whistling sound.

In a few hours' time, dawn would break, and the incipient light of the November sun would gradually light up the plain, lifting the veil of darkness.

What a strange sight it was to behold a lightless city. He could not get used to it. Yet ever since the beginning of the 'Millennial Crisis', as the media had dubbed it, and the war that had ensued, it had all become a commonplace occurrence. Only the dim headlights of the few, still operational cars tore at the nocturnal darkness, triggering a bustling shadowy ballet all around them.

Once again, the man sighed, and then headed towards his armoured command vehicle.

No sooner had he entered than he took off his mask and opened the top of his suit. Casting a quick glance, he found that the data gathered by the reconnaissance units and drones were still pouring in. All around, the electronic map flickered.

As the commander of the 9th 'Rovigo' Artillery Regiment of the Italian Army, Colonel Cialdini had deployed his three batteries comprising four M109L armoured howitzers in such a way that they formed an arc across a vacant lot in the northern suburbs of Milan. Each of the guns had a maximal rate of six 155-mm shells per minute and a range of eighteen kilometres.

Suddenly, the radio began to crackle. Captain Rossi's infantry company, which had been assigned to protect the site, was preparing to make its report. The day before, he and his men had searched all the houses nearby and secured the area.

Once the operator had informed him that everything was normal, the colonel simply nodded thoughtfully. He could not allow any sniper to get close, much less an enemy antitank team armed with its fearsome RPG-29 rocket launcher… No way, he could not run that risk under any circumstances whatsoever — not with the load he was carrying.

Stepping towards the electronic map, the officer made some mental calculations. Seemingly satisfied, he opened his thermos flask and poured himself some coffee, as he glanced through the window.

Dressed in their CBRN suits, his men carefully handled the heavy shells containing the deadliest of agents — VX.

This chemical ammunition had come from the huge stocks that the US Army had left behind on the premises near Aviano while withdrawing from Europe.

Colonel Cialdini gritted his teeth. Without these protective suits, any leak from a defective shell or bad manoeuvre resulting in an accident would mark the end of his unit.

Yet despite the uncomfortable suits, the difficulty in breathing, the fogged masks and the dangerousness of the entire operation, none of his soldiers complained. For everyone understood the seriousness of the situation.

Ten years earlier, the European populations had done everything in their power not to get involved in the war between the United States on one side and China and Russia on the other, but the leaders of the old continent were of a different opinion.

And now, all the former great powers had collapsed. There were no winners, only losers. Chaos now reigned supreme over all of our planet's nations. The populations no longer lived, but were content to survive. The few forces still active, the remaining vestiges of some order or authority, were striving to assemble and make a stand. His unit was one of them. Gathered around a makeshift government, they were doing their best to salvage whatever they could. They were working to restore peace by any means necessary, including the most treacherous methods, such as the one he was about to employ…

The man shook his head. God, how he would have loved for all this to have been no more than a nightmare!

With a thin-lipped smile, he looked at his watch again and placed his hands on the tank's cold steel. Fine drops of bead-like sweat were trickling down his forehead. The time was so near now.

Despite the disgust aroused in him by the method he was about to resort to, he knew deep down that he had no other choice. The new government was so frail that it could neither afford the slightest failure nor the loss of its soldiers' precious lives.

The truth be told, these VX-loaded shells would be very useful to eradicate, once and for all, the insurgents of the Sesto San Giovanni district, renamed 'Sestostan' in 2020.

In the end, though, 'insurgents' was too strong a word. They were rather groups of terrorists or other extremists driven solely by a thirst for power. They implemented their own rules and controlled people through fear, by committing crimes and horrors in the public square.

The military staff could not afford to send infantry units into this urban hell. Too costly in terms of both time and lives. The French example, in the suburbs of Paris, was still fresh in people's memories — months of fighting that resulted in hundreds of deaths on both sides and were crowned with a most disappointing result.

Earlier, in July, the new Italian government had launched its 'Recovery' operation. The reclaiming of Turin had been a bloodbath.

And now it was Milan's turn. The senior officer sighed.

The 9th Artillery Regiment would take action. It would bombard the enemy-held bastion with chemical shells, while the infantry proceeded to encircle the area. The VX agent would do the trick — of this he was certain.

It was a merciless war waged against terrorist groups. There was no room for any sort of negotiation. It was henceforth necessary to eliminate the insurgents using radical means.

One last time, Colonel Cialdini re-did his calculations and checked the coordinates. He knew that his men's pace would not be optimal: an average of three shots per minute for each howitzer, i.e. a total of thirty-six shells per minute. It was, however, more than enough to deal with the enemy stronghold and its entire surroundings.

Sensing that anxiety was now getting the better of him, he looked at his watch once more. H-hour minus five minutes.

Outside, the officers were issuing their last orders, and the men were already in position. Everything was ready — a deadly deluge would soon be brought forth.

The colonel cleared his throat and grabbed his radio.

In a short while, he was going to unleash a fatal downpour on the enemy bastion. In a short while, he was going to betray all of his principles and violate both the Geneva Accords and the Chemical Weapons Convention…

The large hand of his watch shifted and landed across the fateful number, 12. The watch now indicated 6 o'clock sharp.

The officer commanding the regiment pressed the switch on his radio and issued a preparatory order. Finally, in a calm and cold voice, he triggered the shot.

With a deafening noise, the artillery units went into action. Their steady rumble punctuated the next thirty minutes.

Then, everything went silent.

The colonel swallowed hard.

That was it!

He had just used chemical weapons against the terrorist stronghold, on his own territory, in Italy…

He knew that the enemy's men were now rolling on the ground, unable to make the slightest coordinated movement, as their limbs trembled and saliva poured endlessly from their mouths… The nerve agent was unforgiving, and only death would put an end to their suffering…

A long sigh. Then another.

He realised that from that moment on, his name would forever be linked to that morning's attack.

What would people label him? A saviour or an executioner?

During the hour that followed, he spoke very little, contenting himself with giving brief and concise orders, as he remained lost in his thoughts.

More than a thousand shells had been fired. The simulations had predicted that the enemy would suffer losses totalling 95% of their ranks. But what would the outcome really be like?

Then, the long wait began. All that the governmental forces had to do was wait before storming the area. VX vapours and droplets were going to do their work for them. The unfortunate ones impacted by the nerve agent would fall like flies.

It would take a few dozens of minutes longer for those who had only received low doses or those who believed themselves protected by their clothes and shoes to be affected; a sufficient amount of time for the toxic warfare agent to pass through rubber or any other material. Still, the result would be the same. Once the victims become aware of the first symptoms, it's already too late.

Somehow managing to keep his impatience in check, the colonel remained in his command vehicle, waiting for information to be gathered by the observation and combat units.

Around 3.00 p.m., sporadic shots fired from automatic weapons broke the silence. The reconnaissance troops were invading both the enemy stronghold and the entire adjacent neighbourhood, covering an area about two kilometres across.

Finally, in the early evening, the first reports came in, initially brief and then more detailed.

There were only a few pockets of resistance left inside the insurgents' bastion. The governmental forces had now regained control of the neighbourhood.

The mission had been accomplished.

Colonel Cialdini did not know whether to rejoice or cry. From a military point of view, the operation had been a great success. From a human perspective, however, he knew that many innocent people had been sacrificed.

Two days later, complete reports poured in.

The official casualty figure was 7,832 deaths. Surely much more, in fact. Around half of them were among the ranks of local people that had either been kept in a state of slavery or had supported the 'insurgent' forces. He would never find out the exact number of each. As for those armed groups, they had simply been annihilated. None of the different clan leaders had survived the nerve agent.

In the days that followed, the government's troops would carry out further reconnaissance missions. Cellars, granaries and sewers would be searched to ensure there were no survivors or fighters that had managed to escape the gas. Next, the demanding decontamination phase would begin, unless that part of the city was simply to be abandoned. Whatever the case, another unit would take on this task, a unit comprised of engineering specialists employed by the Italian Social Republic.

Suddenly, a sound of engines tore the colonel out of his thoughts.

Tanker trucks were coming to supply the 9th regiment with fuel, a most valuable commodity at the time.

Instinctively, he understood that the government had become aware of the operation's military success.

With a bitter taste in his mouth, he left his command post and walked towards the staff officer that had just arrived.

New orders had been issued—that much he knew. His unit was going to set off on a similar mission. There were so many enclaves still in the hands of 'insurgents'! What city would be targeted next? All of northern Italy needed to be made safe. Next, they would certainly have to push south so as to lend the republics of Naples, Lecce, or even Sicily a helping hand…

The Facts

This story, which is supposed to be taking place in a more or less distant and apocalyptic future, is obviously a work of fiction. Nevertheless, the following question may still be asked: In a crumbling world requiring authorities to restore order by any means necessary, would unconventional weapons, such as chemical shells, biological agents or even atomic bombs be used? Indeed, is history not fraught with a variety of such examples?

From a chemical point of view, did the Greeks shrink from resorting to toxic fumes during the Battle of Sphacteria? And what about the Romans in Ambracia? And the French and Germans who used phosgene, mustard gas and many other agents during World War I? And what about the Egyptians' use of mustard gas and phosgene in the Yemen War? And the United States' resorting to Agent Orange in Vietnam and the atomic bomb against Japan? And what about Iraq and its own use of mustard and sarin gas against Iran and Kurdish populations?

The examples do indeed abound.

Since time immemorial, wars have killed tens of millions of people. Whether soldiers or civilians, the latter were killed with swords, bows/arrows, and rifles; by artillery shells and bombs…

Various rules and conventions[1] have emerged to render these conflicts 'cleaner'. Sometimes, new names were even invented (as was the case with 'surgical strikes') in order to create the illusion that war was not really a massacre. In actual fact, however, what is the difference between killing someone by piercing their bodies, ripping them to shreds or suffocating them? People meet their maker in all of these cases, but it is essentially a certain psychological aspect that comes into play here.

When one realises that most conflicts are accompanied by atrocities such as rape, murder, genocide, and concentration camps, what would the vanquished of the future do if they had nothing left to lose? How would they react when faced with a choice between certain death and the hope of annihilating their enemy using 'forbidden' weapons?

Moreover, even though many states are respectful of the various conventions, some nations, terrorist groups, or unscrupulous leaders could not care less about these treaties. The main obstacle to the use of such substances is the fear of immediate sanctions, including a possible military intervention by the international community. If, however, the powerful were busy bickering amongst themselves, it would be plausible that someone else might take advantage of things and 'settle' internal conflicts or discords with their neighbour by resorting to non-conventional weapons.

What is also interesting to note is that the use of chemical agents does not lead to any heroic feats of arms, since there is no real fighting involved. Those who are not protected cannot evade their own death:

---

1 AN: As early as 1925, a protocol prohibiting the use of choking agents, poisonous gases and the like and banning the employment of bacteriological means was signed. Coming into effect in 1928, it would be known as the Geneva Protocol.

the strong and the weak perish in atrocious pain, decimated by an invisible enemy. The reality is thus somewhat different from what is depicted by Hollywood in its staging of spectacular chemical attacks and its stereotypes of every conceivable kind: bloodthirsty dictators or ruthless terrorists; missiles filled with fearsome gases that bear the most exotic names and cause huge explosions; the impressive deployment of an army whose disembarking or landing is that of a saviour; courageous decisions made by charismatic politicians; and valiant groups of survivors that attempt to leave the city at any cost, despite their own injuries and poignant personal tragedies…

The reality is actually more prosaic than that: chemical agents are designed to quickly immobilise, slow down, injure or kill the enemy. Only silence prevails once the deadliest of them have taken effect, since they eliminate all animal life in the broad sense of the term — which is why they are categorised as weapons of mass destruction. When used under ideal conditions, they can affect thousands of individuals in a matter of minutes.

It must be acknowledged that men have invested a lot of time and money in research programmes to achieve this result. Modern chemical weapons have thus been experiencing continuous improvement. Indeed, the most recent ones have a lethality that is much higher than that of previously used weapons, including those of World War I. As a result, a much smaller amount of the compound in question is necessary to achieve identical or even faster and deadlier effects. A few drops of VX, for instance, can kill an individual in less than a minute when applied to the skin.

On a modern battlefield, chemical weapons can make use of different vectors. They can be sprayed by aeroplanes, or spread by means of conventional rockets and powerful ballistic missiles whose warhead comprises chemical agents, aerial bombardment or, as was the case in the fictional account above, artillery shells.

The handling of such weapons and ammunition requires, of course, specialised personnel equipped with CBRN protective clothing that slows them down and renders their tasks more exhausting. If such operations were to occur, soldiers would have to transport chemical weapons from special warehouses to the battlefield or, at the very least, to launching vectors (planes, missiles, artillery, etc.).

Such handling, which is potentially very dangerous, could lead to disastrous consequences in the event of mistakes or accidents. It therefore requires a long and discerning sort of preparation. Furthermore, deciding to use such ammunition is synonymous with having both adequate logistical support and trained and equipped personnel.

Nowadays, modern armies are equipped with elaborate gas masks and full protective gear. As soon as a chemical threat is predicted, specialised units can also be deployed. The latter have additional equipment at their disposal, including decontamination lines, detection means, CBRN reconnaissance teams, and so on. In addition, some military vehicles are equipped with filtration systems and an overpressure system inside the cabin itself to prevent air and potential toxic agents from getting in. The transported crew and soldiers are therefore safe. Although such protection is highly effective, it remains both expensive and cumbersome.

From a military point of view (excluding the plain and simple massacre of civilians), the use of chemical weapons mainly concerns the following types of actions:

1. Attacks on targeted positions that give the aggressor a tactical advantage in the short or medium term, depending on the agent's persistence. This can be achieved by paralysing a given communication centre, rendering an airfield inoperative, etc.;

2. Attacks on under-equipped or unprepared troops. This type of action is characterised by exceptional effectiveness, in addition to arousing a feeling of dread among the survivors;

3. The protection of crucial sites or the enabling of strategic withdrawal. In this case, one deliberately proceeds to contaminate an

> area with persistent agents. The enemy is thus unable to pursue his offensive, unless his men are wearing CBRN protective clothing, which would render their task more difficult. Moreover, the enemy would have to set up his own decontamination lines, without which he would risk losing some of his soldiers and equipment.

In the event of an operation resembling the scenario above, an attack on 'rebel' forces amidst a civilian population would have tragic consequences. Indeed, civilians have no means of protection, and chemical warfare agents can pass through clothing, shoe soles and even most materials such as wood or rubber. In addition, since the average citizen has not been given the appropriate training to be able to react to this type of threat, the result would be a massacre in the real sense of the word.

Note: During the first Gulf War of 1991, the Israeli population feared an attack involving the use of sarin gas, which could theoretically be loaded into the warheads of Iraqi Scud missiles. The Israeli authorities knew that although such a risk was unlikely, the agent in question could pass through the skin and cause death despite any respiratory protection. This did not, however, stop them from distributing gas masks. This action enabled them to reassure the population and maintain its high morale, a key element that must not be neglected in the event of a conflict.

Analysis

In the fictional scenario detailed above, the world spiralled into utter chaos. The regular army was trying to regain control and restore a semblance of statal structure. The situation was so catastrophic that the new government was compelled to strike both fast and hard, while ensuring minimal losses.

In this context of civil war, it is very likely that the different fractions, ethnic groups or religious groupings that could emerge would bear a fierce, mutual and sometimes justified hatred of one another.

Regardless of the psychological aspect, however, the use of chemical weapons does fulfil the desired criteria of *efficiency and speed.* A street battle, one that is very costly in terms of human lives, could thus be avoided. On the other hand, the death toll among the population is likely to be high. In order to go to such extreme lengths, a set of conditions (such as those mentioned above) would certainly have to be met, namely:

- a chaotic international context;
- the absence of a formal government;
- a willingness to use any means necessary;
- the targeting of groups that are known to commit excesses and atrocities.

In the fictional account above, the compound used was VX, one of the most powerful poisons created by man to kill his fellow humans. A single drop on the skin causes death in terrible agony. Inhaling its vapours produces an identical result. Under normal conditions of pressure and temperature, VX is a compound that remains in liquid form. Its boiling point is at around 300°C. The number of victims mentioned in the story is, of course, subject to discussion. So many parameters must be taken into account that it is difficult to give a reliable estimate, especially when one is unfamiliar with all the data involved. The following is a non-exhaustive list of factors that determine the effectiveness of a chemical attack:

- the number of people likely to be affected;
- the nature of the target populations (civilian, military) and their possible means of protection;
- the area's landscape (valleys, mountain peaks, etc.);

- the type of targeted area (buildings, caves, forts, etc.);
- the targeted surface;
- the climatic conditions (temperature, wind direction and strength, rain, etc.);
- the amount of discharged chemicals (based on the number of shells/bombs and their contents);
- the ability of potential victims to flee.

In our scenario, the shells contained VX. They were not excessively loaded with explosives so as not to destroy the chemical agent that they carried and that they were meant to spread. When the projectile hit its target, the explosion released the VX in the form of millions of droplets. As for the shock wave, it pulverised windows, doors and even some walls, depending on the distance from the impact point. The released heat also turned some of the chemical warfare agent into steam. The intended target was thus impacted by the presence of VX both in the form of vapours and in small droplets scattered all around, amid buildings whose openings had been shattered. A light wind and a relatively low temperature would keep the area contaminated for a long time to come (depending on the amount of VX released and the prevailing climatic conditions, perhaps more than a month).

Since VX can pass through the majority of common materials, it is only a matter of time before survivors are forced to touch contaminated items and are affected as well. In addition, because the liquid residues of VX slowly turn into vapour, they perpetuate the presence of a deadly atmosphere. The soldiers who cordon the perimeter off are, of course, equipped with CBRN masks and protective suits. In the piece of fiction above, their mission, which consisted in closing off the area during the chemical attack, was threefold. Indeed, its purpose was to:

- ensure that no enemy escapes;

- prevent contamination from spreading to nearby allied populations, by hindering the escape of potentially contaminated individuals;

- take charge of any 'allied' or non-military individuals attempting to flee the area.

The reclaiming of the premises was carried out by troops that were both adequately equipped and specially trained. In the absence of an emergency, the mere maintaining of the perimeter allows the toxic agent to perpetuate its lethal action upon the targets. Everyone surely realises that keeping/detaining people in a deadly area (especially if they happen to be civilians) is unnatural and the situation must really be beyond critical for one to resort to such extremes.

Last but not least, the decontamination of an urban zone polluted by VX would require colossal means. It is more than likely that in such a case, the area in question would be abandoned for months or even years. Indeed, none can state with sufficient certainty when the sector will be 'clean' again. A few milligrammes of VX, for instance, could still be left under a stone, in a cellar, etc.

CBRN INCIDENTS: REACTIONS AND PROTECTION

Fortune favours the prepared mind.

— French chemist Louis Pasteur (1822–1895)

1. Familiarise Yourself with the Dangers

A lot of reflection and not a lot of knowledge — that is what one should strive for.

— Greek philosopher DEMOCRITUS (460–370 BC)

Natural hazards such as floods, fires, earthquakes, etc. have always existed. For several centuries now, technological accidents have been added to the list. The latter regularly cause dozens, hundreds or even thousands of deaths. Our current society is therefore never unaffected by this kind of threat. Cities, industries, and governments have all taken steps to prevent and reduce the consequences of such disasters, which can sometimes impact a large part of the population.

While it seems clear that certain administrations or operational units are familiar with the issued instructions, the same cannot always be said of the public. For example, if sirens were activated in an urban conglomeration, how many people would understand their meaning? Would they be able to tell the difference between a genuine alert and the usual monthly test on the first Wednesday of the month? And, if so, could they react adequately?

In developed countries, regulatory constraints change on a regular basis to provide citizens with maximum security. In the case of fixed facilities (factories, nuclear power plants, laboratories, etc.), there are effective prevention and response systems, whether at the level of industry (an internal operation plan, etc.) or that of the government (special governmental intervention plans). These developments are of the utmost importance and include the triggering and spreading of alerts, the behaviour to be adopted by the population, as well as the organisation of rescue and intervention means. In the case of all citizens that reside in sensitive areas (i.e. in the vicinity of facilities that are classified as dangerous under the Seveso directives), the most crucial factor is one's awareness of how the alert would be broadcast and of any specific instructions to be followed.

The situation is more delicate when we are dealing with a CBRN accident in a location that is not usually exposed to any particular risks. Since the population does not feel threatened, it generally has little interest in the various available prevention and protection measures. However, a technological accident could indeed occur almost anywhere in France. An unusual colourful cloud could, for example, spread and obscure the sky. Although such a sight can sometimes relate to natural phenomena, the actual culprits are, more often than not, industrial fires or explosions (oil, chemical products, plastics, paint, etc.). The resulting fumes are, of course, toxic. It is therefore essential to monitor their development and especially the manner in which they are dispersed by natural elements.

Another example is that of a vehicle transporting dangerous goods, which can flip over and discharge its load onto the roadway or into the sewers or a river. The consequences would differ depending on its actual cargo (chemicals, infectious agents or radioactive substances, etc.).

In order to respond appropriately and, perhaps, implement effective protection measures, it is essential for the population not only to become aware of the danger threatening it as quickly as at all possible, but also to be familiar with the nature of the hazards and risks involved.

Although plans and measures do exist throughout our country, it is interesting to note that some situations may, due to their complexity or scale, extend beyond the authorities' control.

In the event of an accident in a nuclear power plant, the alert would most certainly be triggered after a relatively short period of time and the event highly publicised, but what consequences would it all lead to?

In truth, it all depends on the incident's source and size!

The situation could involve a mere minor fire or non-radioactive leak or, conversely, a runaway reactor leading to a Fukushima-type of scenario. The latter example showcases the manner in which the site managers, and then the authorities, tried to play down the consequences until it became too obvious that the event had spiralled beyond their control. In the weeks that followed this major accident, experts found themselves confronted with situations over which they had no control at all: their choices were sometimes unfortunate and the protection/evacuation measures meant to shield the population were sometimes delayed…

In their defence, the complexity and sequence of unexpected events made the situation's development very difficult to predict. Accurately determining the consequences for both the population and the environment had become a real challenge. This is particularly true in a context of radiological contamination. Indeed, radiation is invisible, tasteless and odourless. True enough, simulation software integrating meteorological data and leak estimates is now attempting to provide us with a model of the situation, but nothing can compare to real-time, on-site detection using specialised equipment or fixed sensors.

In addition, the damage caused by radiological contamination is usually delayed and the population cannot know whether it is really in danger or not. In addition to potential ingestion of large doses of radioactivity, which is already possible with tiny amounts of material and results in nausea and vomiting within hours, lower doses are bound to cause noticeable damage several days or weeks later. Very low doses, on the other hand, only serve to increase the likelihood of cancer years later.

The following question thus arises: to what extent can the population trust the information with which it is provided?

The case of Chernobyl, whose cloud was alleged to have stopped at the French border, is an eloquent example. If such an accident were to happen again (in the Ukraine, for example) or an atomic explosion took place in a nearby country, what would the radioactive fallout be? Would you be able to go outside? Would you dare eat your own vegetables?

Some people will thus simply take the initiative and equip themselves with protective or even detection equipment. Although this solution might be of interest, it requires a certain budget, a good knowledge of such equipment and a significant understanding of the risk involved. This topic will be discussed later on in the chapter entitled 'Detection Equipment'.

Alerts and the Dissemination of Information

Both natural and man-made disasters are occurring with increased frequency across the world. In case of emergency, one of the most crucial phases is to make relevant information available at the right time to help people react quickly in affected areas and avoid potential threats.

Alert and warning systems that provide life-saving information need to be efficient and use conventional channels, as well as channels operating with new technologies.

Depending on the country, different systems have been implemented, for instance:

- **Australia:** Standard Emergency Warning Signal

- **Canada:** Alert Ready
- **France:** *Réseau national d'alerte*
- **United States:** Integrated Public Alert and Warning System (IPAWS)

(The Emergency Alert System is now part of the IPAWS)
For more information:
https://www.fema.gov/integrated-public-alert-warning-system.

And the list goes on.

ALERTS AND THE DISSEMINATION OF INFORMATION BY THE AUTHORITIES ARE THE ESSENTIAL ELEMENTS TO TAKE INTO ACCOUNT IN YOUR REACTIONS.

The first reflex you should have is to shut yourself in at home (or the closest and most appropriate place) and turn on your radio[1] to obtain all the information and instructions you require.

Note that *France Inter* stopped broadcasting on its famous frequency of 162 kHz (long wave) on 31 December 2016.

In France

In case of a major nuclear accident, an industrial disaster or a serious terrorist act, the sirens of the national alert network (4,500 in France) will emit an easily recognisable signal, a modulated sound that will rise and fall in pitch and which consists of three one-minute sequences separated by a five-second silence. To listen to the signal, please visit: http://www.iffo-rme.fr/sons. At the end of the alert, the sirens emit a continuous sound lasting thirty seconds.

1 AN: For example, *France Inter*, *France Info*, *Blue France*, *RSR*, *RTBF*, etc. Other local radio stations may also broadcast the authorities' announcements, including *NRJ*, *Virgin Radio*, and so on.

The warning signal and/or information can also be broadcast by means of loudspeakers mounted on vehicles driven by firefighters or the police and gendarmerie. Once the alert is triggered, information should first be sought on the radio. Television and the internet take second place in terms of priority. In Switzerland, a similar plan has been anticipated[2] by the Federal Office for the Protection of the Population (OFPP) in the event of disasters, with a network comprising 5,000 sirens,[3] in addition to broadcasting the announcement on the radio and television and by means of the country's cantonal police intervention centres.

Identifying the Risk

The best way to identify the risk and know what to do is to turn on the radio. The authorities will provide you with specific information and instructions. Nevertheless, in order to understand the situation, each individual can carry out their own analysis by answering the following questions:

1. What—What kind of event is it?

- What is the nature of the risk?
  - Nuclear
  - Radiological
  - Biological
  - Chemical
- What are the risks involved?
  - Irradiation/contamination
  - An explosion/explosive atmosphere

---

2 AN: To find out more, follow this link [TN: in the German language]: http://www.bevoelkerungsschutz.admin.ch/internet/bs/fr/home/themen/alarmierung.html.

3 AN: https://www.admin.ch/gov/fr/accueil/documentation/communiques.msg-id-60485.html.

    - A fire
    - Poisoning
    - Chemical action
    - Infection/contagion

- What are the substances involved?
- In what form are they encountered?

2. How Many/How Much? Scale, amount, number…

- Scale of the leak
- Storage size

3. Where?

- Location
- Distance
- Domino effect?

4. When?

- What time?
- How long ago?

5. How?

- An accident?
- An attack?
- Malice?

6. Conclusion

- What is at stake for me?
- Routes of exposure (inhalation, contact, remote exposure, etc.)?
- How should I react?

Example: While you're working in your garden, the sirens begin to emit a warning signal. If you live in an area where special instructions apply, you simply follow them. If not, you and your family go home and shut yourselves in. Turning on the radio, you realise that a technological accident has just occurred: at 11:30, a truck containing eighteen tonnes of ammonia was overturned in the vicinity of the *Place de la République.* A leak is reported to have generated a toxic cloud. The population is instructed to head to the nearest building and barricade itself. *While continuing to listen to this information*, you check that the house is well-isolated from the outside. You can also connect to the internet or resort to a database on your tablet or phone to obtain further information. Soon, you find out that although ammonia is transported in liquid form, it is released as a gas in case of leakage. This results in a toxic cloud with a strong characteristic odour that can ignite within a certain concentration range (in the presence of a trigger such as a spark). It thus becomes easy for you to answer the previous questionnaire:

1. **What?**
    - Nature of the risk: Chemical
    - Risks: Toxic exposure/explosive atmosphere
    - Substance involved: Ammonia
    - Form: Toxic vapours
2. **How much?**
    - Scale of the leak: unknown
    - Truck transporting eighteen tonnes
3. **Where?**
    - *Place de la République*
    - At an approximate distance of five km (as the crow flies)
    - No high-risk structure located nearby (and thus no domino effect)

4. **When?**
   - 11:30
   - Twenty minutes ago

5. **How?**
   - Traffic accident

6. **Conclusion**

   You now understand the situation well and can remain composed with regard to what must be done. The main risk is to be poisoned by breathing in (inhaling) a dangerous airborne substance. Just stay safe until the cloud passes. If you have any electrical equipment outside (Christmas decorations, etc.), turn off the power to prevent potential gas ignition. Stay informed by means of the radio and await the 'all clear' signal, which will also be broadcast by the sirens.

AN INTERVIEW WITH PIERRE BRENNENSTUHL, THE DEPUTY DIRECTOR OF THE UNIVERSITY HOSPITAL OF GENEVA (HUG) IN CHARGE OF THE ORGANISATIONAL AND ECONOMIC ASPECTS OF THE CARE SECTOR, WHO ACTED AS PRESIDENT OF THE CRISIS UNIT DURING THE 2014 EBOLA CRISIS.

— What organisational steps does the HUG take in the event of a crisis?

Pierre Brennenstuhl: We have our own disaster plan, similar to the 'white plan' in France, which provides for the mobilisation of personnel and resources in case of crises, which are generally of an 'accidental' nature and involve a large number of wounded people being taken to hospital. If what we are dealing with is an infectious disease, we have specific plans in place, particularly in the event of a foreseeable crisis, as was the case with the Ebola epidemic in 2014. Under such circumstances, for example, a crisis unit comprising our main specialists prepares the hospital in a manner allowing it to comply with the required organisational steps to deal with the problem at hand.

In the case of the Ebola epidemic in West Africa, what we did was set up a pre-epidemic unit at the very beginning of the crisis in February 2014, while remaining in touch with any doctors active on site. We prepared ourselves and looked around for places that could accommodate potentially infectious patients for hospitalisation, while also looking into how one could proceed without outpatient consultation and in safety.

Next, we prepared the training plan and protective equipment to be used by our caregivers and coordinated the potential arrival of a patient with the proper airport authorities to ensure that they are given the necessary care on the tarmac and then transported to hospital on board a special ambulance intended for such crises. These elements are provided for and adapted depending on the type of emergency.

The main difficulty lies in our accommodation capacities. The admission of one or two patients can be managed in an optimal way, i.e. by providing isolation, access-control airlocks, power-operated doors, and negative pressure rooms to prevent any and all air particles from escaping, thus avoiding aerial dissemination, even if there is no such risk in the case of Ebola.

If we were faced with a large number of patients, we would be forced to implement what is known as 'cohorting', meaning the grouping of all patients in one single place. Generally speaking, an entire floor or landing is sacrificed to serve as an isolation area in the hospital. This would, of course, put us in an extremely complicated situation, since the staff would work under riskier conditions.

Ideally, an appropriate structure could be built to accommodate any cases involving a high risk of contagion for staff members, but an additional budget of five to ten million CHF[4] would be necessary for this to happen. In addition, the staff would have to be dedicated to this sole purpose for a certain amount of time—indeed, one cannot imagine having a potentially exposed caregiver strolling around the various other units afterwards, considering the risks that this would entail. I should also add that these patients use up a lot of human resources. Every single patient, in fact, usually requires the presence of four people.

Offering one Ebola patient adequate hospital care forced us to abolish ten ICU places. We do not have the capacity, as is the case in the major or military hospitals of large countries such as France, Great Britain or the United States, to provide sizeable infrastructures with hundreds of square metres dedicated to such patients, showers that make use of chlorinated water to disinfect protective clothing, special changing rooms, or water retention and disinfection airlocks. In Geneva, we made do with the means available to us, without having any special premises meant for such a purpose.

4 TN: Swiss francs.

2. Adequate Behaviour

The better path is to go by on the
other side towards justice.[1]

— As stated by the Greek poet HESIOD (8th century BC)

Confinement

When a CBRN alert is broadcast (through sirens, for instance), speed of action is essential to prevent a serious event from escalating into a full-scale disaster for yourself and/or your loved ones.

- **Your first reflex should be to take shelter.**
  In order to be safe, simple containment measures must be implemented:

1. Take shelter (or remain) in the nearest building.
   Note that a vehicle is not an effective source of protection.

2. Close and lock all doors and windows.

---

1 TN: As can be seen in the correct English rendering of Hesiod's words, the authors have unfortunately misunderstood the intended meaning. In French, the word '*justes*' is used, which is the equivalent of both 'just' and 'adequate/correct' actions. Hesiod's statement, however, does not refer to the *adequacy* of one's actions in relation to the circumstances one finds themselves in, but to the moral dimension of *justice* that pervades these actions.

In addition to this, make sure you close all shutters, as long as this action does not endanger the occupants in any way (e.g. rolling shutters, etc.).

3. Carefully obstruct and seal off all openings.
   A well-isolated room greatly impedes any ingress of potential toxic agents. For details, see the section on kits and equipment below.
   Put adequately large and strong adhesive tape along the door to render the interior airtight. Fill the washbasin with water to prevent gas from rising through the pipes. This should be done in all rooms that have their own water point.

4. Keep away from doors and windows.
   In the event of an explosion, being struck in the eyes by fragments of glass would greatly compromise your ability to deal with any unfolding events.

5. Turn off any air conditioners, ventilation systems/devices, and extractor hoods.
   In some buildings, such devices are centralised and thus beyond your control. If this is the case, you will need to obstruct/seal off these air vents and perhaps even isolate all rooms that allow access to them by closing all doors and possibly applying adhesive tape to them so as to ensure that they are completely airtight.

6. Avoid over-consuming air oxygen:
   It is recommended that you do NOT —
   - move unless necessary;
   - light any candles or prepare meals using gas stoves and the like;
   - smoke, especially since the resulting fumes will remain inside;
   - assemble and keep your family in a tight space. A room is a better option.

7. Avoid any flame or spark.

Especially outside (Christmas decorations, an electrical discharge insect control system, etc.). If a cloud of flammable gas is moving around your home, the last thing you want to do is cause an explosion.

8. Stay quiet in order to properly hear any safety-related messages.
Some messages may be broadcast by speakers mounted on governmental vehicles (the fire brigade, civil defence services, the police, the gendarmerie) or industrial ones (EDF, etc.). This information may turn out to be very important when evacuation is recommended, for example.

9. Do not use your landline or mobile phones.
It is crucial to avoid tying up the phone lines, so that emergency services and intervention teams can work in the best possible conditions.

10. Follow the instructions issued by the authorities.
Regular announcements are broadcast through the radio. You will be informed of the actual nature of the danger, kept up to date with regard to the development of the situation and perhaps be given additional instructions to implement.

It is always better to have a battery-operated radio (in case of a power failure).

11. Do not pick up your children at school.
Picking up your loved ones may seem a matter of priority. Nevertheless, doing so is, more often than not, a mistake, and you may find yourself endangered, even mortally so, on the way, without ever reaching your children. Since an alert is usually broadcast at the beginning of an incident, the situation is likely to deteriorate over time. The journey back from school will be even more dangerous, and you will not only endanger your own life, but also that of your loved ones.

In addition to this, each school has its own 'special security plan against major risks'. Teachers are trained to react in case of an alert; the behaviour to be abided by is pre-established, as is the choice of confinement premises.

12. Prepare for evacuation.

There are situations where confinement is simply insufficient. The prefect may then call for and organise an evacuation. Alternatively, certain events such as a fire, an explosion, etc. might not give you any choice.

Note: An ideal confinement zone must offer a minimum space of 3 m^3 per person. It should preferably be located away from fragile openings (such as windows), be free of any hard-to-obstruct areas, and allow easy access to an escape route if necessary. The space consisting of a corridor that connects an office to the toilet is, for instance, a good choice. In the event that the occupants have a survival kit or a bug-out bag with them, the latter must be available in the confinement zone.

Evacuation

There are situations where confinement is insufficient and the best alternative is to escape. In such a context, the term 'evacuation' is generally used. Some natural disasters (fires, floods, etc.) or wars can drive entire families to abandon their place of residence. In case of a total economic collapse, it is quite possible that an outbreak of violent events or the simple need to obtain food will also compel you to leave home.

From a CBRN perspective, a nuclear accident, a pandemic, or a major chemical incident may also drive you to seek safer shelter.

As we have already stated, history is fraught with examples in which a population was forced to 'flee'. Occasionally, only one neighbourhood was impacted, as was the case with the Jiangsu region of China (2005), when an accident involving a tanker truck transporting thirty-five tonnes of chlorine resulted in the temporary evacuation of 3,000 families. At other times, the events led to the final exodus of tens of thousands of people: that's precisely what happened on 27 April 1986, when the inhabitants of Pripyat, an agglomeration close to the Chernobyl nuclear power plant, were only given a few hours to flee. On certain occasions, the very urgency of the situation makes any and all preparation impossible, as witnessed on 3 December 1984, when the population of Bhopal desperately attempted to escape the deadly cloud spreading across the city…

So, in the best-case scenario, you will have a few hours, perhaps a day or two to put your affairs in order. Knowing that an evacuation is only necessary under the gravest and most unexpected conditions, it is better to bear in mind that it will take place under a label of URGENCY! (If you are given more time, you're in luck!) Under such conditions, and especially when your life is in danger, you will not have time to pack your bags and think of everything you need to take. You thus risk losing most of your hard-won possessions. Even if this breaks your heart, it's important not to forget your priorities—to save your own life and that of your loved ones!

Leaving unexpectedly, however, does not mean leaving without preparation. Departing your home at 3.00 a.m. in your pyjamas and travelling barefoot in the hope of finding some kind of shelter is not necessarily a very bright idea. Properly preparing for evacuation requires you:

1. to know where to go;
2. to have adopted adequate operational procedures;
3. to only carry minimal equipment. This aspect is dealt with in the part entitled '*Bug-Out Bags*'.

Where Should One Head To?

During a CBRN incident, the threat may emerge in the form of a toxic cloud or of contamination. If you find yourself compelled to leave the area in all urgency, you must do so in a way that you avoid entering any dangerous areas. If you are already in one, the best option is to leave in a direction perpendicular to the wind and leading in an opposite direction to the place where the incident has occurred. **DO NOT** travel upwind unless you are absolutely certain that the danger is located behind you.

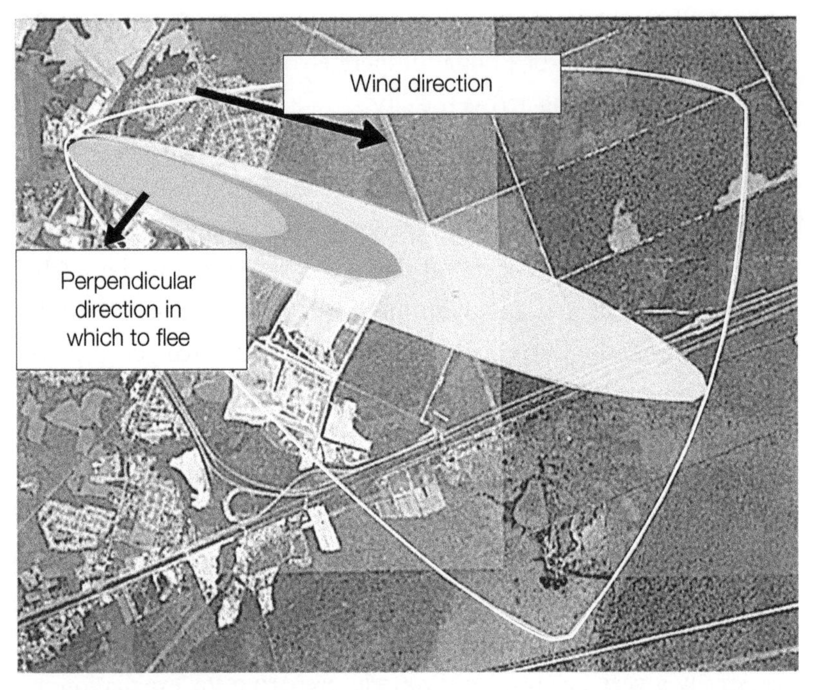

Model situation depicted through the Cameo-Aloha software: the direction in which you flee must be perpendicular to the direction of the wind.

Knowing where to go means having pondered in advance the different possible scenarios and defined a suitable strategy for each single case. In addition, it is preferable to have one or even more places to stay. The latter can be hierarchised according to different levels:

- **An emergency meeting point** (if the family is separated, the instructions are to head there);

- **Temporary or transitional fallback points** (arranged along the itinerary of the 'permanent' refuge, they enable you to spend the night there, drink and eat, etc.);

- **A 'permanent' fallback point** (i.e. a refuge) or a long-term one (i.e. a holiday home, with family, with friends, etc.).

Ideally, all these places should be selected in advance. A few of them may even require some preparation. Regardless of whether it is a transitional one or a main one (and of whether it is actually yours or belongs to friends or family), the fallback point in question can be set up to house 'survival boxes' containing any resources and equipment that you could need. Depending on where it is kept, the equipment could be very basic (a basic map, a knife, a lighter, etc.) or, by contrast, quite comprehensive: drinking water, filters, cookware, food, blankets, clothing, lamps, lighters, batteries, petrol, heating materials, essential medicine, a first-aid kit, hygiene products, weapons and ammunition, copies of official documents (such as passports, birth certificates, deeds of property), etc. Several survival boxes can coexist in different places, depending on your means and strategies. In case of evacuation, they will thus:

- save you from having to carry excessively heavy loads;
- guarantee that you do not forget anything of importance;
- save you valuable time.

Last but not least, this strategy of decentralising some of your equipment will give you the opportunity to move faster, stock up along the way and, having reached your destination, find the necessary means to once again put down roots somewhere. Once you have selected your fallback points, you need to consider two types of factors essential for a successful evacuation, namely:

- the different routes to take (road and topographical maps, compasses, etc.);

- the potential means of transport (on foot, by bike, by car, etc.).

Indeed, it is paramount for you to bear in mind that if you are forced to evacuate your home, then the situation is no longer normal. Your environment could change and even become hostile. A toxic cloud, an area contaminated by radioactivity, and an epidemic that spreads at lightning speed are all examples of situations that may force you to avoid certain places. Depending on the danger involved, it will be necessary to take into consideration the wind direction, the closing of certain roads, the presence of crowds, and so on. Similarly, both natural events and the weather conditions (snow, floods, etc.) are elements one should never overlook if one wants to avoid being stranded as a result of an impassable road or a sector that has become too risky. For all these reasons, it is important for you to plan several possible routes during your preparations. To make the right choices, you need to consider the following aspects:

- feasibility;
- speed;
- discretion;
- the availability of other 'variants' (i.e. the possibility to follow another route should you find yourself stuck).

In order to properly evaluate the above-mentioned criteria, it is necessary to identify, within a radius of at least ten kilometres, any special places that are either useful or should be avoided on the basis of:

- geographical features (streams or water points, lakes, forests, passes, etc.);
- man-built structures (pharmacies, hospitals, petrol stations, police stations, tunnels, bridges, etc.).

Being familiar with the area that surrounds your main residence is therefore essential when it comes to successfully carrying out this task. Although the inspection of road and topographical maps is indeed a must, nothing can replace personal knowledge of the terrain. Take the opportunity to go on family strolls or mountain biking expeditions to explore any interesting areas. Not only will you keep yourself physically fit, but you will also avoid unpleasant surprises if you are ever compelled to implement your evacuation plan.

Operational Procedures

Operational procedures allow one to make use of a well-thought-out organisational system by defining the part that each group or family member is expected to play (i.e. who does what in a general or particular situation). These instructions should be clearly detailed and preferably written in black and white; they should also be kept in an easily accessible place and, of course, be understood by all participants.

The goal is to divide the tasks in advance, especially during a crisis situation. It is clear that during any evacuation, having a pre-established role helps one to keep calm. Regardless of their age, sex and physical or intellectual abilities, each member of the group must therefore be assigned a role.

Implementing such measures, especially within large families, helps maintain psychological cohesion and avoid panic, hysteria or anger. In addition, the group leader does not waste precious time and can focus on his priorities, knowing that the other tasks are being dealt with. Indeed, you must bear in mind that, more often than not, reality tends to differ from theory. In a crisis situation, stress or fear can thus change people's behaviour. Some might become hysterical, while others huddle up in some corner or are left petrified.

The best way to limit such potential departures from adequate behaviour is through training and practice. If you have children, you should regularly organise exercises in the form of games. If you go out on a picnic, for instance, start an imaginary evacuation procedure. Time the whole process and reward the team members so as to encourage them. Write down any weak points and try to do better next time. The following list gives some examples of tasks that can be allocated in case of evacuation:

- Specifying the meeting place (leader);
- Choosing the route to follow (leader);
- Determining the departure time (leader);
- Taking the bug-out bag;
- Preparing the potential means of transport;
- Picking up the children;
- Shutting off the water and gas supply;
- Taking care of the animals;
- Retrieving various tools;
- Closing the windows and doors.

And the list goes on.

Some tasks can be initiated as soon as an alert is broadcast. Others, including the preparing of vehicles, will depend on the route that must be followed. Which is why the leader must gather everyone and briefly explain the group's 'mission' as soon as he has made the necessary choices (destination, route, departure time, etc.). This also offers an opportunity to repeat the instructions: what to do if you are separated from other team members, the behaviour to adopt in case of an attack, and so on. These procedures, which must remain simple, make it possible to differentiate between an orderly evacuation, which is a guarantee of success, and a chaotic one possibly resulting in an uncomfortable or difficult situation that could quickly become dangerous or even fatal.

Military Procedures

Ever since antiquity, armies have been developing command systems that prevent them from forgetting anything and allow them to save time when preparing a mission or engaging in military action. The following three types of command can be adapted and used when evacuating a given group or your own family. You could either write them into a small notebook or learn them by heart. It's up to you, but do try to practice regularly in different situations.

- **The Preparatory Command**

  This is the first thing to do when evacuation is being considered. In French, it is based on the *PATRAC-DR* memory aid (which is a slang term for being ill).[2]

  C: CREW: select the participants.

  R: RADIO: take your mobile phones and walkie-talkies.

  O: OUTFIT: special clothing items to take with you.

  W: WEAPONS: axes, shotguns, tear gas, etc. (if available).

  D: DIET: the food usually included in a bug-out bag.

  C: CAMOUFLAGE: depending on the future mission, a range of both discrete and colourful effects should be provided for.

  A: ASSEMBLING: at such time or within a certain period; at one location or another.

  M: MISCELLANEOUS: turn off the gas, close the doors, etc.

- **The Primary Command**

  It expresses the mission of the group and specifies each person's role. The mnemonic used is *SMEL*.

  S: SITUATION: what's going on?

  M: MISSION: what should we do?

  E: EXECUTION:

  - C—Coordination
  - T—Task assigned to each group member
  - B—Behaviour to abide by in the event of an accident
  - E—External support
  - L—Link (line-of-sight, walkie talkies, etc.)[3]

---

2 TN: In order to create the necessary mnemonic, I have changed the order in which the different points are mentioned so as to obtain an acronym. The result is '**CROWD CAM**'. Although there is admittedly no such thing, it does sound like a camera used for crowd monitoring. And since one is sometimes forced to avoid crowds after abandoning their home during an evacuation process, it all makes perfect sense; and I, for one, am sure to remember this acronym.

3 TN: One way of pronouncing the acronym that refers to the different aspects of execution makes it sound like 'city bell', i.e. an alert.

L: LEADER'S LOCATION — so that everyone may know where to find him.

- **The 'Move Out' Command**

  This command can be issued several times during an evacuation, especially at hazardous locations/times. The mnemonic is *DPIF*.

  D: DIRECTION: general, distant, and not necessarily visible on the ground

  P: TARGET POINT: the exact location to reach; visible on the ground

  I: ITINERARY: the itinerary one must follow

  F: FORMATION: travelling in a column, in a straight line, in intervals, etc.

Note: Commands of the military type have been developed for both peace and war. Transposed onto a civilian context, their greatest value lies in their use in crisis situations. If these mnemonic devices are used, the leader of the group will have to adapt his vocabulary to suit his audience. One does not address former soldiers in the same way one would speak to children (although, at times, the difference is rather minimal!).

Responding to Contamination and Poisoning

Undressing and Decontamination

Decontamination is defined as the complete or partial removal of contamination using deliberate processes of a physical, chemical or biological nature. It encompasses many techniques that vary depending on the material that one wants to decontaminate (people's bodies, equipment, buildings, etc.) and on the type of CBRN agent that one faces. This task is a difficult one and one could indeed write an entire book on the topic of its implementation. Nevertheless, applying some basic rules and a bit of common sense allows us to attain a satisfactory result in the vast majority of cases.

As we have already seen, **contamination can be both external or internal**. The decontamination techniques outlined below relate to the former. With regard to the second type, the ingestion of medicinal substances is necessary (see the paragraph entitled '*Medical Countermeasures*'). The first question you should ask yourself is the following: **has the subject (whether you yourself or someone else) been contaminated?**

This obviously depends on the areas one has passed through and on the type of compound involved. Indeed, **non-persistent agents** (like most gases) **are generally considered not to be a source of contamination**. So, if you only pass briefly in front of a bottle leaking chlorine and hold your breath, you will not be contaminated or poisoned. Of course, your clothes will reek of chlorine a bit, but there will not be any serious consequences. Change clothes and hang those you were wearing on a clothesline outside (provided that there is no risk of contaminating the environment), or put them in your washing machine. If, on the other hand, what you are dealing with are biological agents (saliva, spit, blood, etc.), persistent chemicals (certain acids, VX, etc.), radioactive dust and the like, decontamination becomes mandatory.

Note: Most of the time, you will not actually be aware of having been in contact with the toxic agent and of having thus been contaminated. In this case, it is better to abide by a precautionary principle and proceed with decontamination.

The process of human decontamination can be considered to include two main phases which you can carry out yourself or with the help of a partner — undressing and decontamination per se.

1. Undressing:

The general principle is to remove your contaminated clothes while avoiding re-suspending dust particles in the air or spreading any liquids present on your own person:

- **Avoid sudden gestures**

Clothes are a source of imperfect protection. Most of the time, there will be more contamination on their outer surface than on the inside. When undressing, it is necessary to ensure that the contaminants deposited on your clothes do not fall on your underwear, or worse, on your skin.

- *Undress from top to bottom*
- *Remove your clothes in the following order*:
  - If you have a hood, pull it backwards;
  - Slowly open your jacket;
  - Pull on your sleeves if necessary, so as to free your shoulders;
  - Stretch your arms downwards and backwards and allow your jacket slide to the ground; alternatively, you can place it in a bin intended for such a purpose;
  - Repeat this step with your shirt, etc.;
  - Take a step forward;
  - Open your trousers, allowing them fall to the ground or placing them in a bin intended for such a purpose;
  - Take a step forward;
  - Remove your shoes, leave them on the ground or place them in a bin intended for such a purpose;
  - Take a step forward;
  - Remove your gloves by turning them inside-out in order to keep the contamination inside;
  - If you're wearing a mask, remove it.

Note:

1. Advancing one step at a time allows you to find yourself in a 'clean' area. This is an essential factor when removing your trousers: do not remain there, plodding through the contaminated substance that has fallen off your clothes.

2. If a partner is assisting you, you should preferably use the 'rabbit skin' technique to remove your clothes, i.e. gently roll your clothes with their outer surface facing inwards. Only the inner part of your clothes will thus come into contact with your body.

3. If a person without special protection shows up to be undressed/decontaminated, give them a 'paper' FFP 1-, 2- or 3-type of mask (see the paragraph entitled '*Protective Equipment*'). This will prevent them from inhaling any potential airborne contamination.

4. If someone's shoes prevent them from removing their trousers, merge the two steps: pull down the pants and take them off, removing their shoes at the same time.

5. The cutting of certain garments can sometimes be used to facilitate the process.

6. The removal of one's mask should ideally come last. To do this (once the gloves have been removed), put your fingers under the rubber band (if you were wearing a hood) or directly between your cheeks and the mask itself. Do not touch the contaminated outer part.

   If it turns out to be difficult to take off your mask using this technique, you must reverse the sequence. Keep your gloves on and grab the item from the outside. Hold your breath and turn your head in the opposite direction until you have removed both the mask and gloves.

The picture sequence below, which was made public in circulars 700/SGDN/PSE /PPS (dated 7 November 2008) and 800/SGDSN/PSE/PPS (dated 18 February 2011), gives us an example of an undressing process carried out by a person with adequate CBRN protection.

If the undressing process needs be carried out recurrently (upon entering and exiting your shelter, for example), you must make sure that a specific place is reserved for such a purpose, as it will quickly be left contaminated.

- It is therefore essential to have two different passageways:
  - one enabling people to enter the shelter and requiring an undressing process to be carried out;
  - another allowing uncontaminated persons to enter AND through which one can exit the shelter.

- Providing bins into which contaminated items can be discarded will also be vital. It must be possible to seal off the items afterwards (by using thick bin bags to prevent tearing, for instance).

2. Decontamination:

It is important to consider two main cases:

- **The decontamination of unprotected persons:** If you have come into contact with contaminating agents, especially skin contact, it is essential for you to be decontaminated. Undressing is normally recommended as a preliminary step (unless the contamination is noticeable and localised, in which case it must be attempted immediately). It is recommended one uses soap and mild shampoo to cleanse the body.
- **The decontamination of protected persons and materials:** If you are wearing protective clothing, decontaminate yourself before undressing. If you wish to decontaminate equipment, you can usually use more 'powerful' products than those used for decontaminating the skin. It is better to use a bleach solution (which is more 'powerful' against type-B and type-C agents). The concentrations of the bleach solution are covered a little further on in the text. Professionals can use a BX 24 solution or some Cascad powder in case of biological or chemical contamination, and BX40 in the event of radiological contamination.

Note:

- When the undressing or decontamination phases are carried out with the help of a partner, it is better for the latter to be equipped with a mask, gloves and protective clothing (for more information, see the paragraph on protective equipment);
- Your crewman must avoid getting contaminated by touching or manipulating any parts that may have been soiled

by CBRN agents. If necessary, he can wear a pair of disposable gloves on top of his protective gloves. These will be discarded at the end of the process;

- The products used for decontamination may vary depending on the agent involved and on whether they are to be applied directly to the skin or to one's protective clothing. Examples include: soapy water on the body and face for radioactive dust; diluted bleach for ridding equipment of bacteria, viruses or chemicals, etc.

3. The main decontamination phases for non-equipped/unprotected persons:

- If the contaminated area is localised (a stain on the arm, for instance), the preferable option is to use a sponge (which will be immediately discarded) or wipes (imbued with diluted bleach) to remove as much contamination as possible before continuing;
- Wash your hands with liquid Marseille soap[4] (or, alternatively, with diluted bleach);
- Wash your hair with a mild shampoo by tilting your head backwards so that the water does not run down your face;
- Wash your face with soapy water. If possible, rinse your eyes thoroughly with water once your face is clean;
- Wash the rest of your body with soapy water from top to bottom;
- As for personal items (jewellery, glasses, etc.), they should be left to soak in a bucket containing diluted bleach (in the case of chemical or biological agents) then rinsed thoroughly;

---

4 TN: Marseille soap is a traditional type of soap that has been produced in the Marseille area for six whole centuries now. Its formula is based on the use of vegetable oils.

- Move forward to exit the decontamination area, dry yourself and put on some clean clothes. (For each family member, having a vacuum-packed change of clothes is ideal. The actual set can vary depending on your location and the weather conditions. Generally speaking, tracksuits or old clothes will do the trick.);
- Spray the area with diluted bleach once decontamination is complete.

Note:

- Do not attempt to scrub yourself too much, to avoid damaging the skin;
- Objects contaminated with radioactive substances or toxic warfare agents (VX, yperite…) had better be left behind, as there is still a significant risk of them bearing traces of dangerous elements;
- The undressing and decontamination processes described above should be used in case of emergency or when the relevant professional services are unavailable;
- As is obvious, specialised decontamination lines and trained personnel will be able to provide you with much better care;
- After coming into contact with toxic agents and being decontaminated, you must visit a medical or hospital centre in order to confirm the presence/absence of contamination or receive any necessary treatment.

4. Special cases involving injured persons:

Only general principles are covered below. Indeed, the incredible range of CBRN agents and the large variety of potential injuries or types of poisoning can lead to various often complex cases — which is why it is necessary, if the situation permits, to turn to medical professionals or rescue services for help.

If the subject's injury is not life-threatening (and remaining on the premises will not result in certain and quick death), they must be provided with immediate care, without being decontaminated or moved. Ideally, you yourself should be protected from any surrounding risks. Suppose your friend is bleeding severely from the thigh. Although you are in a contaminated area, you must first remove any debris from the wound and then cleanse it *briefly* with disinfectant-imbued compresses. Last but not least, place a haemostatic bandage on the wound, and without delay. It is always better to ensure that your friend is alive but contaminated than dead and decontaminated!

If the injury is not considered a matter of absolute emergency (involving a mildly poisoned or contaminated person, for instance), take them to a 'clean' area and undress/decontaminate them before proceeding to provide them with care.

5. The decontamination of equipped individuals and items:

For members of the military, law enforcement officers and rescue services, professional solutions are available both with regard to the emergency decontamination of personnel (powdered gloves,[5] RSDL®,[6] etc.) and the decontamination of equipment (Cascad™ powder, BX 24 solutions, etc.).

For private individuals, the best solution is diluted bleach, which is both economical and relatively versatile.

Avoid using packs of concentrated bleach (36° Chl. or 9.6%), which can only be kept for two to three months. Instead, opt for bottles containing a 2.6% active chlorine rate (9° chl.), whose expiry date usually ranges from one to three years after purchase. There are also sodium dichloroisocyanurate dihydrate (NaDCC)

5 AN: Decontamination (powdered) gloves are not normally accessible to the public. The absorbent powder they contain, however, which is known as 'Fuller's earth' (smectic clay), can still be purchased from certain wholesalers.

6 TN: Reactive Skin Decontamination Lotion.

pills or tablets that readily dissolve in water and can be kept for three whole years in a dry and airy place.

To obtain a multi-purpose decontaminant solution, you must have at least 0.8% active chlorine at your disposal. You can mix it according to the following chart:

For packs (250 ml at 36° or 9.6%):

NUMBER OF PACKS	AMOUNT OF WATER TO BE ADDED (IN LITRES)	TOTAL AMOUNT (IN LITRES)
1	2.75	3
2	5.5	6
3	8.25	9

For bottles (1 litre at 9 chl. Or 2.6%):

AMOUNT OF BLEACH (IN LITRES)	AMOUNT OF WATER TO BE ADDED (IN LITRES)	TOTAL AMOUNT (IN LITRES)
1	2	3
2	4	6
3	6	9

For dissolving pills or tablets: 1 gramme of NaDCC releases 0.3 g of active chlorine. Read the NaDCC concentrations listed on the packaging, or ask the manufacturer.

EXAMPLES OF EFFERVESCENT TABLETS	NUMBER OF TABLETS	AMOUNT OF WATER (IN LITRES)
Klorsept 17® (1.67 g of NaDCC)	16	1
Klorsept 87® (8.68 g of NaDCC)	3	1

Note: This bleach solution can also be used for hand decontamination. As for items, they should be left to soak for a period of thirty minutes.

Medical Countermeasures

Medical countermeasures are required whenever a CBRN agent has caused damage or is likely to poison someone or trigger illness. The best solution is to turn to a doctor or hospital. If you do not have the possibility to do so, the following (indicative) information can help you.

- **Cases of radioactive substance poisoning**

  These cases may involve radioactive materials that have passed through the skin or a wound, or have simply been inhaled or ingested. Under these circumstances, it is crucial to administer treatment without delay. The purpose is to eliminate a maximal amount of such internal contaminants as quickly as possible, usually by means of urine. The following products can be used to facilitate the elimination of specific radioelements:

  - DTPA for americium, plutonium, thorium, cobalt, iron and many fission products);
  - Prussian blue (Radiogardase™) for caesium and thallium;
  - As for stable iodine tablets, they are used preventively to avoid any binding of radioactive iodine to the body (the thyroid gland).

  Note: In most cases, it is also recommended that one drinks sufficiently to facilitate urinary elimination.

THE NOTION OF BIOLOGICAL HALF-LIFE AND EFFECTIVE PERIOD

We have already stated that radioisotopes have a half-life (*radiological period* or *Tr*) that corresponds to the average duration after which their activity decreases by half.

In cases where radioactive elements enter the body, mechanisms of purely biological elimination come into play. It is thus possible to define a similar value known as the biological half-life (*biological period* or *Tb*) corresponding to the time required for the body to eliminate 50% of the inhaled or ingested element.

To determine the half-life of a radioisotope in an organism, one must ultimately take into account the radioactive period AND the biological one, thus obtaining ***the effective period*** (Te). This value is always lower compared to the other two, of course.

The calculation formula is as follows:

$$\frac{1}{Te}=\frac{1}{Tr}+\frac{1}{Tb} \quad i.e.: \quad Te=\frac{Tr \ x \ Tb}{Tr+Tb}$$

So as to understand the phenomenon, let us take the example of iodine 131. Its radioactive half-life is 8 days, meaning that its activity will naturally decrease by half after such time. The body will, in turn, eliminate/renew 50% of iodine (which tends to be taken up by the thyroid) in a period of 30 days. In the end, what we get is the following equation:

$$Te=\frac{Tr \ x \ Tb}{Tr+Tb}=\frac{8 \ x \ 30}{8+30}=6.3 \ days$$

The effective period (*Te*) is thus 6.3 days, which means that, as a result of the natural radioactive decay process and the body's elimination mechanisms, the radioactive activity due to the remaining iodine will decrease by half every 6.3 days.

IODINE TABLETS[7]

A nuclear accident can sometimes be accompanied by a massive emission of radioactive iodine. This radionuclide can then enter the body using various routes (the respiratory tract, the skin,

7 AN: According to the interministerial circular of 11 July 2011, which concerns the storage and distribution means of potassium iodide tablets: http://social-sante.gouv.fr/fichiers/bo/2011/11-09/ste_20110009_0100_0052.pdf.

the absorption of contaminated food, etc.) and bind itself to the thyroid gland, an organ that plays a regulatory role and is very important to our bodies. The accumulation of radioactive iodine at thyroid level increases the risk of cancer and other diseases. And foetuses, babies, and young children are more sensitive than anyone else in this regard.

During a nuclear incident, it is sometimes useful to ingest stable (*preventative*) iodine tablets to saturate the thyroid. Any potential radioactive iodine that could enter the body at a later point would thus be unable to bind and would quickly be eliminated by the body. In France, it is only recommended that one take iodine tablets when one is instructed to do so by the local prefect.

What follows is the correct iodine tablet dosage, in the form of 130 mg of potassium iodide:

- **For adults, pregnant women and children over 12:** one tablet to be dissolved in a glass of water;
- **For children aged 3 to 12 years:** half a tablet;
- **For children aged 1 to 3 years:** one quarter of a tablet;
- **For infants (younger than a month):** one eighth of a tablet;
- If necessary, the process can be repeated in the next days in accordance with the prescribed dosage.

Note: It is inadvisable to swallow the tablets separately; they should rather be dissolved in a large glass of warm water. Since the taste is not very pleasant, you can also use milk or fruit juice. The tablets themselves do not easily dissolve: you should thus break them up into small pieces first and then stir them into the liquid.

- **Cases involving people poisoned/infected by biological agents**
  The treatment of victims impacted by biological agents is based on non-specific measures of medical intensive care and on therapeutic

means proper to the agent in question (once it has been identified). Care must be provided by health care professionals, but if this is not forthcoming and you have no other choice but to take over yourself, the advice below can help you cope while awaiting external assistance.

- Isolate ill people to reduce the risk of contagion;
- Protect yourself when visiting the 'patient' by resorting to items such as masks and disposable gloves.

In order to provide specific care, it is crucial to determine which biological agent is responsible. In the event of an outbreak, the name of the bacterium or virus will probably have made headlines. On the other hand, if you have consumed contaminated food or touched unknown powders, things become complicated. The following medical countermeasures are only indicative in nature, since a doctor's opinion is necessary to select the most appropriate medicine:

- **Antibiotics to combat bacteria.** Fluoroquinolones and doxycycline (i.e. broad-spectrum antibiotics that act against a wide range of germs) may be worth a try when the bacterial agent is unknown;
- **Antivirals to combat viruses.** These substances, which are few in number and of low efficiency, are generally active against a limited number of agents. Examples: Ribavirin against Lassa virus and the Crimean-Congo viral disease, but not against Ebola or Marburg virus; Tamiflu against the flu;
- **Serums/antidotes to combat toxins.** These are usually directed against a particular agent. There is, for instance, a serum that contains antibodies that specifically neutralise the botulinum toxin. However, it must be administered as soon as possible (before the toxin can successfully bind) to prevent symptoms from appearing.

- **Regarding vaccination**

  Although there are many controversies surrounding the issue of vaccination, it is certainly the most effective means of protection available; but it must be used as a preventive measure. The principle is that of preparing your immune defences to counter an actual agent if it were to infect you. The main concern is that there are not vaccines against all diseases (and even if this were possible, it would be stupid to administer them all). Plague, anthrax, and tularaemia are all examples of instances in which vaccination is rather ineffective. In other cases, including that of smallpox, a vaccine does indeed exist.[8] However, it is definitely not harmless and the number of adverse effects (sometimes leading to the patient's death) is *relatively* high.

- **Cases involving people poisoned by chemical substances**

  The main treatment is symptomatic, where one prevents, for example, any ventilatory, cardiovascular, neurological or metabolic failure from occurring. There are, however, some specific ways to counteract certain chemical agents.

  - **Against nerve agents:**
    - The Ineurope™ dual chamber syringe is used to counteract the action of nerve agents. It contains a mixture of atropine, contrathion and diazepam, and is not available over the counter. Its use on healthy subjects results in significant side effects;
    - Pyridostigmine. This medicine should be used as a *preventative*. It actually preserves some cholinesterases, thus providing limited protection against certain nerve agents (yet remaining ineffective against sarin and VX). Mestinon

8 AN: Vaccination against smallpox has not been obligatory in France since 1979 and was even suspended after 1984. Booster shots are thus no longer provided.

(produced by Meda Pharma laboratories in 60 mg doses) is sold without prescription and costs 5 euros a box (20 tablets);
- Huperzine. It is derived from a plant native to Asia (moss) and could also serve as a *preventative*, just like pyridostigmine.

- **Against vesicants:**
  - It is essential to carry out the most complete decontamination possible beforehand;
  - There is no specific treatment for yperite. By contrast, armies do have an antidote against lewisite: British Anti-Lewisite (BAL®), which is not marketed in France. In case of skin damage, the simplest option is to act as in the case of a thermal burn. Apply soothing lotions or ointments (such as Biafine®) to the reddened areas. Administer analgesics (e.g. Paracetamol). Avoid any secondary infection of the burns (dressing with sterile gauze, regular cleaning with antiseptic solutions, use of Flammazine™ ointment).

- **Against choking agents:**
  - There is no antidote;
  - Once the person has been evacuated from the danger zone, the treatment is mainly based on respiratory resuscitation and the correction of circulatory disorders.

- **Against cyanides and their derivatives:**
  - In case of gas poisoning (hydrogen cyanide, for instance), evacuate the person from the danger zone urgently;
  - Mouth-to-mouth resuscitation is contraindicated (as the rescuer runs the risk of being poisoned as well);

- Place an oxygen mask on the victim's face; administer Valium™ in case of a seizure; inject adrenaline in the event of cardiovascular failure;
- Hydroxocobalamin (Cyanokit®) is the antidote of choice against cyanide poisoning. It has very few side effects (even in the case of non-poisoned subjects) and brings about a very rapid improvement in the victim's condition. Cyanokit is not accessible to the public. Dodecavit, however, which is produced by SERB laboratories and contains the same active ingredient (but in lower concentrations), is.

Note: Hydrogen cyanide is not necessarily the result of a terrorist act or of a leak in a high-risk business facility. It can also be released by fires involving plastics such as polyacrylonitriles, polystyrenes, polyurethanes, etc.

AUTHORISATION TO USE AN ANTIDOTE AGAINST NERVE AGENTS

Here is what the 'Decree of 14th November, 2015', which authorises the use of an atropine sulphate solution in 40 mg/20 ml injections (PCA) as an antidote against organophosphorus nerve agents, states: 'Considering the fact that the risks of terrorist attacks and exposure to organophosphorus nerve agents are serious health threats that call for the introduction of emergency measures, atropine sulphate injectable solutions, which are manufactured by the Central Pharmacy of the Armed Forces, can henceforth be acquired, stored, distributed, prescribed, dispensed and administered by health professionals within the framework of our medical assistance emergency services to provide anyone exposed to nerve agents with adequate medical care'. In other words, an 'antidote' that is usually reserved for soldiers poisoned by neurotoxic chemical warfare agents has just been authorised in France to treat civilians in the event of an emergency.

Traversing a Toxic or Contaminated Zone

If you find yourself compelled to cross a contaminated or toxic zone, the following tips may be useful in order to avoid complications:

- **Contamination cases**

  A contamination phenomenon can be encountered regardless of the (radiological, biological or chemical) agent involved. Whether the latter is in the form of powder, suspended particles or spilt liquid, the area itself has actually been contaminated. Anyone entering it is thus likely to be contaminated as well. This is by no means inconsequential since, in addition to being impacted yourself, you may also unintentionally/unknowingly:

  - contaminate your loved ones (mainly through contact);
  - bring these harmful substances into your own home and thus contaminate your living space, your possessions, etc.;
  - absorb the agent (through ingestion, inhalation, the skin, an injury, etc.), leading to internal contamination (or the development of an infection in the case of a virus or bacterium).

  To avert such complications, there are several rules to apply:

  - Touch a minimal number of things in the area. The ideal solution is to remain upright (and avoid lying down, sitting or kneeling) and not to use your hands in your interactions with the environment;
  - Try to protect your airways and, if possible, your body. The primary goal is to keep all harmful particles outside your system. If you do not have any adequate equipment, place a tissue or piece of cloth in front of your nose and mouth (for details, see '*CBRN Protection Kits*');
  - You must initiate the undressing/decontamination phase as soon as possible upon leaving the danger zone. However, trust your common sense and assess your priorities: you do not have to take all your clothes off in the middle of winter and in the

absence of a nearby shelter. In most cases, external contamination can wait;

- Another difficulty is that, if you do not have any detection means, it will be difficult to determine the actual boundaries of the contaminated area. Under such circumstances, it helps if one can identify the nature of the agent involved. Indeed, when it comes to dust, for instance, the prevalent wind will give you an indication of how to exit the danger zone as quickly as possible (by leaving perpendicularly);
- Try to locate any medical and emergency services, whose members will certainly have established decontamination lines and other useful facilities in case of such an incident.

Note: The most extensive and long-lasting type of contamination is of the radioactive kind. It may be the result of a major problem in a nuclear power plant or be caused by the explosion of an atomic bomb (radioactive fallout). If the area has been contaminated by toxic warfare agents, you must do anything you can to avoid or bypass it. If you *really* have no choice but to cross it, make sure you evacuate the premises as soon as possible while applying the instructions listed above to the best of your abilities, but with one basic difference: the removal of your contaminated clothes becomes one of your priorities once outside the hazardous location.

- **Cases involving radioactive agents**

If the radioactive elements are encountered in the form of contamination, the advice given in the previous paragraph still applies. In the cases where, on the contrary, you are faced with a sealed source emitting gamma radiation, several means of protection are possible in order to limit the amount of radiation received by your body. There are three main ways to protect yourself against this threat, all of which are very logical:

a) Time-related protection

Imagine for a moment that you are near a large fire and approaching it. Once you have reached a certain distance, you will start to get burnt, and the longer you stay, the more serious the damage to your body will be. In the case of a sealed source emitting radiation, the situation is quite the same, except that you will not actually feel anything (not immediately, at least, until the symptoms surface hours, days or weeks later). If the radioactive source is powerful enough, you will be impacted remotely, without any need to touch the object in question. The principle is simple: *the shorter the time spent near the source, the lower the dose received.*

b) Distance-related protection

The relation that determines the dose rate is inversely proportional to the square of the distance, meaning that the rate will diminish very quickly as the distance increases.

$$(\textit{Distance 1})\ 2 \ \textit{x Rate 1} = (\textit{Distance 2})\ 2 \ \textit{x Rate 2}$$

[With rate 1 (or 2) = measured rate at distance 1 (or 2)]

Therefore, if you know the actual distance and can measure the dose rate, it becomes easy to extrapolate. If you are one meter away, for example, you receive a certain dose rate. Now, if you move:

- 2 metres away, the distance is increased by a factor of 2 and the dose rate is reduced by 4.
- 4 metres away, the distance is increased by a factor of 4 and the dose rate is reduced by 16.
- 10 metres away, the distance is increased by a factor of 10 and the dose rate is reduced by 100.

Distance thus provides very effective protection.

c) Screen-related protection

If we take fire, for instance, the mere fact of hiding behind a wall puts an immediate end to the heat you receive through radiance.

In the case of a radioactive source, the principle is similar, but gamma rays will, generally speaking, not be stopped but only reduced. In addition to this, not all screens are equally effective when it comes radiation. Those made of lead are very effective. Stone or concrete walls offer acceptable protection. Lighter components such as plastic or wood are, on the other hand, less efficient at countering gamma radiation. Ultimately, if you find yourself in the presence of a radiating source, use walls or shelters to protect yourself. Although you will not actually feel that this has made any difference, rest assured it will.

Note: There are also other ways for you to protect yourself, but these can only be implemented in a few specific cases. The idea is to reduce the source's activity mainly by allowing time to elapse or decreasing the amount of radioactive material present.

- The impact of time: if you know what radioactive isotope is involved, waiting becomes a good option. For example, iodine-131 has a period of eight days, meaning that its activity is naturally reduced by half after this time has passed. If you know for a fact that a location has been contaminated with this isotope, simply wait:
  - 8 days for the level of radioactivity to be divided by 2;
  - 16 days for the level of radioactivity to be divided by 4;
  - 24 days for the level of radioactivity to be divided by 8;
  - etc.
- Decreasing the amount of radioactive material: if you are dealing with a gas (e.g. radon), it is enough to aerate the room (or cellar, etc.) for a few minutes to allow it to escape into the atmosphere and ensure that its concentration on the premises is decreased.

Note: Be aware of one important detail, though: *it is not possible to actually modify the activity of an isotope/radioactive element per se.* Indeed, the latter will always have the same decay rate, regardless of the shapein which you encounter. Burning an object or dipping it in concentrated acid may eventually destroy it, for instance, but will have no effect on the radioactivity itself. An item comprised of an ionising element may change physical appearance by being, for example, transformed into vapour, reduced to ashes, or turned into liquid, but the amount of radioactivity present remains the same!

- **Cases involving biological agents**

  Simple hygiene-related measures (such as washing your hands regularly) are essential. If you have to move around a lot (underground railway system, shopping centres, etc.), it might be particularly useful to have some kind of disinfectant gel with you. Although the latter is not effective against all microbes and cannot replace washing your hands with soap, its advantage is that it can be used almost anywhere. In addition to these hygienic measures, plain common-sense tips also apply:

  - Avoid any and all contact with ill people or strangers;
  - Do not touch dead animals;
  - Refrain from placing your hands on/in your mouth;
  - Avoid large crowds by steering clear of shopping centres, markets, public transport, cinemas, schools, etc.;
  - Stay at home and isolate yourself from all external contact as much as at all possible;
  - If you know that an epidemic is raging and have to move about among other people, wearing an FFP[9] (filtering facepiece) and implementing basic hygienic measures can significantly reduce the likelihood of your contracting the disease.

9 AN: For details, consult the section that deals with protection kits and equipment.

- **Cases involving chemical agents**

  Some chemical agents can cause contamination. Toxic warfare agents such as VX or yperite and industrial products (acids, for example) can remain in liquid form for days or even months. If you are contaminated by such elements, you should apply the advice offered in the paragraph above entitled '*Contamination Cases*'. Should the toxic chemicals be encountered in gaseous form (e.g. chlorine, ammonia, etc.), priority must be given to immediate respiratory protection. The rest of the body can be protected at a later point if necessary. Furthermore, your actions will depend on the circumstances:

  - If you find yourself trapped amid a cloud of chlorine, put on a protective mask or place a moist handkerchief on your mouth and nose, and leave the premises without delay. There is no need to linger in a danger zone by putting on some kind of suit to protect the body. Look for a direction that is perpendicular to the wind's and opt for elevated routes rather than ditches. Indeed, most toxic gases are heavier than air and tend to remain at ground level. Those that are lighter, by contrast, will rise and quickly dilute in the atmosphere.
  - In case you find yourself compelled to leave your long-term base and know that the outside environment has been contaminated, it is necessary to put on a mask and protective clothing beforehand and undress/decontaminate yourself upon returning.

Note: In the event of a leak or spill, **contamination** and vapour/gas **toxicity** often occur simultaneously, meaning that you could be faced both with liquids that spread across the ground, flow in the rivers, etc., and with toxic vapours released into the atmosphere.

Reminder: Toxic warfare agents can pass through normal clothing and penetrate the skin.

3. Fallout Shelters

What you are about to watch is a nightmare. It is not meant to be prophetic, it need not happen, it's the fervent and urgent prayer of all men of good will that it never shall happen. But in this place, in this moment, it does happen.

— Opening narration of 'The Shelter', an episode of the American television series (*The*) *Twilight Zone*, which was first aired in 1961

Fallout (i.e. anti-nuclear/anti-atomic) shelters are often stigmatised or mocked for betraying either their owner's paranoia or that of a state overwhelmed by its fear of a nuclear war that will never actually come to pass.

In addition, non-survivability myths are driving many people to believe that any sort of preparation, including one's ownership of such shelters, would be 'useless' if a nuclear conflict or accident were to occur. It is certainly quite easy to convince people who do not actually understand this type of risk that no life would be possible in the aftermath of a nuclear explosion. The examples of Chernobyl and Fukushima, however, each of which released more radioactivity into the environment than any atomic bomb, have shown that even though some areas were indeed evacuated, our world has remained perfectly habitable. In summary, if you survive the thermal and mechanical effects of a nuclear explosion, there is a good chance that you will manage to weather it all as long as you abide by certain simple and logical rules of behaviour.[1] Having a shelter, of course, or even a basic survival kit, for that matter, does make things quite easier.

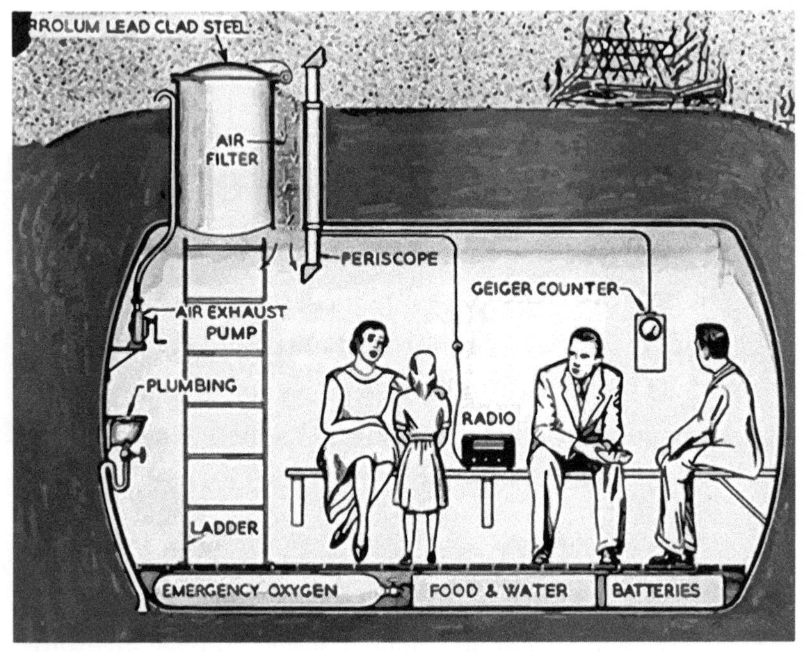

An idealised image of a fallout shelter — USA, 1960s

A fallout shelter can be defined as a partitioned space specially designed to protect its occupants against the effects of an atomic explosion, namely the blast, enormous heat and radiation.

The Legacy of the Cold War

During the Cold War between the United States and the USSR, a large amount of construction work was carried out throughout the world in harmony with the measures recommended by the departments of

1 AN: Here are some useful links:
http://www.besthealthdegrees.com/survive-nuclear/;
http://www.wikihow.com/Survive-a-Nuclear-Attack;
http://www.nytimes.com/2010/12/16/science/16terror.html.

civil defence. During that same period, many prevention campaigns were conducted and various development schemes implemented.

In 1956, for instance, the *National Emergency Alarm Repeater* (N.E.A.R.) civilian programme was developed in the United States to supplement siren warning systems and radio announcements. Considered unsustainable due to the effects of electromagnetic pulses (EMPs) that were purported to scramble or damage communications, however, it was eventually terminated in 1967.[2] President of the United States J. F. Kennedy, who was interviewed by the mainstream magazine *Life* in 1961, urged Americans to prepare, build and make use of shelters.[3] During that same year, *Forbes Magazine*[4] published an article in which personalities such as Nelson Rockefeller, Edward Teller, Herman Kahn and Chet Holifield proposed that a large number of shelters be constructed across the entire country to protect millions of their fellow citizens in the event of a nuclear war.

As part of their 'continuity of government'[5] programmes, some states also planned the construction of gigantic shelters. The objective was to maintain their ability to command their military forces and manage the country in the event of targeted attacks on military command centres and political bodies. These redundancy systems of governmental and military continuity, most of which still exist today, consist of huge bunkers sometimes allowing tens of thousands of people to remain fit to work for months on end. Here are some examples:

- In the United States, one can find them in the Appalachians, in the Ozarks, in the Rocky Mountains, etc.; there are also several

2 AN: https://en.wikipedia.org/wiki/National_Emergency_Alarm_Repeater.

3 AN: *Life Magazine*, September 1961.

4 AN: *Forbes Magazine*, November 1961.

5 AN: https://en.wikipedia.org/wiki/Continuity_of_government.

command centres (Project Greek Island,[6] NORAD, COOP[7] and Site R);

One of these shelters, a very high-profile one that was even featured in some films (*Wargames*, *Stargate*, and others), is the American Cheyenne Mountain command centre,[8] which was dug under 600 metres of granite;

- In Russia: Mount Yamantau and Mount Kosvinsky Kamen;
- In China: Project 131 and the like.

The entrance to the Cheyenne Mountain US Air Force Station

Almost all major powers have constructed such bunkers. France, for example, has some command centres such as the COFAS (Strategic Air Force Operation Centre), which is located near Lyons. In the 1960s, Canada developed its own network of bunkers known as

6 AN: https://en.wikipedia.org/wiki/Project_Greek_Island.

7 AN: https://en.wikipedia.org/wiki/Continuity_of_Operations.

8 AN: https://en.wikipedia.org/wiki/Cheyenne_Mountain_Complex.

Diefenbunkers, named after Prime Minister John Diefenbaker, who authorised their construction.[9]

In addition to building these shelters for military and political purposes, states have also sought to protect their populations. Thus, as part of each country's civil defence policy, a plethora of underground transport systems (tubes, trains), tunnels, and buildings comprising large basements or car parks have been put in place in case of a nuclear disaster. In the USSR, public facilities such as the local metro system were built throughout the Cold War, as were many shelters.

During that same period, Switzerland decided to extend its plans for the protection of its armed forces to include the civilian population. A large number of collective shelters (in schools, hospitals, etc.) were thus built in urban centres. They were designed to withstand a twelve-megatonne explosion at a distance of 700 m from the epicentre and made mandatory in every new building and detached house. In total, Switzerland has more than 300,000 shelters[10] supposedly equipped with beds and with sufficient stocks of food[11] and water to last several weeks, in addition to chemical toilets, air filters and emergency exits. Ever since the end of the Cold War, however, a large number of individual shelters have been diverted from their original purpose and now serve as storage areas, wine cellars, etc. Other countries such as Singapore, Finland, Norway and Sweden have also brought in laws requiring the construction of fallout shelters in multiple public and private buildings.

As for programmes geared towards preparing the population, all industrialised countries have implemented several of them. The most ambitious project was the American *Federal Civil Defense*

---

9 AN: https://en.wikipedia.org/wiki/Emergency_Government_Headquarters.

10 AN: http://www.bevoelkerungsschutz.admin.ch/internet/bs/fr/home/themen/schutzbauten/schutzraeume.html; http://www.bevoelkerungsschutz.admin.ch/internet/bs/fr/home/dokumente/unterlagen_schutzbauten/weisungen.html.

11 AN: http://www.lematin.ch/suisse/Ce-qu-il-faut-en-cas-de-catastrophe/story/31967461.

Administration, which, in the late 1950s, gave birth to an informational programme ensuring public readiness in the event of nuclear warfare. The resulting production of various materials was intense: thirty pamphlets, three films, eleven television and radio commercials, sixty-nine magazine articles and ninety-one official communiqués, for a total of thirty-six million printed and distributed booklets, including twenty-five million which, although the size of a credit card, described the basic measures to be followed in case of an alert. A comic book for children called *Operation Survival* was also designed and printed in four million copies. In addition to these publications, the Boy Scouts of America managed to distribute an awareness booklet to as many as forty million households in a single day of action, on 11 October 1958.

An interesting detail should be noted in all of these documents: the reason for which the public education pamphlets and films mentioned a two-week shelter period was not related to actual radioactive fallout, but rather was related to the average estimated time after which the overcrowding of shelters would begin to cause behavioural problems such as anguish, anxiety, paranoia, aggression and depression!

According to a 1962 report by the National Academy of Sciences entitled *Behavioral Science and Civil Defense*, it was felt that the public was not going to respond adequately to the government's advice. Indeed, during three alert exercises conducted the previous year, it was noticed that a large part of the population was so tied to its habits, daily rituals and social norms that instead of taking shelter and following the instructions, people simply pursued their immediate activities. The same report concluded that a significant portion of the population could not psychologically accept a post-nuclear war world and would end up indulging in alcoholism or drug addiction, developing mental health problems or becoming homeless.[12] Another noteworthy statement is that of Lloyd Omdahl, the governor of North Dakota, who commented on all of these programmes in a single

12 AN: This is what ultimately happened to a part of the population *without there being* any nuclear war.

sentence: 'Our main strategy in the event of an enemy attack can be summarised with one word: run!'

As for the estimated number of casualties, the United States Department of Defense conducted a study in 1967 comprising a detailed simulation of two scenarios of nuclear war between the USSR and the US.

Codenamed BETA I, one of these scenarios envisioned an exacerbation of tensions around West Berlin in 1972, leading to an escalation of the conflict and ultimately to a USSR nuclear attack on the United States, whose government would then reciprocate. The losses were estimated at twenty to thirty million deaths for the US and thirty to fifty million for the USSR.

The other scenario, code-named BETA II, simulated a US surprise attack against the USSR in 1972, in the aftermath of tensions centred around the Cuban situation. This time, the Soviet Union suffered an estimated 100 to 120 million losses, with the US 'only' incurring five to ten million.

Recent Developments

After the fall of the USSR, most civilian programmes were abandoned, particularly in the United States. This phenomenon intensified in the late 1990s, and many of these schemes were completely forgotten... In addition to the abandonment of infrastructures, very few practical exercises were henceforth implemented, and virtually no populational training provided.

The events of 11 September 2001 acted as a powerful backlash. The US authorities suddenly realised that some states or terrorist groups might actually have both the means and the will to strike at their enemies, which could involve the use of nuclear weapons. The US government thus created a new department, the *Department of Homeland Security*, to protect the country from terrorist threats. Its missions include:

- **The surveillance, enhanced control and seizure** of nuclear material;

- **The apprehension** of terrorists preparing for/contemplating a nuclear type of event;

- **The maintaining of governmental continuity**—allowing the government to survive and operate in the event of a nuclear attack.

What is interesting to note is that, this time around, populational protection has not been provided for, whether in terms of the number of shelters available or the implementation of training/educational programmes. This is all the more regrettable since it would have been easy to revive the public education programmes described above after some improvements and updates.

In recent years, Russia has modernised its old shelters and started to build new ones. Unlike the world's leading power, the Russian government has chosen to build shelters for its population as well: 5,000 of them were built in 2012 in the city of Moscow alone.[13] Does this preparation reflect mere foresight or substantial fears in the face of uncertain and turbulent international geopolitics?

On 22 December 2015, sometime after the United States announced the deployment of twenty additional nuclear bombs in Germany, archives were (conveniently?) declassified. They revealed that, in June 1956, the American Strategic Air Command (SAC) had prepared an action plan against the USSR entitled '*SAC [Strategic Air Command] Atomic Weapons Requirement Study for 1959*'.[14] This 800-page document anticipated with great accuracy the use of approximately 10,000 atomic bombs on as many targets across the USSR, Eastern Europe

13 AN: Details can be found here: http://endoftheamericandream.com/archives/russia-has-constructed-massive-underground-shelters-in-anticipation-of-nuclear-war.

14 AN: http://nsarchive.gwu.edu/nukevault/ebb538-Cold-War-Nuclear-Target-List-Declassified-First-Ever.

and China. Cities such as Moscow, Leningrad, East Berlin, Warsaw, and Beijing would simply have been razed. Moscow, for example, was going to be the target of around 180 bombs! In 1962, General Curtis E. Le May, Commander of the Army Staff, presented this scheme to President John Fitzgerald Kennedy, assuring him of his belief that a war with the USSR was inevitable and that it was necessary to launch a massive nuclear attack without prior warning — a genuine act of extermination. Fortunately, President Kennedy refused.[15]

Last but not least, it is interesting to note that, for many years now, literature, television, cinema, video games and even songs have adopted these topics. Examples include novels such as Nevil Shute's *On the Beach* (1959), Robert A. Heinlein's *Farnham's Freehold* (1964), Dean Ing's *Pulling Through* (1983), Walter M. Miller's *A Canticle for Leibowitz* (1960), and David Brin's *Earth* (1990), or the book series *Metro 2033* (2005) by Dmitry Glukhovsky. In 1961, the famous American television series *The Twilight Zone* (broadcast in France as *The 4th Dimension*) included an episode entitled 'The Shelter', which showed that a wrong choice of occupants in a fallout shelter could cause serious problems among neighbours. The cinematic world has produced a very large number of such films as well. Akira Kurosawa's *I Live in Fear* (1955), Stanley Kubrick's *Dr. Strangelove* (1964), LQ Jones' *A Boy and his Dog* (1975), Nicholas Meyer's *The Day After* (1983), Katsuhiro Otomo's *Akira* (1988), Kevin Costner's *The Postman* (1997), Phil Alden Robinson's *The Sum of All Fears* (2002), and George Miller's *Mad Max* series (1979–2015) are among the best-known and most interesting ones.

15 AN: Ernest R. May, Philip D. Zelikow, *The Kennedy Tapes: Inside the White House during the Cuban Missile Crisis*, 1997.

4. Improvised Shelters

I find shelter in this way
Under cover, hide away
Can you hear when I say
I have never felt this way

— British singer Birdy, *Shelter* (2011)

It is not always possible to live in a place that includes a fallout shelter, or to have the means or necessary authorisations to construct one. It is, however, perfectly possible to build improvised shelters at home or in public buildings. These shelters can be quite effective if you take some basics — and a minimal amount of preparation — into account.

The ideal option is to take shelter as far as possible from the actual source of gamma rays and to position as much matter as you can between yourself and those rays, regardless of whether the items/materials are constructions of human origin or simply natural ones. As already stated, the denser the material used, and the thicker the screen, the more effective the protection. The thickness of the materials required to attenuate gamma rays also varies according to the latter's energy. When it is less than 200 keV (kilo-electron volts), for instance, a few millimetres of lead are enough to block 90% of the radiation. For higher amounts of energy, by contrast, such as those encountered in the case of an atomic explosion, thicker screens are necessary: eight centimetres of steel; thirty centimetres of concrete; forty centimetres of earth; sixty centimetres of water; 100 centimetres of wood, etc.

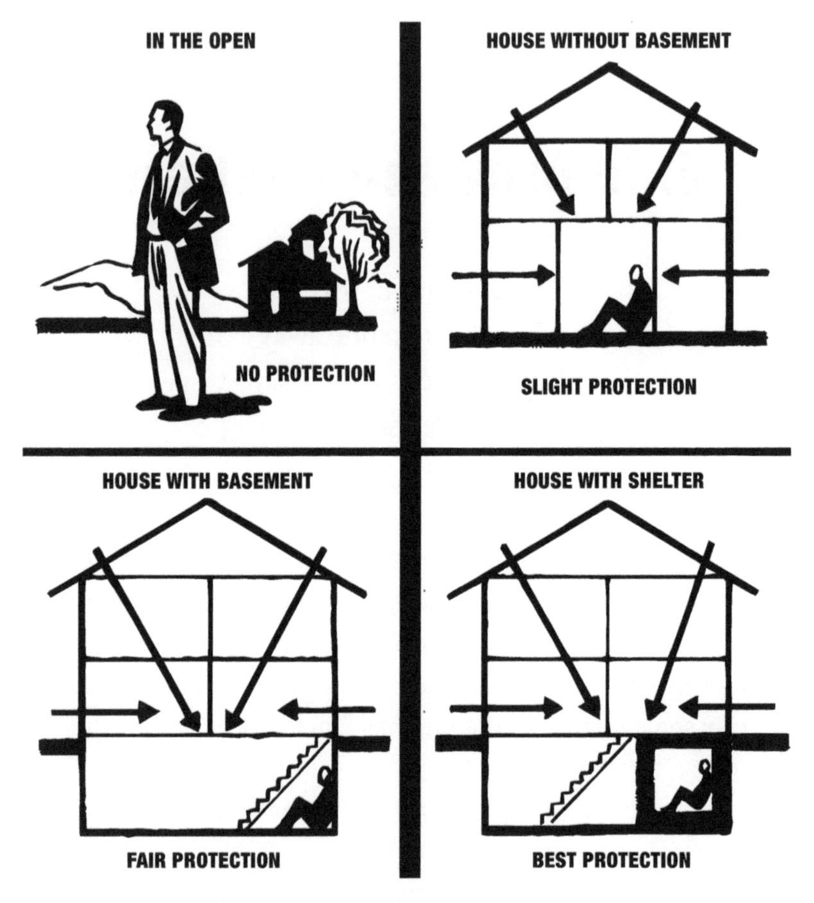

Picture taken from an American brochure released in the 1960s.

A shelter can be created in a quick and simple manner in the basement of a house, a block of flats or a public building. A trench dug in the ground and then covered up can also serve as a shelter. Railway or subway tunnels are also good places, as are the underground car parks of major buildings. In the picture above, originally published as part of an American brochure back in the 1960s, one can find four examples representing four different degrees of protection: no protection (outside), little protection (inside a house), effective protection (in a cellar/basement), and highly effective protection in a specially adapted cellar or basement.

To create an appropriately massive barrier, one can pile up books, furniture, sandbags, bricks, wooden boards, cans and bottles filled with water. An underground shelter can reduce your exposure to radiation by a factor of 100 to 200: if you would have absorbed a dose of 5 sieverts (5,000 millisieverts) outside, which would probably be fatal, you would, by contrast, only absorb 50 millisieverts when hiding in a well-equipped shelter inside a cellar.

Whatever the case, make sure that you keep away from all windows, as they could be shattered by the blast. Position a barrier comprising furniture or other objects between yourself and the windows. If no other options are available to you, protect yourself from the shards of glass using your own clothes, blankets, etc. If you have enough time, it may be worthwhile to place adequately sized wooden planks or panels in front of the windows.

Ask yourself: What would be the best shelter for you in your own city, in your workplace or alongside the roads you take on your regular journeys if a nuclear event were to occur? Is it a tunnel? An underground station? An underground parking facility? A cellar or basement? Is the place easily and quickly accessible on foot (since traffic could come to an almost instant halt)?

Note: One of the authors of this book, Piero San Giorgio, lives in Switzerland with his family and, at times, resides on a mountain farm. Despite the absence of a fallout shelter on the premises, there is a basement there that has been converted into a cellar and can thus be used as a shelter; in the area, the most likely danger is not a nuclear attack, but an avalanche. Piero also lives a few days a week in a small apartment in downtown Geneva, a city that is neither close to a military base nor in the vicinity of other likely targets in the event of a nuclear war. However, since it is one of the headquarters of the United

Nations, a potential strike cannot be ruled out. If one were to occur, he and the other inhabitants of the block of flats could take shelter inside the building's basement. The latter was built at the turn of the nineteenth century and is by no means a formal fallout shelter, but it is still surrounded by thick stone walls. In addition, the basement is partially buried in the ground. As for Piero's parents, they also reside in Switzerland; they live in a house built in 1981 and are therefore under the obligation to have a fallout shelter. This shelter, which is completely underground, remains usable. With its impressive sixty centimetres of reinforced concrete and thirty-centimetre-thick steel door, it serves as a storage room for tinned food, and especially as a wine cellar. It still has a dust-proof air filter, with the chemical toilet, lid and emergency exit giving onto the garden three metres further, allowing one to get far enough from the house should it collapse.

Warning! A shelter inside a cellar or buried structure is useful in case of a nuclear event, but not when it comes to a chemical attack or accident.[1] Indeed, as already said, most chemical gases are heavier than air and thus tend to seep into cellars and underground places.

Necessary Equipment in Shelters

Whatever type of shelter you have planned and set up, it is advisable to equip it with tools and items that will enable you to last there for the necessary amount of time.

The Most Important Thing: Water

Essential for the proper functioning of our organs and tissues, water is the main constituent of the human body and represents 60% to 70% of our total weight. Life expectancy without water is three days on average. Knowing that human beings discharge about two and a half litres of water a day, it is thus important for every person to take in a

1 AN: Unless it is equipped with a CBRN air filtration system.

minimum of two to three litres of water per day, i.e. between twenty-eight and forty-two litres every two weeks! Keeping substantial stocks of bottled water in your shelter is therefore essential. Ideally, you should have two weeks' worth of water for each person. In addition, if the alert is issued in advance, you can also fill all the containers available to you beforehand: bottles, pots, bathtubs, sinks, bowls, etc. — because you cannot predict whether water will always be available after the event or whether it will be left contaminated. There is a wide range of inexpensive and solid containers on the market that can quickly be filled (plastic containers of five or ten litres, jerrycans, etc.), as well as various items that will allow you to purify your water (filters, etc.). It is easy to fill them up regularly and keep them inside the place that could be used as a shelter, alongside some purification tablets that can be used before anyone drinks the water.

Do not underestimate your water needs![2]

Food

Food is much less important than water. Although you can survive for several weeks without food, it is advisable to leave at least some food in your shelter. Psychological considerations must also be taken into account: eating can relieve anxiety, which is inevitable under such circumstances — chocolate will particularly come in handy then. It would be best if most of the food you keep in your shelter is of the non-perishable or long-lasting kind: tins, rationed survival/emergency foods, jars of jam or peanut butter, cereals, dried fruit, and so on. To make the preparation process simpler and limit the necessary tools and equipment, it is preferable to consume food that can be eaten as it is, and that does not require any cooking or preparation. If you want more comfort, however, a gas stove with extra bottles, lighters, matches, and suitable pans and cutlery will have to be provided

2 AN: For further details, see *Rues Barbares — survivre en ville* [TN: Barbarian Streets — Surviving in Cities] by Piero San Giorgio and Vol West, Le Retour aux Sources editions, 2012.

for. Do not forget your tin opener: there have been more than a couple of cases when a person prepared a decent stock of tins only to find themselves in a situation where they lacked the means to open them (there is always another way to do so, of course, but it is definitely not an elegant one!). Last but not least, remember to regularly check your provisions' expiry date and, if necessary, resort to stock rotation.

Information and Communication

Having a radio will be very useful to enable you to listen to news broadcasts, bulletins and reports, especially when it comes updates on the event's consequences, but also in case it all turned out to be a mere false alarm. You need to have some spare batteries, or opt for a hand crank radio that can be started manually. It might be helpful to have a notebook with a pencil, marker and pen at hand so that you can take notes, write a message, or keep a journal. Although this is not essential, it is a good idea to have some ways of 'passing the time' as well, including reading, playing cards, and distractions for your children…

Hygiene

Hygiene quickly becomes crucial in an enclosed area.

To get rid of stool and urine, nothing beats chemical toilets, which dissolve organic matter and are sometimes portable. One can also improvise and create a toilet using a canister with a toilet-rim on top, in which one places small rubbish bags of five or ten litres that are closed after use and placed in a larger rubbish bag to avoid odours. It is therefore necessary to have quite a large number of small rubbish bags available (at least two or three per person for each single day), as well as some bigger ones (about ten will be enough). For matters of personal hygiene, you should have a box of wet wipes, women's tampons, toilet paper, toothbrushes, toothpaste, soap, additional rubbish bags, etc. These are small objects that ensure better comfort, despite not being absolutely essential over such a short period of time.

In hot countries, fly paper and mosquito repellents could offer 'extra' comfort.

Medical Kits

You can keep a small medical kit in your shelter to allow yourself to disinfect and treat cuts, small wounds, burns, headaches, etc. In addition to this, a more elaborate kit can be included in your bug-out bag (see p. 456). If you suffer from a chronic condition that requires specific medication, it would be wise to have a few extra boxes at hand in your shelter.

Specific Aspects of Burn Treatment

You (or someone else in your shelter) may have suffered burns due to the explosion's intense heat. These will have to be treated using whatever is available to you; a first aid kit with special supplies and equipment for such cases would be very useful. There are three degrees of burn severity:

- First-degree or superficial burns are the least serious and most common type of burns only affecting the epidermis. They result in visible reddening. The burned area becomes more sensitive, as is the case with sunburns. First-degree burns do not require special care or treatment, as the skin retains its ability to regenerate;

- Second-degree or deep burns damage the epidermis and, to a lesser extent, the dermis. These burns result in the appearance of blisters on the affected areas. Here again, the skin can regenerate itself as long as the person is careful to avoid any infection;

- Third-degree burns are the most serious ones. They destroy the entire skin (i.e. the dermis *and* the epidermis). The damaged skin then takes on a white, brown or black colour. The affected skin areas become numb, dry and prone to infection. In this third case, there is no possibility of the skin regenerating itself, because all the skin cells have been destroyed. If the damage is very extensive, a skin graft will probably be required.

The basic treatment of any burn is to gently douse the burn with cold water and without generating pressure. Pouring water on a burn decreases its extent, limits the consequences and relieves pain.

In the case of a chemical burn, any and all chemical-soaked clothing should be removed and the burn or affected part washed with generous amounts of water as soon as possible, so as to eliminate the compound. In the event of chemical spraying into the eyes, rinse the latter thoroughly with water. A severe burn will frequently cause cardiovascular collapse, which will lead to noticeable skin pallor (especially at the level of the lips and eyelids), as well as a rapid pulse and a feeling of thirst. If the burn is extensive and the burned person has to be transported, they must be wrapped in a sterile sheet/cloth and immobilised to reduce the pain. If the burn is on the person's back, they must be transported face down.

Temperature Considerations

Earth and concrete are good thermal insulators. The more people there are in a confined space, however, the sooner it will become hot and the faster the air will become stale. Good thermoregulation is therefore important, and opting for clothes that suit the prevailing weather conditions and overall climate is a welcome measure. No need to have a sophisticated wardrobe: some clean underwear, socks, T-shirts, and a variety of items enabling you to go out in all kinds of weather conditions (an anorak, raincoat, walking shoes, etc.). As regards air, the simplest solution is to install a manual fan, meaning a sheet or movable wall (even a mere piece of cardboard) that can be shaken using a string or a rope, thus triggering a fanning effect (through convective motions) and allowing the air to move. The temperature must be tolerable for at least a few hours, perhaps even a few days.

Air-Related Considerations

If the shelter is airtight, filtration systems against both dust and other particles must be available. If no such filtration system is in place, dust will seep in, and it will be necessary to avoid it as much as possible—which is why it is better to seal off all entrances in order to avoid the presence of excessive dust.

Tools

Your shelter could end up being buried under a pile of rubble, and it would thus be useful to have at least some tools that allow you to exit your refuge more easily than if you were forced to use your fingernails. A shovel, pickaxe, saw (with extra blades), hammer or sledgehammer would make it easier for you to clear/move debris. It might also be useful to have some construction materials at hand to improve or repair your shelter if necessary (especially if it is poorly constructed), including planks, nails, bricks, tarpaulin, sandbags, staplers, and so on. Do not forget a pair of high-quality resistant gloves that will allow you to clear debris, etc. without hurting yourself and perhaps also protect you from both cold and fire. Last but not least, having something to put fires out with can be extremely useful. A fire extinguisher remains an inexpensive and highly effective solution in this regard.

Light-Related Considerations

An event of this type is very likely to render the electricity network ineffective: you will therefore need to have your own pocket torches or something to autonomously generate the electricity you need. In a shelter, especially an underground one, there is a risk that daylight may not be able to seep in. You will thus be left in the dark. Having a torch (or more than one), in addition to spare batteries, is a priority.

Candles and a lighter/matches can serve as an alternative if you have a large amount of air or filtration systems in place. In larger, more complex or better prepared shelters, an independent electrical system connected to a generator or solar panels and batteries can be set up.

Means of Contact

It is important to have the necessary means to contact key people (family, friends, etc.) in case of an event. Indeed, something might happen while you are at work/travelling and the children are at school, etc. It would be ideal for you to be able to regroup in time, because the psychological effects of physical separation can be very difficult to overcome. Of course, it is better to know that the children are in the school shelter rather than to have them join you in an exposed and dangerous place. It is likely that telephone networks will be overloaded (during the alert) or perhaps inoperative (because of EMPs). It is therefore crucial to devise a plan and adopt a coordination system before such an event takes place.

For How Long?

This prompts the question of how long one should remain in their shelter. In some sci-fi movies or novels, individuals (or entire civilisations) are forced to stay underground for years, perhaps even generations. This, of course, is completely unrealistic. The best option is actually to stay where you are until help arrives and the area is declared safe. Generally speaking, this should take place twelve to twenty-four hours after the event itself. If no one comes, you are the one who will have to choose the right time. Given that most particles will have fallen in less than three days and short-lived elements will have caused a decrease in the level of radioactivity, exiting your shelter after a week can be considered very reasonable and, after two weeks, safe enough to move to a clean area (nevertheless, do think of putting on a 'dust mask' to avoid inhaling contaminated dust). Whatever the case, since human beings are not meant to exist in a closed environment or in overcrowded conditions among other occupants, you *will* end up leaving, whether you like it or not. This is where those debris-clearing tools will turn out to be useful, as will a bug-out bag, which will provide you with the necessary items to sustain yourself for a few days if necessary, until you have reached a safer area.

Note: In a Fukushima type of incident that does not involve any bombs, the situation is different. Indeed, the quantities of released contaminants can be very significant and trigger long-lasting effects. Under such conditions, it is best to evacuate the area as soon as possible, because medium- or long-lived radioactive elements could render the area 'uninhabitable' for decades, centuries or more.

Once you step outside your shelter more or less unscathed, it is likely that the world will have changed. Although you will not have ended up in a post-nuclear desert (after all, post-nuclear Japan has not fared too badly, has it?), it is still quite likely that the very scale of destruction and the number of fatal casualties will be difficult to accept. Furthermore, the economy, social structures, and lifestyles will remain in disarray for a long time to come. In this scenario, you will have to show mental resilience, pragmatism, courage, etc., and rebuild all that will allow you to start living again. The Germans, Japanese, and Russians have all managed to do this — and so can you.

For further details, here are two very comprehensive guides proposed by the Federal Emergency Management Agency (FEMA):

- http://www.nationalmasscarestrategy.org/wp-content/uploads/2015/10/Shelter-Field-Guide-508_f3.pdf

- https://www.fema.gov/pdf/plan/prevent/rms/453/fema453.pdf

5. Nuclear Attacks

A world without nuclear weapons would be less stable and more dangerous for all of us.

— British Prime Minister MARGARET THATCHER (1925–2013)

In 1945, the United States was the only nation to possess a bomb that could strike virtually anywhere and at any time (as long as the target was within range of its B-29 bombers) and the only one to enjoy complete aerial superiority. The target, usually a city, was likely to suffer significant damage both on a human and a material level. This 'sword of Damocles' hanging over enemy populations could guarantee a quickly won war.

In 1949, after a long and arduous development effort, and thanks to the help of a pro-Communist spy network in the Manhattan project, the USSR completed the construction of its first A bomb. Baptised *Pervaya Molniya* ('First Lightning') and characterised by a power of approximately twenty-two kilotonnes, it was successfully tested on 29 August in the military zone of Semipalatinsk (in present-day Kazakhstan). From that moment on, one began to speak either of a 'balance of terror' or of deterrence and dissuasion. This meant that every nation that attacked another in the knowledge that the latter had the necessary weapons and vectors to strike back and reciprocate ran the risk of suffering very heavy losses (especially with the development of intercontinental missiles by the late 1950s). Paradoxically, nuclear weapons have helped to make the world more peaceful and reduce the number of wars.

Officially, all powers claim that they would not want to be the first to resort to such weapons or, if necessary, would only resort to them in a defensive situation. And the balance is thus maintained: conflicts are limited to diplomatic battles, and wars remain indirect by supporting rebel factions, organising and financing terrorist groups, or instigating ideological subversion. Even the development of so-called 'tactical' weapons (designed to only have an impact on the battlefield) or *mini-nukes* meant to destroy high-priority targets (command bunkers, buried research centres, etc.) has contributed to keeping this reality unchanged.

Attack Plans

Since 1945, various plans have been developed to create doctrines for the use of weapons. Here are some examples:

- *Plan Totality*, established under the Truman administration in 1946, anticipated attacks on twenty to thirty major Soviet cities. In fact, the US only had nine bombs at the time and this plan was part of a 'bluffing' strategy whose purpose was to dissuade the Soviet Union from any attempt to violate the Yalta treaty;

- From 1961 to 2003, the United States adopted a plan known as the *Single Integrated Operational Plan* (SIOP), which constantly selected targets considered to be of strategic importance and assigned adequate and available means of attack for each of them (bombers, missiles, artillery, submarines, etc.). In 2003, the plan (which had been repeatedly modified to improve flexibility) was re-baptised *OPLAN8022*. What makes this plan special is that it does not exclude the possibility of using nuclear weapons as part of preventive strikes;

- 'Seven Days to the Rhine River' was the name of a Soviet plan dating back to 1964 and maintained under several aspects until 1984–1986. It anticipated a response to a potential attack carried out by the United States and NATO against the Warsaw Pact. The plan comprised a general offensive across Europe, supported by 131 tactical strikes against major concentration points of NATO forces, bases, and command and communication stations (which would have wreaked substantial destruction upon many European cities, including Brussels, Amsterdam, Cologne, Bonn, Frankfurt, Stuttgart, Munich, Nuremberg, Vienna, Vicenza, Brest, Toulon, Aalborg, and others). The goal was to reach the Rhine in seven days, then Lyon within nine, before advancing further to the Pyrenees. The plan took into account the fact that NATO would have used many tactical weapons (totalling twenty-five) on all possible Vistula crossings from Gdansk to the Slovak border;

- The United Kingdom has had an atomic weapon since 1952 and deploys about 200 warheads, mainly in its Trident II missiles aboard Vanguard class nuclear-powered ballistic missile submarines (Indeed, its 'V-type' long-range bombers — Valiant, Vulcan, Victor — have now been deactivated). Its doctrine has been rooted in maintaining its major power status through these weapons, all of which are intended to be used in defensive situations, especially in the (now past) event of a Soviet attack against its territory;

- Having risen to the rank of a nuclear power on 13 February 1960, thanks to its first atomic test nicknamed 'Blue Jerboa', France has been developing its own strike force known as 'Deterrent Force' since 1961, and has both the necessary vectors to use it (*Vautour IIB*, *Mirage IIIE*, *Mirage IV*, *Jaguar*, *Super Etendard*, *Mirage 2000N*, and *Rafale F3* bombers; *S2*, *Pluto*, *Hades*, and *S3* ballistic missiles; *Le Redoutable*- and *Le Triomphant*-class ballistic missile submarines equipped with M1, M2 , M20, M4, M45, and M51 missiles) and a very clear policy: if any enemy forces were to forcibly enter our national territory (which, in a Cold War context, meant that Soviet forces had successfully vanquished the French and German field armies and were preparing to cross the Rhine), an adequate response would be triggered;

- The doctrine espoused by the People's Republic of China, which has had its own nuclear weapons since 1964, is a purely defensive one, especially since its deterrence policy had long failed to establish credibility due to a lack of efficient vectors (indeed, China's bombers were obsolete and limited in their combat radius, the accuracy of its missiles very random, and so on). Thanks to the modernisation of its armed forces, however, its capabilities now match those of the American, British, French and Russian army in terms of precision, reliability and range, especially when it comes to ballistic missiles and cruise missiles. China has thus become both an efficient and credible power.[1]

Despite these plans, and in spite of the 'balance of terror', very serious tensions did indeed surface during the Cold War, almost leading to a nuclear war that the world only narrowly evaded:

- During the Korean War (1950–53), US General MacArthur proposed using atomic bombs against the Chinese army. His proposal was, however, rejected by President Truman,[2] and MacArthur was sacked;

- During the Cuban missile crisis of 15–28 October 1962, the sequences of events and bluffs that took place between the United States and Soviet-supported Cuba reached its peak on 27 October 1962. On that day, an American reconnaissance aircraft was shot

1 AN: http://www.ucsusa.org/sites/default/files/attach/2015/03/chinese-nuclear-strategy-full-report.pdf.

2 AN: Here — https://www.airspacemag.com/military-aviation/how-korean-war-almost-went-nuclear-180955324.

down on Cuban soil, exacerbating tensions so much that the American naval forces maintaining a blockade on Cuba initiated a fictitious anti-submarine depth charge attack (which was not, however, interpreted as such) on the Soviet submarine *B-59*, which then nearly launched a nuclear torpedo attack against the American aircraft carrier *USS Randolph;*[3]

- During the Yom Kippur War (6–25 October 1973), tensions between US and Soviet naval forces in the Mediterranean were so high that the United States, fearing a Soviet attack, actually went on alert;

- On 9 November 1979, the NORAD[4] command centre's computer screens revealed that a massive attack had been launched by the USSR. After a few minutes of intense panic during which a retaliatory strike was considered, someone pointed out that no missiles had been detected by any radar. It had all just been a computer error;

- On 26 September 1983, it was the computer screens of Soviet control centres which, this time, warned the Soviets that American missiles had been launched. As a retaliatory strike was being prepared, someone suddenly realised that it was a false alarm due to a computer error;

- During NATO's *Able Archer 83* exercise (2–11 November 1983), the Soviet army's General Staff became convinced that these exercises were actually a diversion intended to conceal a real surprise attack. Warsaw Pact forces, including strategic ones, were put on alert. Fortunately, tensions decreased once the exercise ended, because the Cold War was peaking at the time and the USSR had armed

---

3 AN: For details, see http://www.theguardian.com/commentisfree/2012/oct/27/vasili-arkhipov-stopped-nuclear-war.

4 AN: North American Aerospace Defense Command.

itself with 45,000 warheads of every conceivable kind, as compared to the United States with its total of 31,175;

- On 25 January 1995, a Norwegian scientific research rocket was launched from *Andøya* base. This triggered an alert on Russian radars, which identified the rocket as a ballistic missile. The situation was exacerbated to the point where Russian president Boris Yeltsin, awakened from his usual ethylic coma, found himself holding the briefcase that contained the launch authorisation codes. Luckily, he returned to bed once the Norwegian rocket had fallen into the ocean.

It is also important to note that some nations did not wait for an external threat to surface to carry out nuclear tests on their civilian population or military forces for evaluation and study purposes. Although such behaviour is difficult to imagine in today's highly mediatised and crowded world, such tests did take place in both the US and the USSR…

For example, the exercise code-named *Snezhok* ('snowball') conducted by the USSR on 14 September 1954 in Totskoye, near Oldenburg, comprised various operational manoeuvres involving more than 45,000 soldiers, standard and armoured vehicles, and both air support and logistics across an area that had previously been impacted by a nuclear strike. The troops had no special protection and thus served as guinea pigs to study the effects of a 'nuclear battlefield on combat forces'. Although the number and scope of casualties, including the surrounding population, remain unknown, several thousand deaths were witnessed in the months and years that followed the event, a fact that was meticulously concealed by the Soviet authorities…

Proliferation and Threats

Due to nuclear proliferation (a term describing the worldwide increase in the number of countries possessing nuclear weapons), the risk of potential conflagration is now growing.

A nuclear conflict between India and Pakistan, for instance, two countries characterised by an extremely nationalistic discourse, remains possible. Indeed, the two are involved in important territorial, religious and political disputes, especially with regard to the strategic Kashmir region, and have many scores to settle both in connection to the 1947, 1965, and 1971 wars and the Pakistani support of Islamist terrorism. In addition to this, they have displayed, in a highly official manner and in full view of the world, a large number of warheads, medium-range ballistic missiles and planes capable of carrying sophisticated bombs. Pakistan is, moreover, traditionally supported by China, the United States and Saudi Arabia (with the latter having financed the Pakistani nuclear programme and allegedly signed a secret agreement to quickly obtain bombs and missile launchers if necessary). For its part, India traditionally enjoyed Soviet support before succeeding in developing its own independent programme, while maintaining good diplomatic and military relations with most countries, including Russia. A conflict between these two states, which could be sparked off in many different ways,[5] could quickly degenerate into an atomic conflict. Indeed, India has at least 90 to 110 bombs in its possession, and Pakistan no fewer than 100 to 120 bombs... The effects of such a conflict would thus be considerable from a human, economic and geopolitical perspective.

Another case is that of the Democratic People's Republic of Korea, which has tested several bombs and medium-range ballistic missiles. However, no one really knows how large its arsenal really is, nor how

---

5 AN: In order to gain a better understanding of the possible scenarios, visit: http://warontherocks.com/2015/11/the-pink-flamingo-on-the-subcontinent-nuclear-war-between-india-and-pakistan/

serious its leaders are in their threats of using it. If it ever chose to use its offensive or defensive weapons, these could cause serious damage both to neighbouring South Korea (the thirteenth largest economy in the world and a major source of electronic components) and Japan. An attack on a more distant target, especially the United States, is unlikely, because the country does not currently have any missiles with a sufficient range—although placing a bomb in a classic container or transporting it on board a standard private plane could indeed constitute a very effective way to impact any part of the world with a highly destructive strike force.

The fact that Israel possesses a total of 75 to 400 nuclear weapons is an open secret. The country has the ability to use bombs to either defend itself or attack someone by means of its *Jericho II* medium-range missiles; *F-4*, *F-16I*, or *F-15I* planes; or cruise missiles launched by *Dolphin*-class submarines. The latter bestow upon Israel a second-strike capability of global extent. As for Israel's willingness to resort to these weapons, its successive governments have repeatedly stated that they would not hesitate to pulverise any enemy that threatens the country's existence, even at the risk of self-destruction.[6]

The greatest threat, however, is currently embodied by the highly aggressive US policy. Since 11 September 2001, US administrations have declared themselves ready to use their weapons,[7] including first-strike ones,[8] against any country defined as an enemy: Iraq, Iran, North Korea, Libya, Syria, etc. This doctrine seems to have been extended to include genuine military powers such as the People's Republic of China or Russia. These two nations have at least 400 and 5,000 bombs respectively, in addition to various global-scope

6 AN: This scenario is known as the 'Samson Option'. For more details, see https://en.wikipedia.org/wiki/Samson_Option.

7 AN: http://www.i24news.tv/fr/actu/international/67309-150411-les-usa-prets-a-utiliser-des-bombes-anti-bunker-contre-les-sites-iraniens.

8 AN: As confirmed by the following article: http://www.washingtonpost.com/wp-dyn/content/article/2005/05/14/AR2005051400071.html.

vectors—bombers, intercontinental missiles, cruise missiles, and missile-launching submarines.

Targets

What are the potential targets of a nuclear attack? The doctrines that emerged between the 1950s and the mid-1980s only replicated, but with more destructive weapons, the strategic bombing concepts of the Second World War, targeting all military and economic infrastructures (including the civilian population) in order to force the enemy to submit, in what was meant to be a total war. Powerful atomic bombs could also have a strictly military use when it came to the destruction of concentrated military forces: a fleet in a harbour, concentration points of troops and equipment, and so on.

It was from the 1960s onwards that the idea of using nuclear weapons for tactical purposes on the battlefield began to emerge, using smaller bombs this time. Nuclear warheads small enough to be used in short-range missiles (*Tochka*, *Scud*, *Frog*, *Scaleboard*, *Scarab*, *Spider*, *Little John*, *Lance*, *Pershing*, *Pluto*, *Hades*, etc.) or fired by artillery guns (152 mm, 180 mm, 203 mm, 240 mm, and 280 mm) began to be manufactured. Although these tactical weapons still exist, the development of small nuclear warheads has, especially since the 1990s, been oriented towards *mini-nukes* intended to be placed on high accuracy precision-guided cruise missiles or bombs. Indeed, these can be particularly effective against highly protected targets such as command bunkers, communication centres, military bases, airfields, underground factories or research centres, a government leader's residence or any other 'hard' target. Despite their small power, none of these nuclear *bunker busters* has officially been used in a theatre of military operations. The explanation proposed by some analysts is a simple one: if a small atomic bomb were used, people would notice that the resulting destruction was not as great as they had imagined (which is, however, the main point behind the development of these new weapons in the first place!). This would shatter the terrifying myth surrounding such weaponry and break down the psychological barrier to its use, thus paving the way for increasingly frequent usage.

With the advent of new doctrines regarding the use of nuclear weapons, it would seem that the idea of wreaking mass destruction upon enemy cities so as to exterminate the population is no longer favoured by the military staff of the world's great powers. Indeed, why take the moral and historical responsibility of destroying famous metropolises laden with emotional symbolism when targeted strikes would suffice? Moreover, the destruction of military targets (of which there are so few left in many Western countries) would certainly have such a psychological and demoralising impact that only few mini-nukes would be necessary to achieve objectives that would have required an entire downpour of nuclear bombs back in the 1960s. Last but not least, the dependence of modern countries on the proper functioning of all their networks and supply chains[9] is such that any attack would cause virtually unmanageable and utterly crippling economic and social chaos. Furthermore, such de-structuration could also be achieved without causing any deaths thanks to the electromagnetic pulses that result from a powerful detonation in the upper atmosphere (see below).

It is also obvious that it is preferable not to live or, in the event of an alert, not to remain in the vicinity of military bases and airfields, command centres (unless they are secret, of course), strategic arms plants, important research centres, the headquarters of NATO-type organisations, or significant policy-making centres.

---

9 AN: See Piero San Giorgio's *Survive the Economic Collapse*, Le Retour aux Sources editions, 2011.

Reacting to a Nuclear Attack

Before the Event

Considering the fact that a nuclear attack can occur without prior warning, and in view of its consequences, it is better to prepare in advance both physically and in terms of knowledge, behaviour and any actions suited to the situation you might face.

During the Event

A nuclear explosion is immediately visible. The resulting light is very intense and can even blind you, burn your skin and ignite objects. The shock wave that follows and the deafening sound stemming from the explosion itself make it impossible not to notice. Radiation, by contrast, is not felt (except in very high doses) and cannot be seen, as damage will only occur at a later point.

- **If you are outside:**
  - Every second counts!
  - Take shelter immediately by throwing yourself to the ground, into a hole (if possible) or behind a solid structure (a low wall, kerb, etc.);
  - Cover your head and face with your arms and curl up to limit the area that will come into contact with the flash of heat and the explosion's blast;
  - Do not attempt to look at the explosion and keep your eyes closed;
  - Close your mouth and do not breathe in when feeling any possible heat.

  Once the heat wave and blast have subsided, the circumstances will guide your actions. You can choose to flee immediately (in an opposite direction to where the explosion has taken place and preferably one that is perpendicular to the wind's direction); to collect your bug-out bag first and join your family somewhere; or to confine yourself to your long-term autonomous base (or a fallout shelter if available), as long as the latter's location is suitable and it has not been badly damaged.

- **If you are inside:**
  - Stay away from all windows and take cover as soon as possible by throwing yourself to the floor in the corner of a room or under a table;
  - When the blast has passed, take the necessary time to think the situation through. If your Long-term Autonomous Base (LAB) is still in good condition, you can consider shutting yourself in: cut off any ventilation, air conditioning, and/or heating; close all shutters and windows. Try to seal off your abode by using the contents of your containment kit. The information broadcast through the radio should enable you to determine whether evacuation is necessary;
  - If your LAB has been destroyed or damaged, you must head for one of your other refuges (your friends', family's or any other shelter) as quickly as possible.

- **Whatever the case:**
  - You must assume that radioactive contamination is already present. A part of it has been brought about by the blast and other dust particles will be added to this as a result of the fallout;
  - It is imperative for you to take the necessary precautions to protect yourself and avoid internal contamination (use plain tissues, pieces of cloth or FFP masks if available) and to furthermore prevent any contamination of your home, shelter, etc. on returning to your dwelling;

- Beware of any rains that may follow, as these will be severely soiled with radioactive elements: do not drink this water and avoid any and all contact with it.

Note: In the event that an alert announcing a nuclear attack is issued, you have but a few minutes to search for the safest possible refuge: a fallout shelter, basement, cellar, underground railway station, etc.

After the Event

If you live far from the location where the attack has taken place, you will not be affected by the thermo-luminal or mechanical effects. On the other hand, the fallout can still reach you, and rain can intensify it. Unless an alert is given or you possess your own detection capabilities, the presence of radiation will become evident only when you witness the arrival of personnel with specialised CBRN equipment. Under these conditions, you must understand that radioactive contamination is your main enemy. It is therefore in your own interest to apply the appropriate instructions, use adequate items and equipment and implement the undressing/decontamination procedures described in the previous chapters. Keep listening to any broadcast information if available, but always with a bit of scepticism, especially if it sounds too optimistic (e.g. 'The radioactive cloud has not crossed our border'; 'Everything is under control', etc.). And, as in all other cases, think for yourself. If the authorities ask you to evacuate, and you are in a perfectly safe shelter, will you comply only to find yourself in a refugee camp? It's up to you. Do not venture close to the explosion's or accident's epicentre.

ELECTROMAGNETIC PULSES

An electromagnetic pulse (EMP) is an abrupt and sudden electromagnetic wave emission. This type of phenomenon was discovered during the first nuclear tests as a side effect of a nuclear explosion.

Physicist Enrico Fermi had foreseen it and had even predicted that it was likely to destroy many electrical or electronic device components and scramble telecommunications.

It should be noted that there are also electromagnetic pulses of natural origin that are caused by the Sun or by lightning, but which do not have the same characteristics or the same effects as a nuclear EMP. Last but not least, some weapons can generate this type of pulses without the use of nuclear weapons (NNEMP). They are, however, very limited in their scope. A nuclear EMP comprises three different pulses: *E1*, *E2* and *E3*.

- **E1:** The *E1* type of pulse is produced when the gamma rays of a nuclear explosion eject into the upper atmosphere a very large quantity of electrons out of their atoms, at relativistic speed (more than 90% of the speed of light) and in less than one millisecond;
- **E2:** The *E2* type of pulse results from the gamma rays produced by the neutrons which the explosion generates for a longer amount of time, ranging from one millisecond to a second after the explosion;
- **E3:** The *E3* type of pulse is, on the other hand, very different. It is a very slow pulse that can last for dozens of seconds, or even more than a minute. It is triggered by the temporary displacement of the Earth's magnetic field as a result of the nuclear explosion, and then by its return to normalcy. An *E3* pulse resembles a geomagnetic storm caused by the Sun.

Several important factors influence the ability of a nuclear bomb to produce EMPs, including the altitude of the detonation, the distance between the detonation and the target, the geography, or the power of the Earth's magnetic field at the time of the explosion. A nuclear bomb will, for instance, trigger a much greater EMP if it explodes at high altitude (HEMP in English) than if does so near the ground. From zero to 4,000 meters, the effect is devastating for electrical and telecommunication infrastructures because the distances travelled by the pulses are short (only a few kilometres). From 4,000 meters to thirty kilometres, the effect is more limited than at higher or lower altitudes, because the atmosphere absorbs the radiation. There are thus few EMP effects in this case. At an altitude of more than thirty kilometres, the effect of such an explosion is optimal when it comes to inflicting a maximal amount of damage to electrical and telecommunication infrastructures.

Consequences:

- An E1 pulse can cause very high voltage in electrical and electronic conductors, which then exceed their breakdown voltage and burn out. It can destroy computers and communications equipment and disrupt radio signals for a long period of time;
- Although less powerful, an E2 pulse shows many similarities with lightning. It is usually absorbed by surge arresters and has, in general, little effect;
- Akin to a solar storm, an E3 pulse can produce geomagnetic currents in large electrical conductors, which can then damage components such as in-line power transformers.

In the event of targeted nuclear attacks, EMPs can therefore lead to a loss of power supply means and communication tools for several months. This can cripple a nation's defences by scrambling radar readings and decreasing the effectiveness of armies, which are nowadays highly dependent on computer and communication technologies as well as on electrical equipment. The pulse can also incapacitate certain command[10] and logistics centres, paralyse

10 AN: Since the army has been aware of the consequences of an electromagnetic pulse for several decades now, most command centres have taken adequate measures to protect themselves.

industries that depend on electric power, and render some intelligence and communication satellites inoperative.

In the absence of electricity and communication means, an entire segment of the economy collapses, marking the end of the modern era: no more light, no more communications, no more refrigeration, no more water or gas pumping, no more sewage treatment, no more operating theatres or life-saving/life-support machines (incubators, dialysis machines, respirators) in hospitals, etc.

Together with some authors of anticipation novels such as William R. Fortschen (*One Second After*), films like *The Day After Tomorrow* and the TV series *Jericho* describing such attacks/events and their consequences, it is interesting to read what the US army has been publishing on the topic.[11]

Mandated by the United States Congress in 2001, the *Commission to Assess the Threat to the United States from Electromagnetic Pulse (EMP) Attack*[12] convened a meeting of scientists and technologists to publish a report—the *Critical National Infrastructures Report*—in 2008, which describes in great detail the likely impact of a nuclear EMP on civilian infrastructures, which are crucial for the proper functioning of the country's armed forces.[13] This EMP commission has furthermore determined that protection means are almost completely absent in US civilian infrastructure and that even certain major sectors of the US military had some very severe deficiencies. History also tells us that electromagnetic storms of solar origin occur on a regular basis:

11 AN: Miller, Colin R., Major, USAF, *Electromagnetic Pulse Threats in 2010*, Air War College, Air University, United States Air Force, November 2005.

12 AN: http://www.empcommission.org.

13 AN: http://www.globalsecurity.org/wmd/library/congress/2004_r/04-07-22emp.pdf.

- The solar flare of 1859 triggered many auroras that were even visible in some tropical regions and greatly disrupted electric telegraph telecommunications;
- On 10 March 1989, following a solar flare, a powerful cloud of ionised particles was expelled by the Sun towards the Earth. Two days later, the first voltage variations were observed on Hydro-Québec's transmission system, whose protection systems were triggered on 13 March at 2:44 am. A general blackout plunged Québec into the dark for more nine hours;[14]
- In 2003, a similar storm caused a power outage in Sweden and damaged several electrical transformers in South Africa;[15]
- In an article dated July 2014, NASA announced that a very powerful solar storm had passed very close to the Earth on 23 July 2012. If it had occurred a week earlier, it could have caused large scale damage and 'sent civilisation back to the eighteenth century'.[16]

How to protect yourself against EMPs:[17]

EMP protection has been very well-mastered from a theoretical point of view, and a number of military and civilian facilities, devices or tools are thus protected. One solution is to construct the hardware (circuits, cables, etc.) entirely from solid materials inside which electrons, or other charge carriers, remain completely confined. A second option, a much simpler and less expensive one, is the use of a *Faraday cage* around sensitive devices.

---

14 AN: http://www.hydroquebec.com/comprendre/notions-de-base/tempete-mars-1989.html.

15 AN: http://www.nasa.gov/topics/solarsystem/features/halloween_storms.html.

16 AN: Here it is: http://www.lemonde.fr/sciences/article/2014/07/25/la-terre-a-echappe-de-justesse-a-une-gigantesque-tempete-solaire-en-2012_4462546_1650684.html.

17 AN: Details can be found here: https://fr.wikipedia.org/wiki/Impulsion_électromagnétique.

A Faraday cage, named after its inventor Michael Faraday, is a metal enclosure usually made of aluminium and connected to the ground so as to maintain its potential unchanged. This 'Faraday cage' is often used when one wishes to make precise measurements in the field of electronics, electricity or electromagnetic waves. It enables one to protect equipment against external electrical and (on a secondary level) electromagnetic issues or, conversely, to prevent equipment from polluting its surroundings. A Faraday cage is thus impermeable to electric fields. It operates by means of a simple difference of potential, without any current being necessary, regardless of whether the disturbance source is inside or outside the enclosure. This structure can also have an indirect protection effect against any disturbances of electromagnetic origin due to the presence of a current. This is then known as electromagnetic shielding. When used for such a purpose, it is no longer necessary for the structure to be grounded, but its efficiency will depend on the disturbance's frequency, the conductivity and the magnetic permeability of the material in question.

A Faraday cage must, in principle, be completely closed, but it can also consist of a wire mesh: the orbs of the mesh are a few centimetres across and act as a mirror on an ultra-high frequency range. The higher the frequency (and the shorter its wavelength, therefore), the smaller the orbs must be.

The performance and cost of a Faraday cage depend largely on its accessories: doors, windows, honeycomb vent panels and fluids (for attenuation). Any conductors entering and leaving the Faraday cage must be equipped with radio filters—otherwise, they behave like antennas and greatly reduce the overall performance of the cage. There are three main techniques for producing industrial Faraday cages:

- Modular cages: they are made from steel trays or wooden panels coated with a steel sheet on both sides. The trays are held together by bolts, and the wooden panels by steel profile connectors. The advantages of using steel trays include insensitivity to both moisture and hygrometric variations and a decent geometrical performance in the long run. The main benefit of using wooden panels is that they can be re-cut and the overall dimensions modified. Modular cages enable a performance above 130 dB at 1 GHz;

- Architectural copper cages: They comprise a 0.2 or 0.3mm copper strip which is put on top and continuously brazed with tin. This technique is used in large rooms and allows one to adapt to complex geometries (corners, recesses, beams, pillars). There is no waste of space, since the copper is applied directly to the walls. Copper Faraday cages provide performances in excess of 120 dB at 1 GHz;
- Architectural cages made of a metallised fabric: in this third case, the cage is constructed using a metallic tapestry held in place by glue. This technique has the same advantages as copper cages. The performances achieved are above 60 dB at 100 MHz and sufficient for some implementations.

As part of the three main techniques, it is possible to add windows to the cage. These windows, however, have limited efficiency (approximately 80 dB), because they reduce the overall performance of the Faraday cage; which is why armoured video cameras are often installed instead of windows in order to preserve the cage's overall shielding level.

Cars act as Faraday cages as well. However, the use of non-conductive composite materials and the presence of glazed openings render it flawed.

The metallic cases of computers are also Faraday cages. If this housing happens to be non-metallic (plastic), it is lined, at key locations, with a thin metal sheet connected to the device's electrical ground, so as to meet the standards of radio-compatibility.

In general, many household appliances are equipped with internal shielding forming a Faraday cage, at least in the case of sensitive parts. Due to construction cost imperatives, metal shielding sheets are often replaced with a layer of a conductive material applied through spraying onto the interior of the body, which is made of insulating materials. For example, a microwave oven comprises a Faraday cage in which food is placed for heating, while the door is equipped with a wire mesh that is fine enough to hold the waves in and large enough to allow people to see inside. This, at least, is theoretically true, because practically speaking, tests carried out using high-frequency radiation-measuring devices have revealed leaks in many currently marketed models.

MRI devices are surrounded by a Faraday cage to isolate any waves that could interfere with the radio frequency waves emitted by the radio wave generator.

Electrophysiology equipment is always surrounded by a Faraday cage to keep the interference noise low, thereby increasing the signal-to-noise ratio.

Single-family houses are sometimes equipped with a Faraday cage protecting all those inside. This protection against external radiation, however, does not allow the electromagnetic fields generated inside the house to be normally discharged.

6. Alimentary Protection

The best way to detoxify is to stop putting toxic things into the body and depend upon its own mechanisms.

— Andrew Weil, naturopathic doctor

Obtaining Information

Nowadays, the population of 'developed' countries seems to have become aware of its own interests. People are thus looking for better foods, i.e. buying organic products, growing vegetables in their own garden or even raising small-sized livestock. Some choose to leave the city and return to the countryside. This approach is, of course, accompanied by a desire for more autonomy and less stress. But what happens in case of a nuclear accident or war? Would one still be able to cultivate one's own land or buy fruit and vegetables without having to worry?

In order to protect yourself, you must first be informed.

We all hope that the government authorities and media will inform the population in the event of contamination, regardless of its cause. The Chernobyl and Fukushima accidents have, however, brought to light a certain sluggishness and lack of transparency. This can be understood as a desire to limit potential panic effects. And yet, this can also have a negative effect on the protection of populations, particularly by delaying the initiation of containment or evacuation procedures — all the more reason to prepare in advance, both in terms of equipment and the behaviour to adopt. But what would happen if no information were broadcast, either because it was intentionally suppressed or as a result of the country's inoperative or destroyed means of communication?

Even though radioactive gases and very fine particles can travel the Earth's width several times, most of the dust that is expelled into the atmosphere by the explosion of a nuclear bomb, for instance, will eventually fall back. The levels of radioactivity characterising the deposits thus formed obviously depend on their size and composition.

The following chart,[1] based on the results of US nuclear tests conducted between 1950 and 1970, provides us with a calculation of fallout-related dose rates over time:

TIME ELAPSED AFTER EXPLOSION (HOURS)	DOSE (R/h)	TIME ELAPSED AFTER EXPLOSION (HOURS)	DOSE (R/h)
1	1000	36	15
1.5	610	48 (2 days)	10
2	400	72 (3 days)	6.2
3	230	100 (4 days)	4.0
5	130	200 (8 days)	1.7
6	100	400 (17 days)	0.69
10	63	600 (25 days)	0.40
15	40	800 (33 days)	0.31
24	23	1000 (42 days)	0.24

It can be seen that the decrease in radioactivity levels is fast. Thus, in less than two days, radioactivity will have diminished by more than

1 AN: Adapted from *The Effects of Nuclear Weapons* by Samuel Glasstone et Philip J. Dolan, DoD, ERDA, 1977 — http://www.deepspace.ucsb.edu/wp-content/uploads/2013/01/Effects-of-Nuclear-Weapons-1977-3rd-edition-complete.pdf.

half; and after two weeks, its rate will have been divided by approximately 1000.

In case of environmental contamination resulting from a Fukushima-type accident, things are a bit different. First of all, the quantities of dispersed radioactive material are much larger than in the case of a nuclear explosion. Secondly, the closer to the damaged reactor you are, the higher the proportion of heavy elements. Radioisotopes such as plutonium and uranium will therefore be present in large amounts. As for strontium-90 or caesium-137, they could be transported over greater distances. In truth, all these radioisotopes are extremely dangerous and some of them have incredibly long lifetimes... The areas located in the proximity of the plant will therefore remain off-limits for at least several hundred years.

However, the further one moves away from the accident site, and thus from the source of contamination by heavy radioisotopes, the more rapidly the danger decreases.[2] In addition, this decrease in activity will be intensified by natural phenomena such as the seeping of radionuclides into the soil, their dispersal through runoff, and their expulsion by waterways and rivers, until they have 'disappeared' into the seas and oceans. Most of these areas are likely to have acceptable levels of radioactivity to enable the return of human activity quite quickly.

This does not, of course, mean that we are to neglect the presence of contamination. On the contrary, the purpose is to learn to live in conditions of radioactive contamination *if necessary* and to minimise the effects of radiation on the body...

---

2 AN: Be warned, though: this is not always the case. If the activity of a source is really high, it could remain dangerous even after diminishing by a factor of one thousand.

Health-Related Effects

More than thirty years after the Chernobyl nuclear accident, the populations inhabiting the Ukrainian and Belarussian territories that were contaminated by the fallout are still absorbing the radiation emitted by the radionuclides deposited on the ground. Some of these, being short-lived, have already disintegrated, but others continue to generate radiation, i.e. to bombard everything around them with particles/waves. This external irradiation, however, which is quite low, does not embody the main health hazard for the inhabitants. Much more serious are the consequences of internal contamination, which is caused by the penetration and settling of radionuclides into the human body.[3] Indeed, the latter will keep on causing damage as long as they remain inside one's body, i.e. until they have disintegrated or been discharged from it. Internal irradiation is all the more dangerous because it acts directly on a person's cells, without there being at least some protection offered by the skin. It is interesting to note that these radioactive elements can have particular affinities and be prone to binding themselves to specific parts of our organism:

- With its half-life of about eight days, Iodine-131 is easily absorbed by the body and binds itself to the thyroid gland. It is a genuine sort of danger, especially in the early stages (i.e. a few weeks to a few months). Its short half-life causes it to disappear quickly;
- Caesium-134, whose half-life is 2.06 years, and caesium-137, whose half-life lasts thirty years, are alkaline metals similar to potassium. Just like the latter, they tend to invade the whole body.

---

3 AN: As seen in the analysis carried out by the Belrad Institute in Belarus (www.belrad-institute.org), which has shown that almost 90% of the total irradiation rate absorbed by the populations that were affected by the contamination resulting from the Chernobyl accident is actually internal (relation between the caesium contamination rate of a territory and the amount of caesium accumulated in the inhabitants' organism, based on data gathered from 100 localities near the Gomel region).

They can dissolve in water and spread very quickly throughout the surrounding environment. Of the two, caesium-137 is by far more dangerous;

- With its half-life of 28.1 years, strontium-90 has similar properties to calcium and is mainly deposited into the bones and bone marrow.

Effects

According to the studies carried out in Ukraine and Belarus by Dr Vladimir Babenko,[4] the main source of irradiation in the aftermath of the 1986 accident was due to the ingestion of radioactive particles with locally produced foods, which are heavily contaminated by caesium-137 and strontium-90 radionuclides. This internal contamination accounted for 70 to 90% of the total irradiation rate received. The higher the concentration of ionising elements, the more the person's organs were damaged, of course.

It is important to realise that children and adolescents are more vulnerable to radiation than anyone else. Indeed, their fast growth — and thus the rapid proliferation of their cells — makes them more susceptible to mutations and other harmful effects.

The health-related consequences of internal contamination can be very serious and lead to:

- an increase in the number of patients afflicted with diabetes;
- chronic diseases of the gastrointestinal tract;
- diseases of the respiratory tract;
- autoimmune diseases;
- allergies;

4 AN: Vladimir Babenko, *After the Nuclear Accident. How to Protect* [TN: *Yourself*] *Against Radiation — a Practical Guide*, Tatamis editions, 2012.

- various types of cancer (thyroid gland, leukaemia, etc.);
- childhood tuberculosis;
- heart dysfunction;
- hypertension;
- cataracts, the degeneration of the eye's vitreous humour, the degeneration of the retina, blindness;
- congenital malformations, miscarriages;
- etc.

In addition to this, the immune system is very sensitive to ionising radiation and can be significantly affected, which sometimes causes a drastic reduction in the body's defence capabilities. The sooner one becomes aware of internal contamination, the sooner one can take measures to facilitate the discharge of radionuclides from the body. To measure the body's radioactivity, a specially designed measuring device called the *Spectrometer* of *Human Radiation* (SHR) must be used, which is harmless to humans. In the event that someone ingests radioactive elements, it is necessary to attempt to eliminate them as quickly as possible in order to minimise their harmful effects.

The Elimination of Radionuclides

The human organism discharges radionuclides (just as it does with many other substances, whether toxic or not) through its haematocrit organs, including the kidneys, liver and gastrointestinal tract.[5] To reduce the harmful effects of contamination, it is necessary to enable an accelerated discharge of radionuclides from the body. For this purpose, Ukrainian and Belarusian doctors use pectin-based products.

---

5 AN: For further information, read Christopher Vasey's *The Detox Mono Diet: The Miracle Grape Cure and Other Cleansing Diets*, Jouvence editions, 2014.

Pectin is a large molecule found abundantly in apples and characterised by its ability to absorb certain heavy metals and radionuclides during its passage through the digestive tract. It seems to be able to help the body eliminate any caesium-137 it contains faster and without the side effects of chemical chelators.[6]

In actual fact, pectin is used as a dietary supplement for children living in the areas that have been exposed to Chernobyl fallout: these children suffer from pathological states relating to the accumulation of caesium-137 ingested with food or drinks. Belarusian professor Vassili Nesterenko of the Belrad Institute[7] (an international institute fighting against the effects of this radioactivity) mentions[8] an experiment involving sixty-four children in the Gomel district, a zone heavily contaminated by Chernobyl fallout. These children spent a month in a sanatorium where they only consumed uncontaminated food. One control group took pectin in the morning and evening, the other a placebo. One month later, the children of the pectin group saw their caesium-137 levels decrease by 62.6%. In the other group, the decline was only 13.9%. These results justified the Belrad Institute's development of a powder enriched with vitamins, trace elements and pectin and known as *Vitapect*.[9] This is now administered to children residing in heavily contaminated villages as part of treatment courses lasting from three to four weeks (the prescribed adult dose is one or two teaspoons, two to three times a day, in a quarter-full glass of water, tea, compote, juice or any other drink; the dose for children is one teaspoon twice a day). About 200,000 children from Belarus have been given this preparation, and the level of caesium-137 was checked before and after treatment. Professor Nesterenko has demonstrated that three to four annual pectin-based treatment courses lasting four

---

6 TN: Chelators are tiny molecules that bind very tightly to metal ions.

7 AN: http://www.belrad-institute.org/UK/doku.php.

8 AN: http://www.liberation.fr/week-end/2004/05/08/la-pomme-contre-l-atome_478694.

9 AN: http://www.vitapect.eu.

weeks each and distributed to children attending school in highly contaminated villages successfully kept the caesium-137 rate below the threshold of 50 becquerels per kilogramme of weight (Bq/kg), which is the threshold at which irreversible damage to the heart, eye, immune and endocrine systems or other organs occurs.[10]

The ACRO[11] (Association for the Control of Radioactivity in the West) also found that children who had received pectin during their stay in France saw their caesium-137 contamination decrease by 31% on average, compared to only 15% in the case of those who had not received any (except for the part naturally comprised in their diet). Still according to the ACRO, pectin increases and accelerates the elimination of caesium, yet less quickly than the studies advertising it have claimed.[12]

Note: In the absence of *Vitapect* (which can sometimes be difficult to obtain), *Naturactive* sells capsules containing a concentrated extract of apple pectin. *DB Pharma laboratories* also offer a powder for oral suspension called *Gelopectose.* For an even more favourable effect, these two products can be supplemented with a multivitamin cocktail (B2, B6, B12, C, beta-carotene, folic acid) enriched with minerals (selenium, zinc).

How to Manage Your Food

As shown from the experience gained from the Chernobyl accident, radionuclides enter the human body mainly through the food chain. They can also penetrate it, but to a much lesser extent, through inhalation (breathing) and contact (i.e. through the skin and mucous membranes). Last but not least, a cycle of radiological contamination

---

10 AN: Y.I. Bandazhevsky, Chronic Cs-137 incorporation in children's organs, Swiss Med. Wkly., vol. 133, no 35–36, 2003, p. 488–90.

11 AN: http://www.acro.eu.org.

12 AN: Bandazhevsky Y.I. & Bandazhevskaya G., *Caesium-137 Myopathies*, Cardinale, 2003;15(8): 40–3.

can sometimes be restricted to a radioactive source 'polluting' the surrounding environment, and therefore man's entire food chain:

- The source of contamination can come from the explosion of an atomic bomb, a nuclear accident, or the dispersion of radioactive waste;
- Following environmental contamination, soil and water are soaked up by plants and ingested by animals, which thus absorb the radionuclides. Next, the contamination can pass from plants to man (through the consumption of fruit and vegetables), from plants to animals (including seaweed to fish), from animals (including fish) to man, from the milk of animals to man, from cheese milk to man, from (plant) firewood to humans (through the ashes), etc.;
- Food is logically the main source of radionuclide intake, and one single meal can sometimes suffice to contaminate a person.

The radionuclide content of foods can be substantially reduced through proper culinary treatment. However, the proposed methods can only be implemented if the food product's content in radionuclides is two to three times higher than normal, not more. Indeed, *if the contamination is too extreme, no treatment can make the food edible* and one had better not consume it nor use it to feed animals or livestock.

Water is very useful for getting rid of radionuclide-laden dusts. In addition to this, since caesium-137, one of the main sources of contamination, is actually soluble and does not bind to fats, water is also an effective means of dissolving and removing some of the contaminating elements. In the aftermath of the Chernobyl accident, in fact, the Soviet authorities rinsed the roads and facades of buildings, including the windows and facades of all the buildings in the city of Kiev. So, if your home has been contaminated by radioactive fallout, rinsing the surface to reduce radioactivity may well be worth it. Other measures can also be applied:

- Roads and walkways/driveways can be rinsed;
- The upper layer of garden soil can be removed and buried;
- Trees and bushes can be removed, as their leaves will have collected a large amount of dust;
- Dead leaves can be cleared, buried or disposed of in distant locations. Be careful not to burn them (nor any dead wood), as the fumes (which are actually ash) will be loaded with radioactive elements.

Figure 21. The Contaminated Territories Around Chernobyl

The Vegetable Garden

Caesium-137 has chemical properties very similar to those of potassium. If a plant lacks potassium, it will tend to take in more radioactive caesium. Following this same pattern, strontium-90, whose chemical properties resemble those of calcium, will be absorbed in greater quantities by any plant that lacks the latter. It is thus a wise choice to add products rich in both potassium and calcium to nourish plants grown in a vegetable garden that may end up being contaminated:

- dolomite powder, which is rich in calcium and magnesium, once every four to five years at a rate of 40–50 kg per 100 m^2;
- a complete artificial fertiliser for vegetable gardens, at a rate of 40g/m^2 for green vegetables, pumpkins, squashes, etc; 60g/m^2 for cabbages; 90g/m^2 for cucumbers;
- 1–1.5kg/100 m^2 of carbamide and 2–3kg/100 m^2 of potassium chloride annually;
- generally speaking: manure, compost, organic peat, etc.;
- Hemp, rapeseed, reeds and sunflowers seem[13] to be plants with an ability to cleanse soil and stagnant water of heavy metals. This phenomenon, known as *phytoremediation*, is carried out by the roots of these plants, which, characterised by their highly filtering properties, absorb 'waste' of any kind, thus helping to decontaminate the soil in your struggle against radioactivity. Warning—these plants must not be consumed, and their physical elimination must take place in another location, so as to prevent heavy metals and other contaminants from returning to the soil.

In the event of contamination, one must no longer use as fertiliser any ashes stemming from the incineration of wood in the contaminated area. Indeed, these ashes would thus be a concentrate of the forest's radioactive elements. It is therefore preferable to bury them in a special place intended for such a purpose, at a depth of approximately one meter, once you have rendered the bottom of the hole impermeable using a *polyane*[14] plastic film.

---

13 AN: As has been stated in studies carried out in Japan, Belarus and Ukraine: http://lesbrindherbes.org/2013/03/26/cannabis-tournesol-colza-plants-depolluantes-contre-radioactivite.

14 TN: *Polyane* is a range of insulating and waterproof plastic films sold in France and used by both building and public works professionals. At times, the term is applied rather loosely to refer to any sort of plastic film.

Plants

The relation between the number of radionuclides present in the soil and the amount absorbed by plants depends on the actual type of soil and the plant species itself. Plants growing in marshy, peaty, sandy and podzolic soils (which absorb and retain most water) are those that assimilate them most. Lichens, mosses, mushrooms, legumes and grasses thus tend to tap radionuclides most intensively. The leaves of trees also retain a lot of radionuclide-laden dust.

To treat vegetables, it is necessary to start by removing any parts where radionuclides have accumulated, i.e. those that are on the surface. For instance, when cabbages are cleared of their large outer leaves, their radioactive contamination could be rendered forty times lower. The removal of the green leaves of beets, radishes, turnips, carrots, etc., reduces their contamination by a rate of fifteen to twenty times. Next, of course, you must clean the parts that are to be consumed. If you brush the soil off thoroughly and wash the potatoes, tomatoes, and cucumbers meticulously, you can reduce their contamination by a rate of five to seven times. You can also reduce the dangerousness of root vegetables by removing the upper part of the root (1 to 1.5 cm from the top). A washed potato also becomes twice less radioactive after being peeled. Wheat, once threshed and separated from its bran, is ten to fifteen times less radioactive. All this must be done very carefully if one is certain (or if one suspects) that one lives in a contaminated area.

Because the roots of trees, especially fruit trees, plunge deep into the soil, their fruit may be radiologically clean even if the area has recently been contaminated. It is, however, necessary to wash the fruit before preparing it and imperative to peel it if the contamination is recent.

As for the conservation of vegetables and fruit (through lacto-fermentation, in vinegar, etc.), it allows us to further reduce their content in radioactive elements, provided that we do not to consume the salting layers or marinades in which they have been kept.

Meat

Radioactive caesium is deposited in the soft tissues of animals and is particularly concentrated in their liver and kidneys, which act as endocrine filters. Offal must, therefore, as a preventive measure, be avoided. Caesium almost never binds itself to bones. Radioactive strontium, by contrast, does settle in the latter and is practically unextractable once there. For this reason, it is better to refrain from consuming bone-based broth. As for bacon and fat, they accumulate few radioactive elements. Be wary of the meat of wild animals, though, as it usually contains more radionuclides than that of farm animals.

To reduce the radionuclide concentration in meat, the most effective solution is to soak it in a 2% salt solution after cutting it up into small pieces so that the contact area is as large as possible. You must then leave it there for at least twelve hours while replacing the brine regularly. The longer the meat is left to soak (and the more often the saline water is changed), the fewer radionuclides remain in it. Most of the caesium will be removed along with the salt-water. If the meat is boiled for about ten minutes, around half of the radionuclides will be dissolved in the broth, which thus becomes unfit for consumption and must be discarded.

Milk

Milk can have a fairly high concentration of radionuclides, including whey, which must not be used or consumed. On the other hand, it is possible to treat the milk and rid it of a large part of the radionuclides polluting it. Once we separate it, what we are left with is a cream that is four to six times less contaminated than the milk was. White cheese (or *fromage blanc*) made from contaminated milk also contains a four-to-six-times lower rate of radionuclides, with cheese comprising eight to one hundred times less, butter eight to ten times less, and clarified butter ninety to one hundred times less.

Forest Products

Forest products are the most dangerous food category. Indeed, radioactive fallout is deposited on the forest floor's litter, which acts as a screen and prevents it from seeping deeper. Most of the radionuclides thus remain in the forest's three to five cm superficial layer. There is also a high concentration of radionuclides in tree bark, dead wood, mosses, lichens, berries, and mushrooms. The latter are the riskiest to consume, considering that they have the unfortunate property of concentrating caesium. If you really want to eat them anyway, they must first be boiled in salt-water, with the water frequently changed and both vinegar and citric acid added to accelerate the transfer of radionuclides from the mushrooms to the water. Likewise, berries growing in contaminated forests may also be contaminated as a result. Based on people's experience in Ukraine, Belarus and Russia and on the measures taken there, the most contaminated ones are bilberries, cranberries, and blueberries, with raspberries and wild strawberries among the least affected. Nonetheless, all of them are to be avoided as much as possible. Game from contaminated forests is also a source of danger due to the higher concentration of radionuclides. Last but not least, remember that drying does *not* reduce radioactivity in the case of meat, mushrooms and fruit. On the contrary, it *increases* it.

Miso

According to a 2001 study[15] conducted by Doctor Masayuki Ohara of Hiroshima University, miso appears to have properties that allow it to act against radioactivity, since it contains dipicolinic acid, an alkaloid that has the particularity of chelating[16] heavy metals such as strontium, lead, mercury and cadmium while facilitating their elimination through the urinary tract. Miso is a beige-to-brown paste obtained

15 AN: http://www.ncbi.nlm.nih.gov/pubmed/11833659.

16 AN: Chelation is a physicochemical process that leads to the formation of a stable, inactive, non-toxic and soluble complex that is easily eliminated by the kidneys.

from a mixture of fermented soybeans, sea salt and a fermentation-starting enzyme called *kōjikin* that contains *Aspergillus oryzae*, a noble mould also used in the production of sake. When faced with radioactive pollution, dissolve the paste in a bowl of hot water and take one teaspoon of it four times a day.

Note: Piero San Giorgio visited the Chernobyl site and its surroundings in 2010. With a diameter stretching approximately thirty kilometres around the site of the former power station, the exclusion zone can be visited under certain conditions. The one-day visit takes visitors to a Ukrainian army barracks where they are given a short historical rundown of the accident that took place on 26 April 1986, as well as some basic notions about the risks involved in exposure to radiation. The zone is relatively safe and, in most places, the radiation level does not exceed a dose rate of 1 μSv (one microsievert) per hour. The guide accompanying visitors carries a Geiger counter that measures the level of radioactivity, which is now normal almost everywhere. Mosses on trees and on the ground are, however, clearly more radioactive, because they concentrate radionuclides. All visits to the power plant itself are obviously forbidden. Visitors are taken to a building 100 m from the power station, where models displaying what it looked like before the accident can be examined and where one can look through the window and see the pile of metal and concrete of Reactor 4, expeditiously covered with a tin plate infrastructure (a new protective dome is under construction and will be put in place in 2017). The visit then continues to the city of Pripyat, which was evacuated most urgently following the accident on 26 April 1986. Since the city has remained abandoned since then, the spectacle is an impressive one — a modern city, very well-equipped in terms of public infrastructures (including sports centres, theatres, schools, playgrounds, and so on), blocks of flats, parks, etc., now totally overgrown with vegetation that covers each and every space in an opportunistic fashion, shrouding and weakening structures and, at times, causing their destruction. It is a genuine ghost town in which everyday objects ranging from trolleybuses to school notebooks have remained where they were left (with furniture and other personal belongings stolen a long time ago). The feeling aroused in visitors within this context enables them to envision the effects of a nuclear accident on the daily lives of the families and individuals that were compelled to flee the town without delay. In 2016, Ukraine maintained fifteen nuclear reactors in operation at four different sites, and given the country's economic and financial situation and the level of corruption afflicting its successive governments, it can only be hoped that the measures necessary to avoid any potential accident have indeed been taken…

Piero San Giorgio and a Ukrainian friend in front of the Chernobyl power plant in 2010.

KITS AND EQUIPMENT

Make a kit, make a plan, get informed.

— Slogan of the American Red Cross

Two sureties better are than one; and caution's worth its cost, though sometimes seeming lost.

— Jean de La Fontaine, Fables, *The Wolf, the Goat and the Kid*, 1668

1. General Kits and Equipment

The foolish learn sense through misfortune, not words.

— Greek philosopher Democritus (460–370 BC)

Containment Kits

In the event that you are forced to confine yourself to your home (house/flat), it would be useful to have the following kit at hand:

- 1 pair of scissors;
- 1 cutter or carpet knife;
- several rolls of strong and waterproof adhesive tape (duct tape);
- 1 DIY stapler;
- several plastic protective films or, failing that, large-sized rubbish bags to insulate doors, windows and ventilation/air-conditioning grids;
- some cloths (when moistened, they can be put to effective use by blocking various openings such as vents, etc.);
- 1 pair of leather gloves (in case any windows are broken, etc.);

- 1 flashlight (as power failure cannot be excluded).

Note: If you believe that more than one person will be present, do not hesitate to increase the number of items (e.g.: 2 pairs of scissors, 2 pairs of gloves…).

Bug-Out Bags

A bug-out bag is, above all, a tool whose purpose is to answer all of your physiological needs when you're evacuating a high-risk place to reach a stop-off point, preferably in less than seventy-two hours. It is not meant to be a tool enabling indefinite survival. It is ideally comprised of a hiking backpack whose size and weight depend on how much load you can carry and the distance to be covered. As previously stated, you might have to use a vehicle (a car, motorcycle, bicycle, etc.) when evacuating. There are, however, many factors that could drive you to abandon your means of transport, including:

- impassable or congested roads;
- mechanical breakdown;
- the requisitioning of your vehicle;
- the final stretch of your journey towards your fallback point only being accessible on foot;
- etc.

BAGS

The ideal option is to have a waterproof hiking backpack in which you can place a trash bag to house the various items. This technique will guarantee outstanding waterproofness in the event that you have to cross a stream. Your backpack can then act as a life ring to help you swim. Having a bag/backpack with several pockets will also allow you to divide the objects you are carrying into several groupings to facilitate access to the different items.

When filling the bag/backpack, it is important that you arrange your items in a particular order. The things that you will not need urgently will thus be left at the bottom of the bag (e.g.: a change of clothes). Other items such as first aid kits, by contrast, must be readily accessible.

CLOTHING OPTIONS

Adequate clothes represent your first line of defence against the cold (hypothermia) and certain kinds of external injuries (cuts and abrasions, for example). Make waterproof clothes and shoes your first choice in order to remain dry.

- 1 scarf/shemagh/bandana/cloth/handkerchief/etc.;
- 1 pair of socks (made of wool or any other material);
- 1 change of underwear;
- 1 T-shirt;
- 1 head protection item (hat, balaclava, woolly hat, etc.);
- 1 hand protection item (gloves to fight the cold);
- 1 set of thermal underwear (top and bottom) against the cold.

On your person:

- 1 pair of sturdy hiking or walking shoes, ideally made of leather and waterproof (Goretex®);
- 1 rugged pair of trousers (a pair of jeans can also be a good choice if they are not too tight, since they are plain and can be worn both in town and the countryside);
- 1 pair of work gloves.

If you are forced to move about in cold weather conditions, it is wise to opt for a so-called 'three-layer' outfit and avoid any materials that retain perspiration (do not wear cotton clothing).

- **First layer (comfort)**
  - 1 long-sleeved T-shirt (thermal or merino wool) to keep your skin dry;
- **Second layer (intermediate or insulating)**
  - 1 hooded fleece jacket or the like to transmit moisture to the next outer layer and keep body heat unchanged. This layer can be doubled in case of extreme cold;
- **Third layer (protection)**
  - Warm and breathable clothes (Gore-Tex®, Windstopper, etc.) that also offer protection against the rain, wind and various injuries (cuts, scratches, etc.).

CLOTHES FOR 'SLEEPING ROUGH'

Depending on the place, environment and climate, sleeping outside may turn out to be easy and pleasant or, alternatively, very uncomfortable or even dangerous. If you evacuate during the winter, one of your priorities will obviously be to keep your body temperature unchanged and avoid potentially fatal hypothermia. Be sure to have the following items at the very least:

- 1 protection item against the elements (rain, wind, sun, etc.): a tarpaulin sheet, tent or anything similar, the size of which must suit your needs. A tent is a good solution if you have children, especially when you consider the wide range of very light and compact models available today;

- 4 black and thick 120-litre (or, better yet, 200-litre) rubbish bags. These bags can be used for the construction of shelters or the harvesting of rainwater, or as insulation blankets (when filled with fallen leaves or newsprint paper), etc.;
- 1 sleeping bag or, failing that, a 100% wool blanket;
- 1 Bivvy Sack™ by Adventure Medical Kits® or, if unavailable, an aluminium survival blanket.

THE 'WATER' STRATEGY

The purpose of this strategy is to allow you to recover, store, sanitise and drink water during evacuation. Be sure to have:

- 1 box of coffee filters to pre-filter the collected water;
- 1 mechanical filtration system: Aquamira® Frontier Pro™ emergency straw filter (with an autonomy of 200 litres). For large families: 1 MSR® Miniworks EX™ pump filter (autonomy: 2,000 litres!);
- 25 Micropur® pellets for the chemical purification of water (1 tablet/litre);
- 1 army surplus-type aluminium gourd, with its own quarter-litre to boil water, soup or any other freeze-dried beverage;
- Your water transport and storage capacity must be at least 3 litres per person. Depending on your preferences, you can for example use 'Nalgene'-type bottles with a wide neck (1 litre) (one of them can also serve as a collection container/drip pan, and therefore as a 'dirty bottle' to avoid contaminating all the systems), or Collapsible and light bottles of the Platypus® type (1, 1.5 or 2 litres) and/or MSR water bags (2, 4 or 6 litres) for transport and storage.

Note: During your preparation process, and particularly when examining special places that may turn out to be useful in case of evacuation, do think of cemeteries. Indeed, these always have their own water point!

THE 'FOOD' STRATEGY

The purpose of the food strategy is to provide you with sufficient energy during an evacuation process of about seventy-two hours, while enabling you to remain as light and low-key as possible. Since evacuating will require a lot of effort (forced march, cold weather conditions, heavy weights, etc.), a basic minimum of 2,000 calories per adult per day should be provided for.

It is not important to have a refined sort of diet, but one should seek rather to have foods that are both energy efficient and space-saving. Freeze-dried meals are ideal in this regard: indeed, they are light, do not require a lot of water to be prepared and offer a highly adequate caloric intake in view of their size and weight (more than 2,000 calories for a weight of less than 500 g).

- Freeze-dried meals (e.g. Backpacker's Pantry);
- Freeze-dried soups.

To complement them:

- Dried meat;
- Dried/dry fruit (nuts, hazelnuts, almonds, apricots, and so on);
- A tube of honey, a packet of peanut butter;
- Cereal/energy/protein bars.

THE 'FIRE' SYSTEM

Fire is absolutely crucial for survival. It allows us to regulate the surrounding temperature, boil water, prepare food, make tools, illuminate dark places, reassure people and even cauterise a wound… Ouch!

Make sure you have:

- **Three means of ignition:**
    - Firesteels (i.e. fire strikers, known as 'rifles' in good French);
    - Lighters;
    - Storm-proof matches.

    (Considering their small size, it is strongly recommended that you take all three.)

- 1 barbecue firelighter — or cotton wool that can be moistened with Vaseline (take 1 tube of the latter) — to increase efficiency when starting a fire.

THE 'LIGHT' SYSTEM

Although this could be perceived as a luxury (since our ancestors managed to survive without a flashlight), having the means to illuminate your surroundings in case of evacuation remains a significant asset. Many situations may require some brightness or, at least, they can be managed in a much more effective way in the presence of light. Examples include administering care in poor light conditions, advancing through a tunnel, looking for an object in the dark, etc. There are many kinds of lamps for such a purpose. If possible, opt for those that are waterproof and diode-based (and thus characterised by low power consumption).

- **Headlamps:** it is preferable to choose those that can produce white and red light and are thus less conspicuous (e.g. Petzl's Tikka XP3).

- **Torches:** their qualities have evolved enormously in recent years. It is now easy to find powerful and low-consumption models that are also space-saving (Fenix E35, for instance). Some, like Frendo's TR 150, can be recharged through a USB port.

- **Electric lamps** that can be hung somewhere or placed in position (e.g. Black Diamond MOJI 100), or a bulb-lamp (of the ambient light type) for your children. This kind of lighting is ideal when you are 'travelling' as a family. Indeed, one single lamp can provide a small group with sufficient light.

- **Waterproof dyno torches** (e.g. Topoplastic) — thanks to this accessory, you are guaranteed light whenever you need some.

Note: Opting for a lamp with a compact design is a definite advantage, since it not only affects the weight of the lamp itself, but also that of spare batteries. Having a mini-lamp (e.g. Inova Microlight) on your key ring at all times can occasionally be useful.

THE 'ADMINISTRATIVE' SYSTEM

This system is important because it allows you to orient yourself, to communicate, to obtain information, to bribe, but also to identify yourself when necessary.

Be sure to have:

- The keys to your stop-off point;
- Road maps (of your country, region, etc.);
- Topographic maps of the city/ region/stop-off points;
- Maps of the underground or railway station;
- A detailed map of the city;
- Special maps of places such as catacombs, sewers and the like;
- 1 compass;
- 1 small battery-operated or dynamo-powered AM/FM radio;
- Your identity documents — birth certificate, passport, identity card, driver's license, family record book, etc. These must be kept in a sealed envelope;

- 1 small notepad with a pen/pencil;
- Your mobile phone — mobiles have now become essential tools. Having one allows you to instantly contact your loved ones or get in touch with emergency services. In the case of a smartphone, you can also connect to the world's largest database, namely the Internet, and thus obtain the latest information, maps, and so on. Be careful, however, not to rely entirely on your phone, as it could suffer irreparable damage or the network might be inoperative;
- If you are not travelling alone, a pair of walkie-talkies may prove to be an appropriate choice to maintain your communication capabilities should you get separated. To ensure that you are able to communicate over greater distances, you could go for a pair of walkie-talkies such as Motorola's T40 and T80 models (which are water resistant and offer a range of four to five km and ten km respectively);
- Carrying cash (banknotes) or metal (gold/silver) coins is the best way to purchase equipment or goods when necessary. On some occasions, they can also be used to bribe people. History has shown that even in those countries most affected by crises, gold and silver coins have continued to be used as transaction currencies;
- Gold/silver coins (to be resorted to only in the event of a major crisis).

THE 'ENERGY' SYSTEM

Keeping a low profile can sometimes be worthwhile. There are circumstances in which it is out of the question for you to light a big fire. A small camping gas stove is the best way to boil water or cook your meals without being spotted at a distance of up to ten kilometres. There is no need to have thirty gas cartridges, as one will suffice for a period of seventy-two hours.

- 1 hotplate/stove — MSR's PocketRocket™, for instance, is a compact, lightweight and inexpensive option;
- 1 USB/smartphone dynamo charger — it is very useful to recharge your phone when necessary during the evacuation. Opting for a USB-rechargeable lamp will double the benefit.

THE 'HYGIENE AND CARE' SYSTEM

The ideal solution is to have a first-aid kit at your disposal, with the following contents:

- Skin antiseptics (spray and wipes);
- A box of adhesive bandages of various sizes;
- A box of sterile compresses (10 × 10 cm);
- Stretchy crepe bandages;
- Adhesive plasters;
- Self-adhesive elastic bandages (for strapping);
- Instant cold compresses (e.g. Instant Cold Pack);
- Disposable gloves;
- A pair of round-tipped scissors;
- Tweezers;
- Some hydroalcoholic gel (to disinfect the hands);
- An emergency haemostatic cushion (in case of haemorrhaging);
- A painkiller and antipyretic (e.g. paracetamol);
- Your usual anti-inflammatory medication (e.g. ibuprofen);
- An ointment to treat insect bites/stings;
- An ointment to treat burns (e.g. Biafine);
- An antihistamine to fight allergies (e.g. Humex Allergy Cetirizine).

As regards the hygiene kit:

- Soap;
- A toothbrush, toothpaste, dental floss;
- A disinfecting gel/sanitiser;
- Baby cleansing cloths;
- A mini sewing kit.

THE 'TOOLS' SYSTEM

The tools system's purpose is to enable you to influence your immediate environment as efficiently as possible. Having a decent fixed-blade knife and some string is a minimal requirement.

- A fixed-blade knife;
- Some string.

Depending on your load capacity and destination, you can also take other accessories with you:

- An adhesive roll made of waterproof plastic cloth (duct tape);
- A multi-purpose knife including a screwdriver (of the Leatherman, Gerber, Decathlon type, etc.);
- A wire cutter;
- Some wire;
- A wire saw of the Commando type;
- Some kind of machete (an item that can also be regarded as part of one's defence system);
- Binoculars.

In cases where you can carry out the evacuation by means of a motorised vehicle, other materials can be added to the list, including:

- A sledgehammer;
- An axe or hatchet;
- A pry bar;
- Large pliers.

THE 'DEFENCE' SYSTEM

According to prevalent legislations and within the framework of legitimate self-defence, try to use every trump card to your advantage. Different weapons will thus be allowed, depending on the country you reside in. Chaotic situations can quickly degenerate into panic and some individuals or groups may resort to violence, especially if the authorities are overwhelmed or absent. In case of evacuation, bear in mind that the use of weapons is your last resort. It is much better to avoid being seen than to get involved in a confrontation.

THE 'MENTAL/EMOTIONAL' ASPECT

Remember to pack a small item to help you remain determined. It can be a lucky charm, a talisman, a religious necklace or pendant, a picture of your family, or any object of sentimental value… A 'cuddly toy' is not to be overlooked if you have small children.

Undressing/Decontamination Kits

In order to carry out basic decontamination in the best possible conditions, the following items can be included in your kit:

- 1 pair of 'Mapa'[1] gloves;
- 1 box of disposable gloves (preferably made of nitrile);
- 1 set of 'paper' masks of the FFP 1, 2 or 3 type;
- A few pairs of splash goggles;
- 1 pair of boots;

1 AN: http://www.mapa-pro.fr.

- 1 set of type 5/6 protective suits[2] or, failing that, a disposable paint coverall;
- Several large rubbish bags;
- Some sponges or cleansing cloths;
- 1 solar shower with a water bag (to be hung);
- 1 garden sprayer — preferably worn on the back and with a minimum capacity of 12 litres (e.g. Tecnoma's Pulsar 1200) — or, if unavailable, a garden hose connected to a water supply;
- 1 set of towels for drying purposes;
- 1 set of spare clothes or tracksuits for each family member;
- Some gentle shampoo (ideally in the shape of small samples);
- Liquid Marseille soap;
- Several bottles (2.6%) or, if unavailable, packets (36°) of bleach.

Some items such as boots, goggles, a sprayer, etc. are meant to be used by a potential partner that assists you in the decontamination process — the sprayer can, of course, also be used by a person carrying out the process on their own: after self-decontamination, the person uses this piece of equipment to spray the area with diluted bleach. If your budget allows it, you can also opt for items or materials that enable you to decontaminate/neutralise the effects of acid and base substances. You are, however, more likely to use them when dealing with problems that relate to household products (hydrochloric acid contained in descaling agents or chemical strippers, soda lye, etc.) or other substances (battery acid, etc.) than during an actual 'CBRN attack'.

- a 100-ml Diphoterine® spray (to be used on the skin);
- Diphoterine® eyewash (to cleanse the eyes).

2 AN: See '*CBRN Protection Equipment*'.

2. Protection Equipment

The tiles that protect from the rain have
all been laid in good weather.

— Chinese proverb

Personal Protective Equipment (PPE) is essential if you are going to enter, or escape from, a contaminated or toxic area. There are two main categories in this regard: items used for the protection of the respiratory tract and those providing protection for the skin and body.

Respiratory Protection

Its role is not only to prevent harmful substances from entering your body when you breathe, but also to prevent any and all ingestion. Since each CBRN agent has its own distinctive properties, it is clear that some types of respiratory protection will be better suited to certain uses, while some will be of no use at all. There are many mask categories on the market. The masks can basically be grouped according to their shape and the type of air supply mechanism involved (filtering cartridges, assisted ventilation, pressurised bottles, etc.):

TYPE OF MASK	PICTURE	REMARKS
'Dust' Mask	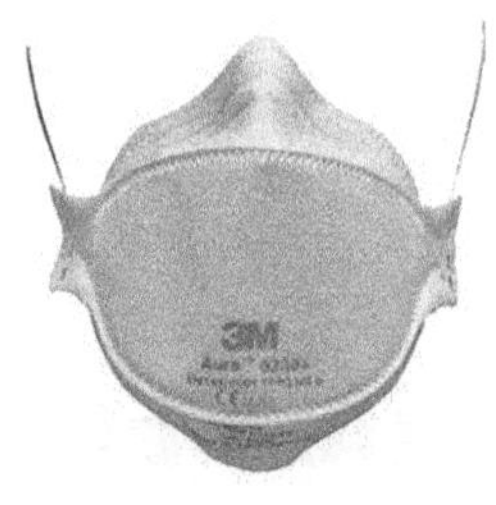	▪ Does not protect the eyes and the face. ▪ Lacks efficacy with regard to chemical threats.
Half-mask dual cartridge respirator		▪ Does not protect the eyes and the face. ▪ Protection depends on the type of filter used.
Full-face dual cartridge respirator		▪ Uncomfortable (compared to other types of masks). ▪ Protection depends on the type of filter used.
Assisted ventilation protective apparatus		▪ A turbine connected to a filter pumps air through the mask.

TYPE OF MASK	PICTURE	REMARKS
Self-contained breathing apparatus		▪ One inhales air from pressurised tanks (as in the case of SCUBA diving).

We will now focus on the two types of masks that seem most useful for any individual facing a CBRN threat, namely the 'dust' mask and the full-face dual cartridge respirator.

A. 'Dust' Masks

Technically, these are FFPs (Filtering facepieces) that cover the mouth, nose and part of the chin. They are an excellent solution with regard to biological risks and radioactive dust. The various available models are classified according to their efficiency, as defined by European standards (EN 141 and EN 14 387):

CLASS	FILTRATION	LEAKAGE RATE	USES
FFP1	Stops at least 80% of all particles	Less than 22%	Particles, smoke/ fumes, fog…
FFP2	Stops at least 94% of all particles	Less than 8%	
FFP3	Stops at least 99% of all particles	Less than 2%	

The FFP3 mask type is recommended in case of radioactive contamination.

The available models may or may not comprise a valve and can either be semi-rigid or more pliable and elastic.

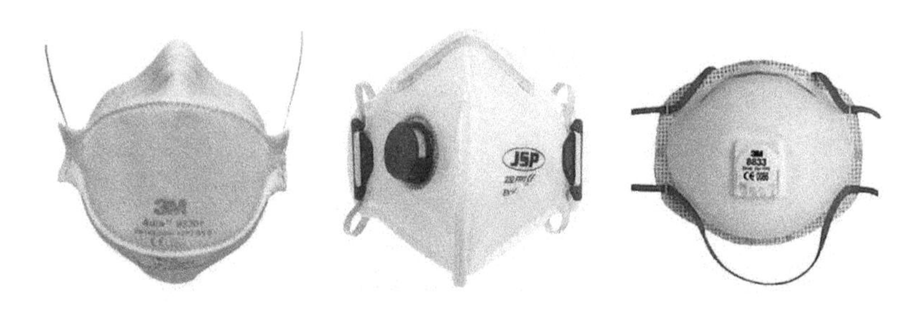

The Aura FFP3 mask (3M)	The FFP3 pliable Olympus (JSP)	The FFP3 Cool Flow (3M)
Pliable, valveless	Pliable, comprising a valve	Semi-rigid, comprising a valve

There is a wide range of suppliers for this kind of equipment. In addition to the three possible classes (FFP1, 2 or 3), the appropriate abbreviation (NR, i.e. 'Non-Reusable', or R, meaning 'Reusable') must also be provided.

Note: The main function of a medical or surgical mask is to minimise the number of droplets that reach the surroundings and environment. It offers a lower degree of protection compared to FFP masks and remains unsuitable for CBRN use, especially in the case of radioactive contamination.

B. Full-Face Masks

In France, both purchase and ownership of full-face military gas masks that offer protection against CBRN agents are prohibited, since such masks are considered to be third-class weapons of war under the old legislation and A2-category warfare weapons/equipment according to the legislation which entered into force on 30 July 2013 (Decree number 2013–700).[1] This is not the case in Switzerland, for instance,

1 AN: Decree number 2013–700, dated 30 July 2013, states: 'The warfare equipment and weapons that one is prohibited from acquiring and possessing and which fall into Category A are: (…) Item 2: All weapons pertaining to warfare materials; equipment intended to be carried or used in combat; firearms; and

nor in other European countries, where such equipment is not subject to any sort of ban.

Nevertheless, there is a certain range of equipment that can be used by everyone, without any legal restrictions. Such masks are intended for private or industrial use.

Most full-face masks available to the general population offer similar characteristics. Your choice must therefore take into account:

- the intended use (how long it will be worn, the type of risk involved, etc.);
- the price (of the mask and cartridges);
- the different filtration systems available;
- the visual field;
- the comfort level (which depends on the mask's shape and material);
- the earphone/microphone assembly (which improves speech transmission);
- the presence of a drinking mechanism (connected to a camel bag, for instance), etc.

Bear in mind that some masks are available in different sizes! The purpose of a model that matches your facial shape is to limit leaks and thus offer better protection. This can be important for people with very fine or strong facial features.

---

protective equipment against chemical warfare gases are classified into category A2 as follows: 17° materials specially designed for military use, detection and **protection against biological or chemical agents as well as radiological hazards.**'

Advantage 3200 (MSA) **Advantage 3100 (MSA)** **Optifit (Honeywell)**

In general, screw filter cartridges (the Advantage 3100 and Optifit masks pictured above) offer more options in terms of filtration and performance. There are, of course, other models such as the SGE 400/3 (MSA), which is available in the United States, and many manufacturers have put up more or less equivalent products for sale (e.g. Dräger, Matisec, 3M, etc.). Various European standards govern the various possible protections and their level of efficiency.

<table>
<tr><th>COLOUR</th><th>TYPE</th><th>USE</th><th>CHARACTERISTICS</th><th>NORM</th></tr>
<tr><td>Brown</td><td>A</td><td>Organic gases and vapours
(Boiling point > 65 °C)</td><td rowspan="4">Maximal gas concentration not to be exceeded according to the cartridge class:
▪ Class 1: low
0.1 vol. % (1,000 ppm)
▪ Class 2: average
0.5 vol. % (5,000 ppm)
▪ Class 3: high
1.0 vol. % (10,000 ppm)</td><td rowspan="4">EN 141
(EN14387)</td></tr>
<tr><td>Grey</td><td>B</td><td>Inorganic gases and vapours (except CO) such as chlorine, H_2S, HCN…</td></tr>
<tr><td>Yellow</td><td>E</td><td>Sulphur dioxide and acid gases and vapours</td></tr>
<tr><td>Green</td><td>K</td><td>Ammonia and ammoniated organic derivatives</td></tr>
</table>

COLOUR	TYPE	USE	CHARACTERISTICS	NORM
White	P	Particles	▪ P1 class: low efficacy Filter stops at least 80% of aerosols ▪ P2 class: average efficiency Filter stops at least 94% of aerosols ▪ P3 class: high efficacy Filter stops at least 99.95% of aerosols	**EN 143**
Brown	AX	Organic gases and vapours (boiling point <65 ° C) = low boiling point substances (groups 1 and 2)	▪ Gr.1 : 100 ml/m³, max. 40 min. ▪ Gr.1 : 500 ml/m³, max. 20 min. ▪ Gr.2 : 1000 ml/m³, max. 60 min. ▪ Gr.2 : 5000 ml/m³, max. 20 min.	**EN 371**
Blue	NO	Nitrogen oxides NO, NO2, NOx	Maximal time of use: 20 minutes	**EN 141** (EN 14 387)
Red	Hg	Mercury vapours	Maximal time of use: 50 hours	**EN 141** (EN 14 387)
Black	CO	Carbon monoxide	Local legislations	**DIN 3181**
Orange	Reactor P3	Radioactive iodine	Local legislations	**DIN 3181**
Violet	SX	Specific compounds (gases and vapours)	Only classified in one type and one class (SX)	**EN 372**

Bodily Protection

Protective Clothing

WARNING! These suits do not protect against ionising radiation (external irradiation)

Their purpose is to protect your body from external impact, especially from corrosive, toxic or infectious agents, and to stop any and all contamination from entering your body or coming into contact with it.

Many models are available on the market, but the safest choice is any suit that complies with European regulations for 'Personal Protective Equipment' (PPE) or US regulations (OSHA, NFPA) in the case of people residing in the United States.

The chart below categorises the types according to the different European standards and presents the various options relating to biological risks and radiological contamination.

TYPES	DESCRIPTION	PICTOGRAM	PROTECTION LEVEL
Type 1	Gas-tight		
Type 2	Limited gas-tightness		
Type 3	Liquid-tight		
Type 4	Aerosol-tight		
Type 5	Particle-tight		
Type 6	Limited tightness against splashes		
Radioactive contamination	Protection against radioactive contamination in the shape of particles.		

TYPES	DESCRIPTION	PICTOGRAM	PROTECTION LEVEL
Biological agents	Protection against contaminated liquids		

Type 1 suits correspond to the heavy and cumbersome CBRN suits that can be seen in some disaster films. These models are, of course, prohibitively expensive and do not suit the needs of individuals at all.

For private use, it seems reasonable to limit oneself to types 5 or 6 (which may also include further options against biological agents and radioactive contamination) or even to type 3 in case of an identified chemical risk.

The chart below shows some examples of suits:

TyvekClassicXpert suit (DuPont)		Type 5 + type 6 + radioactive contamination + biological risk
Suit 4535 (3M)		Type 5 + type 6 + radioactive contamination

The Proshield 10 suit
(DuPont)

Type 5 + type 6

Accessories

In addition to the suit, it is imperative for you to protect your extremities (feet and hands) and eyes, which is why the 'accessories' category includes shoes, gloves and goggles.

- **Shoes**

  In case of contamination, one solution is to put on overboots. For outdoor use (in a forest, on gravel, etc.), however, this method is not ideal. Indeed, the basic models offer very little resistance to abrasion and tearing; as for the 'sturdiest' ones, they are actually a financial black hole, as they must be discarded after each use (if contaminated).

  Although there is no magic bullet, there are several paths you might take:

1. If you are merely attempting to evacuate an area that has become contaminated or toxic, you can opt for waterproof walking shoes. Although they are not intended to provide chemical protection, they will still limit any such effects and isolate your feet from any biological agents and radioactive dust, while simultaneously allowing you to easily progress through the wilderness when necessary. In most cases, they will unfortunately have to be thrown away once they have been contaminated.

2. If you have to travel back and forth across a contaminated area (from your long-term autonomous base, for instance), a pair of quality rubber boots (or, ideally, any material that can resist chemicals) is a good alternative. All you have to do is clean/decontaminate them after each use, which will enable you to wear them repeatedly.

- **Gloves**

  As in the case of boots, it is important for you to know what your gloves are intended for — crossing a contaminated area, driving, climbing, etc. For tasks that do not require resistance to tearing or puncturing, the ideal option is to have disposable nitrile gloves. You can wear them directly on your hands or on another pair of gloves, and they can be replaced whenever necessary (if you think you have touched something contaminated, for example). Before entering a non-contaminated location, all you have to do is throw them away. If you intend to perform a more demanding task, you will need tougher gloves, which will probably have to be discarded afterwards. Thick nitrile gloves or, to a lesser extent, plain dish-washing gloves are a good enough alternative.

Box of 100 disposable nitrile gloves		10€ a box
Nitrile work gloves	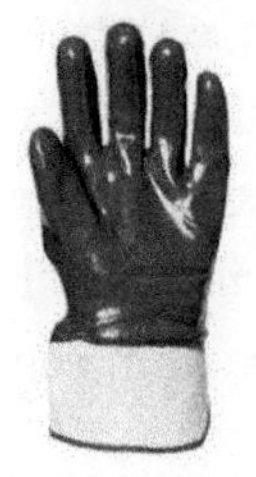	25–30€ for each ten pairs

Nitrile-coated cut-resistant gloves		8€ a pair

- ## Safety goggles

The eyes can act as entry points for potential contamination. The ideal solution is to use goggles whose contours stick to your face as closely as possible.

SAFETY GOGGLES	PICTURE	CHARACTERISTICS
Hobygam LM60580		▪ Low cost ▪ Price: 3€
Dexter AD60970		▪ Mechanical resistance ▪ Anti-fog ▪ Compact design ▪ Price: 16€
3M 2890		▪ Can be used as OTG (over-the-glasses) goggles ▪ Impermeable version for better protection against gases and liquids ▪ Limited protection against chemicals ▪ Price: 19€

3. Detection Equipment

To foresee things is to both ponder the future and shape it; foresight is, in itself, synonymous with action.

— French musician François-Joseph-Marie Fayolle (1774–1852)

People's fear of an impending economic collapse, their dread of nuclear attacks, their mistrust of public authorities or perhaps the mere fact of residing in a danger zone may lead some to invest in detection equipment. Owning such equipment could indeed turn out to be advantageous should a dangerous situation develop critically, or might serve simply to ensure that the alert is raised. Furthermore, possessing certain kinds of detectors would be prudent in special cases such as the exploration of caves or other underground sites.

Be careful, however, as many of these items are very costly and their misuse or misinterpretation could lead to erroneous conclusions that would endanger you. Such devices are basically meant for people with a large budget or those that have identified the presence of a specific risk. The following list is not an exhaustive one, and although the suggested pieces of equipment are a safe bet, other brands could also offer items with similar characteristics.

Radioactivity Detectors

Given the plethora of detectors available for purchase, making a choice can sometimes be complicated, and the item could indeed fail to meet your expectations. Drawing up your budget and defining your needs (i.e. what you would like to use the device for) must be your main priority, even if other factors must also be taken into account:

- The item's size/weight/ergonomics (compact, equipped with sensors, etc.);
- The detection technology involved (a Geiger counter, scintillation, semiconductors, etc.);
- Sensitivity (the sensitivity of most state-of-the-art devices is more than sufficient for personal, individual use);
- The operational range (some devices may overload quite quickly and stop working as soon as dose rates reach a certain threshold, leading one to believe that everything is alright when, in fact, the very opposite is true);
- The radiation's energy ranges (since some excessively low or high ones are not taken into account);
- The response speed (the faster it is, of course, the better);
- The nature of the elements to be detected (gas, dust, etc.);

The following (non-exhaustive) examples are intended to help you make a choice, while taking into account the criteria listed above, including the detected energy range and the sensitivity/operational range.

Neutron Detection

Since such radiation is hardly present in nature at all, ***it does not seem appropriate to invest*** in a detector intended for this type of purpose. If neutrons were to emerge as a result of human action, they would either not remain there for too long (e.g. an atomic explosion) or be accompanied by other kinds of easily detectable radiation (the gamma rays of plutonium[1] waste, for instance).

The Detection of Human/Object Contamination

Contamination detectors are known as *contaminometers.* These devices are designed to detect alpha, beta and gamma radiation. Basic models do not differentiate between the various types of radiation, since *their purpose is merely to indicate the presence or absence of radioactive contamination.* Most of the time, their measurement unit is in counts per second (some also offer the Bq/cm²). In addition to having a digital display, most can produce sounds (such as crackling, etc.) to help locate the contamination source.

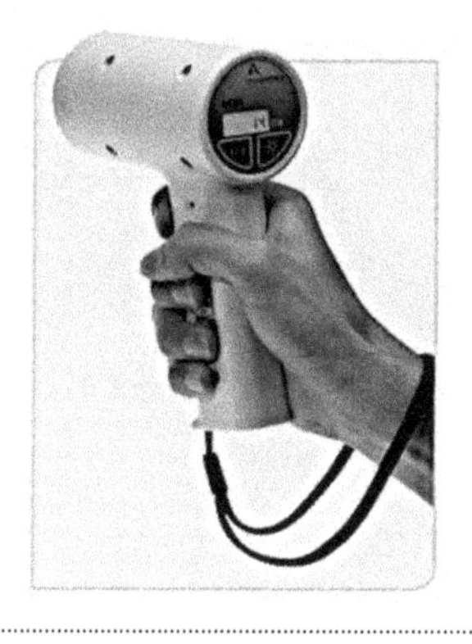

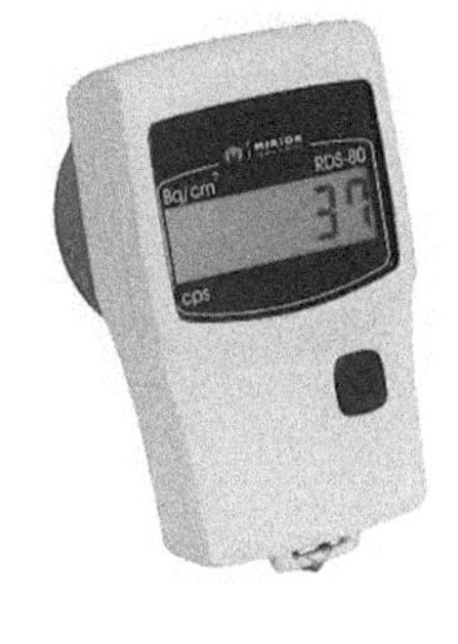

MCB2 (Canberra) RDS-80 (Mirion)

1 AN: The gamma rays that stem from the different plutonium isotopes are characterised by their low energy and small number. On the other hand, those released by its descendants (which are present in the waste) are generally much easier to detect.

The Detection of Atmospheric Contamination

Although well-made, the detectors presented above are not meant to measure very low airborne contaminations. They would not, for instance, allow you to detect a radioactive cloud in Europe if a new Fukushima-like event occurred. For this kind of capacity, it is usually necessary to use special components that draw in and filter the air. Contamination thus slowly accumulates on a strainer and is measured in real time. Another option is to have an atmospheric sampling device and then perform the measurements yourself using a radioactivity detector. This type of equipment is very expensive and requires regular maintenance. It is mostly reserved for use by technicians working in the field of radioactivity or for specialised services. The following chart shows two examples of atmospheric contamination detection monitors:

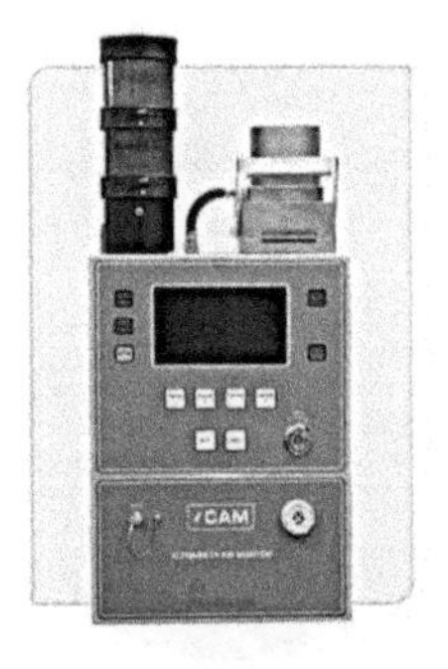

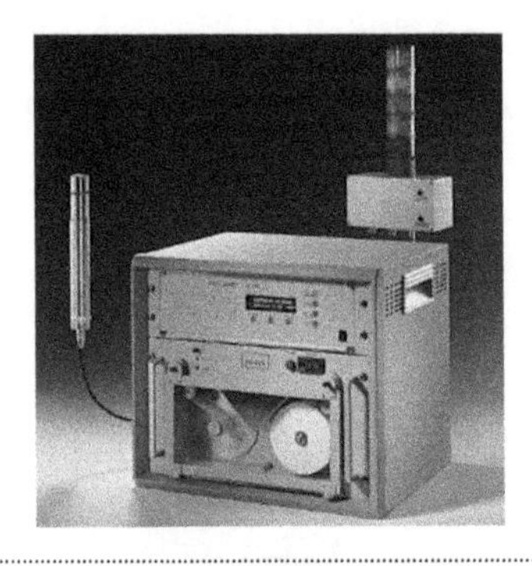

iCam (Canberra)	BAI 9128 (Berthold)

Dosimeters (for Gamma—or Sometimes Beta—Radiation)

This type of equipment is used in the field of radiation protection to monitor the doses received by staff members over time and to alert them when a dose or dose rate has been exceeded. Its presence constitutes a legal obligation for anyone working in the nuclear field. In a context of personal/individual use, the role of a dosimeter is limited to that of an alert device. Simply put, you take it with you and forget about it. Should there be an increase in radioactivity levels around you, it will emit a beeping sound. By checking the monitor, you will then be able to ascertain what has triggered the alarm: an excessively high dose rate or an above-normal cumulative dose.

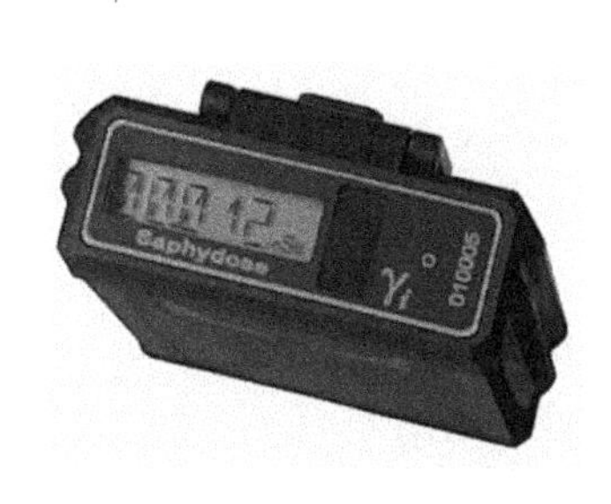

Saphydose (Saphymo)

PM 1208M (Polimaster)

Radiameters (Gamma Radiation Dose and Dose Rate Indicators)

These devices allow one to measure gamma radiation 'dangerousness' in real time. They usually display their measurements in either grays or sieverts (or their submultiples, rather — otherwise, you're in big trouble!). This type of detector is generally quite easy to use.

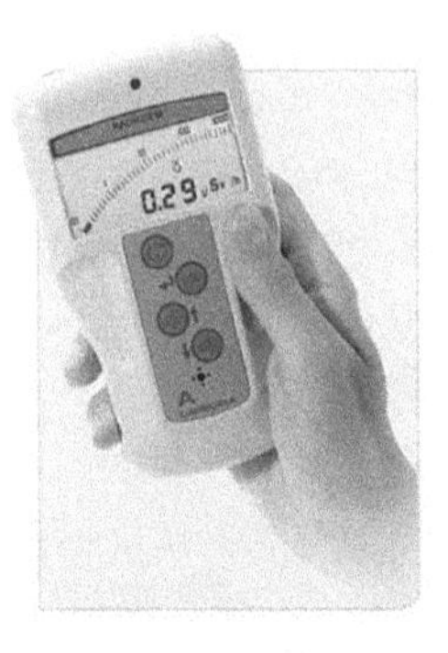

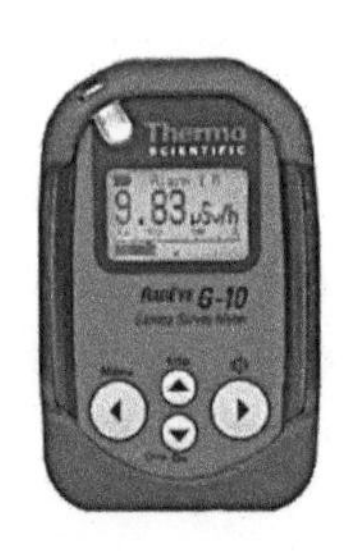

Radiagem 2000 (Canberra) RadEye G10 (Thermo Scientific)

Phone Applications and Accessories (for Gamma Radiation)

There are two ways for you to convert your smartphone into a radioactivity detector:

- **Option 1:** Install an *application* that will transform the camera into a gamma ray sensor (the lens must be covered to prevent the passage of any and all light).
- **Option 2:** Add a genuine *detector* to your phone.

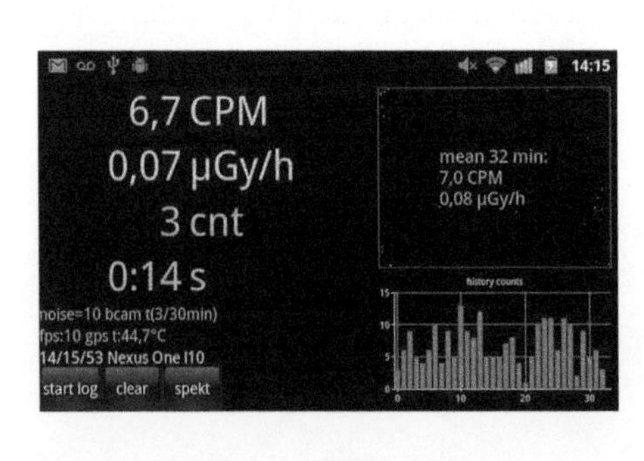

Option 1

Radioactivity Counter (Android)

Option 2

PM 1904 (Polimaster)

Note: Phone apps cannot provide you with the same accuracy as a specially made detector. You thus have to be careful with the results. They can, however, serve as a low-cost warning tool. There are, of course, other alternatives to those presented here, but you should always be careful with regard to your choice of application, as some are nothing more than scams.

Spectrometry (for Gamma Radiation)

This is **a device capable of identifying radioactive elements** through gamma radiation analysis. There is an entire plethora of models that mainly differ in terms of the detector's size and the type of technology used (NaI, Germanium, etc.). The price range stretches from a few hundred euros to tens of thousands of euros. Most models can also be used as radiameters and display sub-multiple values in grays or sieverts. A small number can also provide you with optional neutron detection.

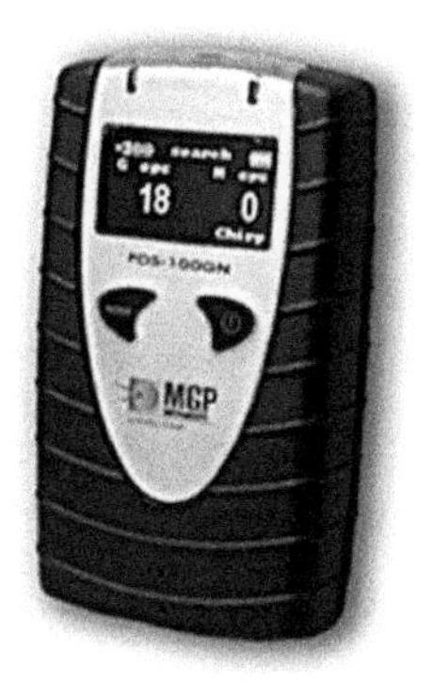

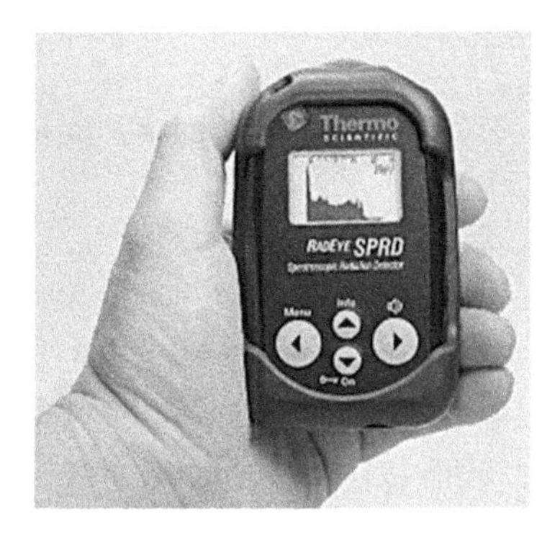

PDS 100 G id (Mirion)

RadEye™ SPRD (Thermo Scientific)

Chemical Agent Detectors

There are millions of chemical compounds. Attempting to detect them all in real time is quite a challenge, even for professionals equipped with the latest equipment. Hundreds of different devices are, however, available on the market. These detectors make use of a large variety of technologies ranging from mass spectrometry to electrochemical cells, through flame spectrometry, ion mobility, etc. Private individuals lack the necessary knowledge and budget to be able to afford such detectors, many of which cost more than 10,000 euros, with some blithely priced at more than 100,000 euros... The following list thus focuses on simple and affordable items that can be used for the detection of credible threats such as the presence of toxic gases.

(Colorimetric) Detection Tubes

This is one of the oldest technologies. Using a pump, you draw in air through a reactive tube. A chemical reaction occurs inside, which either leads to a colour change or not. With regard to its use by a private individual, the main drawbacks of this technique are the following:

- The tubes are disposable (and you must therefore have several of them);
- The tubes are specific to a given chemical type or agent.[2] If, for example, you have a tube that detects chlorine, it will usually remain unresponsive to other products.[3] You thus need as many different tubes as there are gases you seek to detect.

---

2 AN: Tubes with a wider spectrum can be used, but the result remains too imprecise. To identify the gas, one would have to resort to different tubes successively.

3 AN: The situation on the ground is a little more complex than that, since there may be positive reactions to compounds containing chlorine or even to chemical elements similar to the latter.

- There is **no (continuous) response in real time**, but only when you take a measurement.

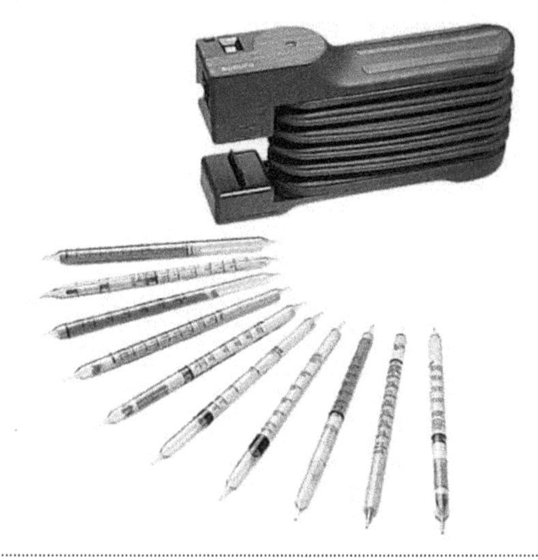

A sampling pump and some colorimetric tubes (Dräger®)

Single-Gas Detectors

As suggested by their name, these are small devices whose purpose is to detect a single gas. In most cases, however, it is possible to change the sensor and thus choose to detect a different gas. These detectors can perform a measurement in real time and therefore represent a decent means of alert and detection (when the detected gas is actually the source of the threat). The most important ones for us to detect are, for example, carbon monoxide (CO), chlorine (Cl_2), ammonia (NH_3), hydrogen cyanide (HCN) and hydrogen sulphide (H_2S). Depending on your needs and the intended use, you can also opt for a *waterproof* model such as the Pac 7000.

ToxiRAE® II (RAE System)

Available sensors for: H2S, CO, O2, NH3, Cl2, ClO2, HCN, NO, NO2, PH3, SO2

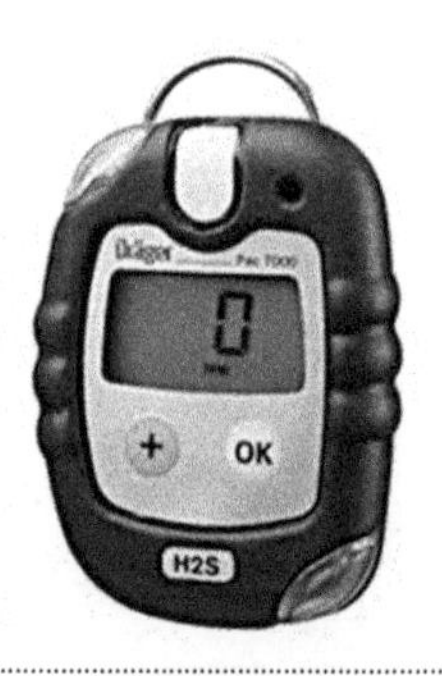

Pac® 7000 (Dräger)

Available sensors for: H2S, CO, CO2, NH3, Cl2, HCN, NO, NO2, SO2

Multigas Detectors

These detectors use the same type of technology (i.e. electrochemical or infrared cells) as the single-gas types we have just covered. The main difference is, however, that these devices can measure several gases simultaneously. This type of detector is therefore more versatile. Depending on the model in question, it can measure a variety of things, such as:

- the oxygen level (O2): this capacity makes the device particularly useful in a confined environment (when under the ground, confined to your home, etc.);
- the concentration of carbon monoxide (CO): this is essential in the event of a fire;
- explosive gases and vapours (leakage of flammable products, etc.);
- the concentration of hydrogen sulphide (H2S), a deadly gas resulting from the decomposition of organic matter (sewers, etc.);

- the concentration of a specific toxic gas (chlorine, ammonia, and so on).

In the case of some detectors, the configuration is fixed and cannot be altered (many 'four-gas' devices, for instance). As for other types, the choices range from two to six detection cells. Depending on the type of use considered, one is advised to pay special attention to the detector's impact resistance, waterproofness and autonomy.

Note: Beware of the maintenance costs, as these devices must be checked and calibrated on a regular basis!

ALTAIR 2XT (MSA)	ALTAIR 4X (MSA)	X-am® 5600 (Dräger)
Two sensors to be chosen from the following: H2S, CO, NO2, SO2	Fixed configuration: CO, O2, H2S and explosive gases	Up to six detections: O2, H2S, CO, CO2, NH3, Cl2, HCN, NO, NO2, SO2, PH3, COCl2, explosive gases and organic vapours

Detection Systems

These devices enable a continuous and permanent sort of detection so as to potentially trigger a sound alarm and warning light. In anticipation of a CBRN incident, they should ideally be placed outside.

Although they can be plugged into the mains, most can run on batteries when required. Some models function wirelessly and can communicate with your home computer from a distance. You will thus receive, in real time, alarm notifications indicating the responsible agent (depending on the installed sensors) and an estimate of its concentration. Such equipment is, of course, quite expensive and many models are reserved for professional use. Two main types of devices should be considered:

- Those that comprise a removable detector (e.g.: X-zone 5000);
- Those that can truly be categorised as dedicated detection systems (e.g.: Safesite).

X-zone® 5000 (Dräger)

For use with detectors of the X-am series (X-am 5600, for instance)

The sensors used are those of the X-am detector: O2, H2S, CO, CO2, NH3, Cl2, HCN, NO, NO2, SO2, PH3, COCl2, explosive gases and organic vapours.

Safesite detector (MSA)

This detection system can comprise up to six sensors enabling one to detect a range of agents and concentrations from the following list:

- Chemical warfare agents (GA Tabun, GB Sarin, GD Soman, HD/HN-3 sulphurous and nitrogenous yperites);
- TICs (CO, H2S, Cl2, CLO2, SO2, NO2, NO2, HCN, HCl, PH3, AsH3, Br2, NH3);
- Oxygen concentration, flammable gas concentration, etc.

Biological Agent Detectors

The detection of biological agents is a real challenge, especially when one resorts to field equipment. Most devices use immunological techniques or DNA replication (*Polymerase Chain Reaction — PCR*) to enable the identification of a limited number of viruses, bacteria or toxins.

In addition to their high price, some detectors are not authorised for sale and are reserved for governmental departments or professionals (including hospitals, laboratories, etc.). The possession/use of biological agent identification devices is thus inadvisable in the case of private individuals.

4. CBRN Kits

One must be wary of engineers — what starts with a sewing machine ends with an atomic bomb.

— French writer and filmmaker Marcel Pagnol (1895–1974)

Given the wide range of scenarios and risks to be considered, the setting up of a CBRN kit is a difficult task. The cost and possible maintenance of such kits are essential elements to take into account, as is the environment in which you live (in the countryside, near a chemical plant or a nuclear power station) and the desired use (the evacuation of a given area or round trips from your long-term autonomous base). The set-up and contents of your CBRN kit can therefore vary considerably depending on your goals and budget. Your kit should, however, primarily be focused on respiratory protection.

Non-CBRN-Dedicated Protection Kits

In case you have not dedicated a part of your budget to such a purpose or do not want to invest in a CBRN kit, you will simply have to use some items you already own. Do not, of course, expect any protection against aggressive chemicals. This kind of kit should, however, allow you to limit internal and external contamination:

- For respiratory protection: a water-soaked (or urine-soaked[1]) handkerchief.
- For bodily protection:
  - K-way style of clothing/hooded windcheater;
  - rubber boots or waterproof walking shoes;
  - leather work gloves.
- Detection means: none.

'Basic' CBRN Kits

Such kits are useful when CBRN risks are not your top priority or when you want to limit your investments in CBRN equipment. As with the previous type of kit, you can use your own walking shoes or rubber boots, as well as your K-way jacket and other windcheaters. In contrast to the first kit, the protection of your respiratory tract, eyes and hands is somewhat improved in this case. The following list itemises the objects that can make up 'basic' CBRN kits. The different examples are listed for informational purposes only, so as to provide you with an idea and overview of the necessary budget. You are obviously free to choose the same type of equipment at the best possible price.

- **Fold flat disposable FFP 3 masks**

Fold flat valved FFP 3 mask
(SilverLine — 457043 series)
(5 euros/item)

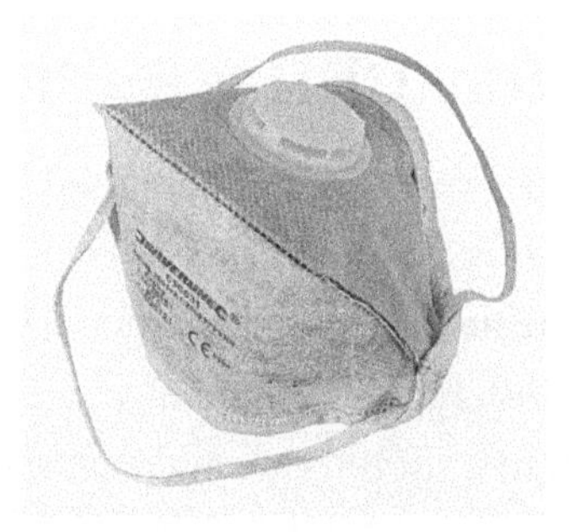

1 AN: This practice was used during World War I to provide people with limited protection against certain chemical agents.

Offers effective protection against biological agents and radioactive dusts. In addition to this, the thin layer of carbon it contains provides limited protection against certain chemical compounds.

▪ Safety goggles

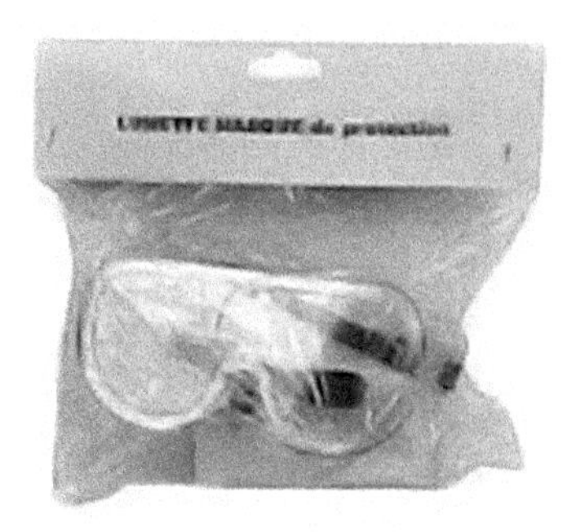

LM60580 (Hobygam)

(3 euros)

These goggles protect the eyes, which represent a potential entry route for toxins and contamination.

▪ Disposable *nitrile* gloves

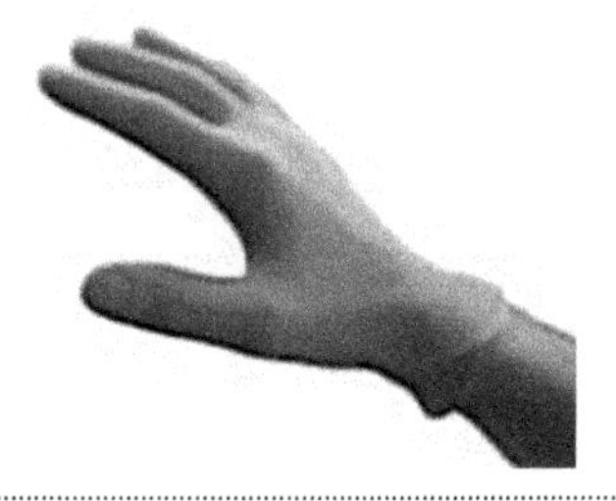

Various brands

(10 euros per 100 pairs)

These gloves are usually sold in boxes containing 100 pairs each. They offer effective protection against biological agents and radioactive dusts, as well as limited protection against chemical agents.

'Intermediary' CBRN Kits

▪ Type 5/6 protection suit

TyvekClassic Plus — Type 4B/5/6

Green

(10 euros)

These suits offer effective protection against both particles and splashes, and therefore against contamination. They also provide limited protection against certain chemicals.

▪ Protective half-masks

X-PLORE 3300 (DRÄGER)

(20 euros)

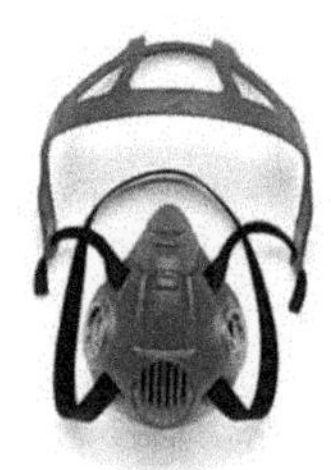

2 pieces of type ABEK1HgP3 filtering cartridges

(25 euros)

With its two cartridges, this mask is very effective against particles (P3) and provides moderately-efficient filtration in the case of many chemical components.

▪ Safety goggles

Model 2890 (3M)

(18 euros)

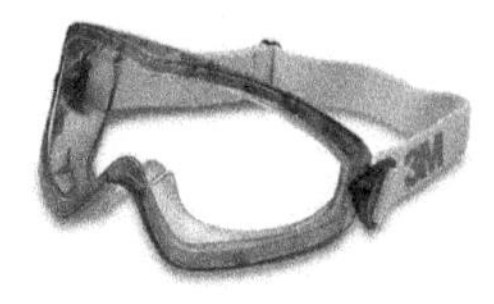

These goggles offer greater comfort and resistance than those included in the basic kit.

▪ Protective gloves

Disposable nitrile gloves

(10 euros per 100 pairs)

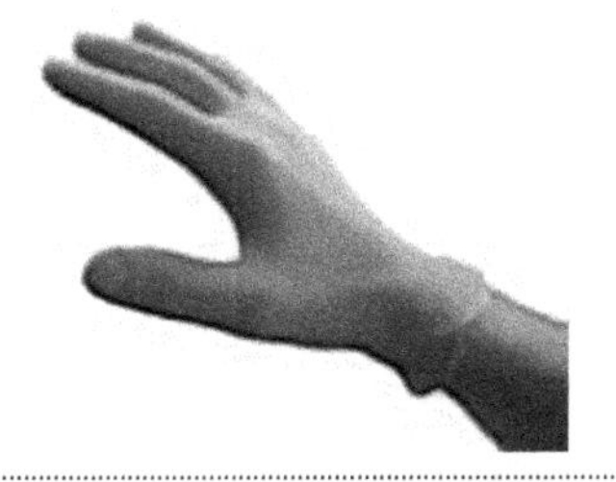

These gloves are usually sold in boxes containing 100 pairs each. They offer effective protection against biological agents and radioactive dusts, as well as limited protection against chemical agents.

Ultranitril 492 gloves (Mapa)

(3 euros a pair)

These gloves are more or less resistant to physical impact and damage (perforation, abrasion, etc.) and offer decent protection against toxic chemical agents.

A set whose contents approach those of an 'Intermediate' CBRN kit can also be obtained at a lower cost through the purchase of a protection kit against phytosanitary products. The one sold by Protect Nord for about 60 euros, for instance, includes:

- **1 green TYVEK suit** (type 4B-5-6), XL-sized
- **1 protective half mask** (X-PLORE 3300)
- **1 set of 2 A2P3 filters**
- **1 pair of Nitri-solve 330mm chemical protection gloves**
- **1 pair of Chemglass safety goggles** (safety mask)

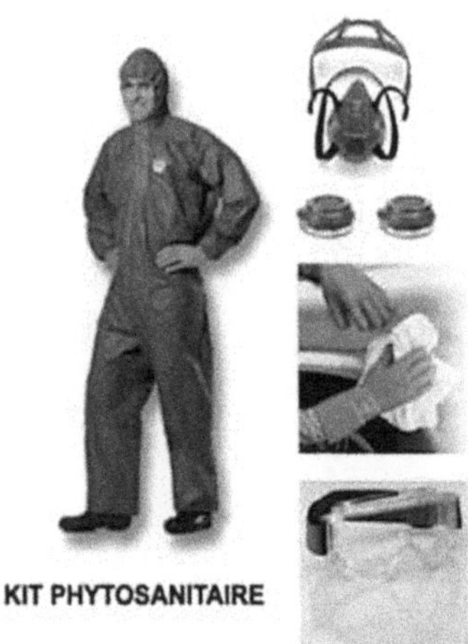

As for detection, considering the fact that the necessary items are rather costly, it seems wise to content oneself with:

- **Gamma ray detection**

Radioactivity Counter for Android systems
(5 euros)

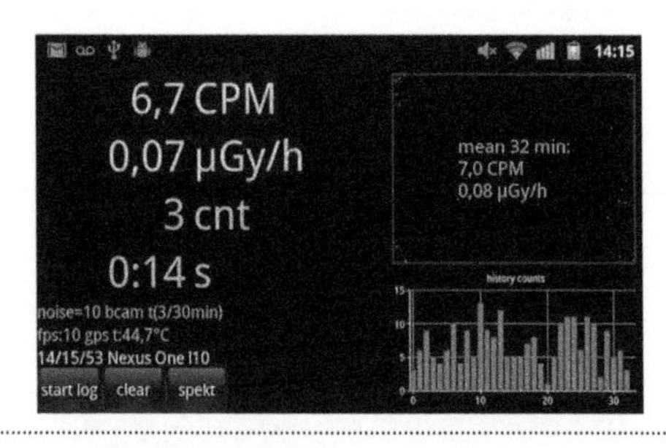

This smartphone app allows one to detect gamma rays through their mobile phone camera.

'Advanced' CBRN Kits

If your budget permits, and you are worried about a particular risk and interested in CBRN issues, you can opt for this kind of kit to provide yourself with additional protection and detection capabilities.

Type 3 protective suit

Grey TYCHEM 'F' suit

(35 euros)

Type 3 protective suit – offers effective protection against biological agents, radioactive dusts and most industrial chemical products.

Full face mask and filtering cartridges

Full face Optifit mask

(140 euros)

A2 B2 E2 K2 P3 cartridge

(30 euros a piece)

This cartridge is highly effective against particles (P3) and enables a thorough filtration of many chemical compounds.

Protective gloves

Disposable nitrile gloves

(10 euros per 100 pairs)

These gloves are usually sold in boxes containing 100 pairs each. They offer effective protection against biological agents and radioactive dusts, as well as limited protection against chemical agents.

MaxiChem 56-633 cut-proof nitrile/neoprene gloves

(15 euros a pair)

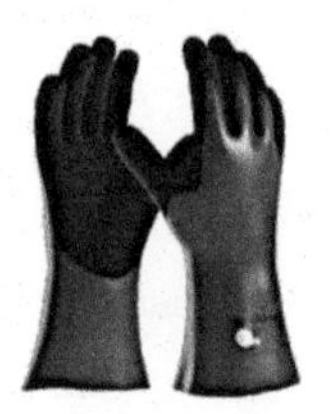

These gloves offer decent protection against both physical damage (perforation, abrasion, etc.) and toxic chemical agents.

▪ Boots

1 pair of chemical protection boots (made of nitrile/rubber, nitrile/PVC or hypalon).

Bronze S5 PVC security boots (Netco)

(15 euros a pair)

Made of PVC/nitrile, these boots offer moderate protection against aggressive chemicals.

Chemical-resistant safety boots (Hypalon SA)

(150 euros a pair)

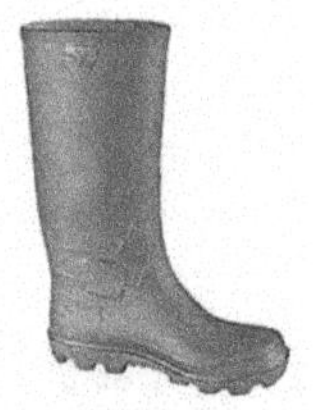

Designed for professional use, these hypalon boots are highly resistant to chemicals of all types. They also come with an anti-perforation sole and a safety toe cap (the latter could, however, become a disadvantage in case of a long walk).

▪ Radioactivity detection devices

The purpose of the models presented below is to raise the alert or detect radioactivity. Radioisotope identification must, by contrast, be carried out using other, more expensive equipment (for further details, see the section dealing with detection).

PM 1904 (Polimaster)

(250 euros)

This is a genuine Geiger counter that you can plug into your smartphone, enabling you to detect gamma rays in real time.

One of the advantages of using this system is that it allows you to have a map displaying your route and the different levels of encountered radiation.

PM 1208 (Polimaster)

(550 euros)

This watch comprises a real gamma ray detector, thus allowing you to measure radiation levels in real time.

RadEye B20-ER (Thermo Scientific)

(1,600 euros)

This small device can measure the surrounding gamma radiation dose rate while also detecting any alpha, beta and gamma ray contamination.

It can be particularly useful if your home or long-term autonomous base is located in an area that has been contaminated by radioactive elements.

- **Chemical gas detection devices**

  Chemical detectors must be chosen in accordance with the risk you are likely to face or the use that you would like to make of them. If a chlorine plant is located nearby, for example, use a single-gas unit to detect the compound. If you think the actual risk is more extensive than that or are involved in such specific activities as the exploration of a confined space, you can opt for a more complete detector (of the X-am 5600 type).

Single-gas ToxiRAE II detector with built-in chlorine sensor

(expect at least 600 euros price-wise)

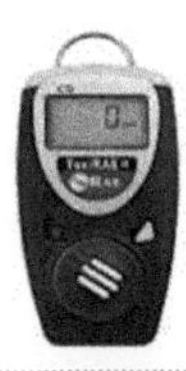

X-am® 5600 (Dräger)

(Variable price depending on the configuration. Expect at least 2,500 euros)

You could, for instance, opt for the following configuration: the detection of oxygen levels, carbon monoxide concentration, explosive vapours and gases, and three toxic agents (such as chlorine, ammonia, and hydrogen sulphide).

CBRN Kits: Conclusion

The items presented above are just examples among many others. Once you have understood the principle, strive to find the best possible prices for various pieces of equipment whose performances are, at the very least, equivalent. If, for whatever reason, you decide to invest in only one category of equipment, choose the one that **protects your airways**. If your budget is very limited, opt for a disposable FFP type of mask. Although it offers you virtually no protection against toxic or aggressive chemical gases, it will, on the other hand, be very useful with regard to avoiding internal contamination through biological agents or radioactive dusts. Choosing a full-face mask and an A2 B2

E2 K2 P3 cartridge will provide you with effective protection against most chemical toxic agents.

Warning! It is crucial to be aware of the limits that apply to the use of filtering masks, namely:

1. The air's oxygen concentration must total at least 17% (with 19% the recommended rate). It is therefore advisable to use filtering masks in an open environment.

2. Some compounds are not filtered. For example, A2 B2 E2 K2 P3 cartridges cannot stop radioactive iodine or carbon monoxide (a deadly and very common gas in cases involving a fire or incomplete combustion).

3. If the air concentration of a toxic agent is too high, the cartridge quickly loses its effectiveness. It is unfortunately impossible to give an estimate of how long it takes for it to stop working properly, since this depends on both the compound encountered and its concentration. If you enter a room saturated with a toxic gas, your cartridge may become ineffective after merely a few breaths.

   From a more general perspective, durability is mainly determined by your physical activity; the environmental conditions (the humidity, etc.); and, as previously stated, the agent involved and its concentration.

4. All cartridges have an expiry (or storage) date.

   As regards bodily protection, the main question that influences your choice—in addition to any financial considerations—is the following: what is the intended use?

If a CBRN kit is part of your bug-out bag, a set of disposable nitrile gloves, FFP3 masks and some safety goggles should enable you and your loved ones to leave the area under favourable circumstances (especially in the case of biological or radiological contamination).

If you decide to keep a CBRN kit in your long-term autonomous base in order to remain in a contaminated environment or to leave it at a later point, other items might be worth considering, starting with a pair of boots (*Landes*[2] boots made of PVC/nitrile and costing 12 euros, Bronze S5 PVC boots for 15 euros, or *Ardèche* S5 boots for 55 euros, all of which are characterised by superior thermal resistance, thermal insulation and the presence of an anti-perforation sole). The purchase of disposable suits, a genuine mask and perhaps even some detection tools is also worth considering.

2 TN: In France, many safety boots are named according to the country's different departments, including *Landes* boots, *Ardèche* boots, etc.

Conclusion

Mankind will derive more blessings
than blight from new discoveries.

— French physicist Pierre Curie (1859–1906)

We fly over cities,
freeway wastelands,
lost diagonals,
and straight lines at random.
Faceless women
upon landing —
let us be casual
and wear an impassive expression.

— *Tostaki* (1992) by Noir Désir (a French music group)

As we have already seen, there is a considerable number of scenarios that can give rise to an event of a nuclear, radiological, biological or chemical nature. It might turn out to be a technological accident (ranging from a truck transporting hazardous materials that overturns near you to the failure of a nuclear power station or a natural disaster that damages a high-risk plant); a malicious action or an attack (stretching from the dispersion of a toxic warfare agent to the explosion of a bomb containing radioactive elements or the sabotaging of a chemical factory); or an armed conflict (a civil war or hostilities between countries with the potential use of nuclear bombs).

Look around you. How do you see the world evolving?

Ever since the economic crisis of 2007/2008, the richest people have increased their fortune by 40%, while others have seen the value of their assets decrease by the same rate.[1] And the cumulative property of the richest 1% of the world's population has since been in excess of that of the remaining 99%. The world's sixty-two richest people thus own as much as the poorest 3.5 billion.[2]

Global debt has never been greater. In January 2016, it amounted to more than fifty-seven trillion dollars; and it is still growing… How far will it all go before causing a genuine economic and financial collapse? And what will the consequences then be for the complex and fragile supply systems which we all depend on,[3] especially in cities?

The Earth's population is now more than seven billion people. According to current estimates, the number is expected to total ten billion by 2050. And yet, the forecasts made in the early 2000s were based on a maximal limit of seven billion people, which no one thought would be exceeded…

Conflicts and wars are ubiquitous:[4] in Syria, Yemen, Iraq, Afghanistan, Mali, Chad, Nigeria, the Central African Republic, Uganda, Somalia, Congo, Libya, and, closer to us, Ukraine. Considering the case of this country, which is now on the brink of financial collapse; who will guarantee the maintenance and proper operation of its nuclear power plants once no part of the budget is

---

1 AN: http://www.gauche-anticapitaliste.ch/old/?p=12436.

2 AN: As seen in an article published in the French newspaper *Le Monde* on 18 January 2016, which cited a study by the British NGO OXFAM: https: //www.oxfam.org/fr/rapports/une-economie-au-service-des-1.

3 AN: Especially with regard to food, medicines, etc. For details, see Piero San Giorgio's and Vol West's *Barbarian Streets—Surviving in Cities*, Le Retour aux Sources editions, 2012.

4 AN: http://borgenproject.org/global-peace-index-offers-critical-poverty-insights.

allocated for this purpose? These conflicts, which spread far beyond traditional battlefields to reach our streets, are increasingly taking on the shape of a 'molecular civil war', to use the words of the Swiss historian Bernard Wicht,[5] a war that will be more akin to the Thirty Years' War (1618–1648)[6] or the Lebanese Civil War (1975–1989) than to the Kursk Battlefield of 1943.

As for terrorism, recent events have shown that Africa, the Middle East and Afghanistan are not the only places to be impacted by it...

Are all the elements of a major crisis not already in place? How are we not to anticipate a disaster in such a world, where inequalities are constantly being reinforced, tensions over water, food and natural resources keep worsening, and man has initiated the sixth extinction of life?

Admittedly, none of us can predict the future. However, when one considers the worsening tensions, the growing everyday use of hazardous materials, the increasing population density and the ever-larger transportation networks, the risk of a CBRN event does seem to be on the rise.

Which catastrophe will hit the jackpot in this macabre lottery, then? A sudden, lightning-fast pandemic? An atomic war between enemy states? A new Fukushima? A CBRN attack? Chronic chemical spills?

But in the end it really doesn't matter.

Even if luck can also have a say in this (you could, indeed, find yourself in the wrong place at the wrong time, and that's that), it is obvious that a well-prepared person can overcome such an event more readily than anyone who remains ignorant of the risks involved.

So, no, it really does *not* matter in the very long run.

---

5 AN: Wicht himself adopted this notion, which was originally proposed by German poet Hans Magnus Enzensberger and is explained here: http://www.theatrum-belli.com/note-prospective-de-lete-2015/.

6 AN: https://en.wikipedia.org/wiki/Thirty_Years%27_War.

99.9% of the species that once inhabited our planet have now disappeared. The Earth has already witnessed several mass extinctions, yet each and every time, life has gone on, persisted and flourished, and so it shall again, whether in human form or not.

None of us could, however, look our loved ones in the eyes, nor even lead the life of a citizen, if we embraced such fatalism, such a philosophy.

At the very least, we have a moral obligation to obtain information, if not to actually prepare ourselves, so as to increase our chances of individual and collective survival should a CBRN type of event come to pass.

In this book, we have attempted to gather as much useful information as possible to facilitate people's understanding of these threats and their operational (i.e. response-based, procedural) and material preparation for them.

Our aim is not, of course, to turn everyone into a CBRN survival 'expert', but simply to allow you, our reader, to improve in this field.

Having read this book, you have certainly become aware of new dangers and broadened your knowledge. As for those who would seek further involvement in CBRN matters, they should now have all the necessary details to guide them in their choices.

One day, perhaps, the information contained in this book will help you or your loved ones survive a nuclear, radiological, biological or chemical event. It might even enable you to avoid serious trouble without you even realising it...

In the end, none of us can foretell the future, but everyone can prepare on their own personal level and in harmony with their own convictions, since few people have the necessary means to actually build a fallout shelter...

Indeed, as already stated, protecting oneself does not entail any major expenses. In addition, although governmental entities do have the means to respond to CBRN events, it is still important for you to depend on yourself and your own efforts so as to take on the role of a responsible and committed citizen.

There are two key and inseparable steps that allow you to do this: the first is to acquire sufficient knowledge, and the second to take action.

The ball is in your court now!

OTHER BOOKS PUBLISHED BY ARKTOS

Sri Dharma Pravartaka Acharya	*The Dharma Manifesto*
Joakim Andersen	*Rising from the Ruins: The Right of the 21st Century*
Winston C. Banks	*Excessive Immigration*
Alain de Benoist	*Beyond Human Rights* *Carl Schmitt Today* *The Indo-Europeans* *Manifesto for a European Renaissance* *On the Brink of the Abyss* *The Problem of Democracy* *Runes and the Origins of Writing* *View from the Right* (vol. 1–3)
Arthur Moeller van den Bruck	*Germany's Third Empire*
Matt Battaglioli	*The Consequences of Equality*
Kerry Bolton	*Revolution from Above* *Yockey: A Fascist Odyssey*
Isac Boman	*Money Power*
Ricardo Duchesne	*Faustian Man in a Multicultural Age*
Alexander Dugin	*Ethnos and Society* *Ethnosociology* *Eurasian Mission* *The Fourth Political Theory* *Last War of the World-Island* *Political Platonism* *Putin vs Putin* *The Rise of the Fourth Political Theory*
Edward Dutton	*Race Differences in Ethnocentrism*
Mark Dyal	*Hated and Proud*
Koenraad Elst	*Return of the Swastika*
Julius Evola	*The Bow and the Club* *Fascism Viewed from the Right* *A Handbook for Right-Wing Youth* *Metaphysics of War* *The Myth of the Blood*

OTHER BOOKS PUBLISHED BY ARKTOS

	Notes on the Third Reich
	The Path of Cinnabar
	Recognitions
	A Traditionalist Confronts Fascism
Guillaume Faye	*Archeofuturism*
	Archeofuturism 2.0
	The Colonisation of Europe
	Convergence of Catastrophes
	A Global Coup
	Sex and Deviance
	Understanding Islam
	Why We Fight
Daniel S. Forrest	*Suprahumanism*
Andrew Fraser	*Dissident Dispatches*
	The WASP Question
Génération Identitaire	*We are Generation Identity*
Paul Gottfried	*War and Democracy*
Porus Homi Havewala	*The Saga of the Aryan Race*
Lars Holger Holm	*Hiding in Broad Daylight*
	Homo Maximus
	Incidents of Travel in Latin America
	The Owls of Afrasiab
Richard Houck	*Liberalism Unmasked*
A. J. Illingworth	*Political Justice*
Alexander Jacob	*De Naturae Natura*
Jason Reza Jorjani	*Iranian Leviathan*
	Novel Folklore
	Prometheus and Atlas
	World State of Emergency
Roderick Kaine	*Smart and SeXy*
Peter King	*Here and Now*
	Keeping Things Close
	On Modern Manners

OTHER BOOKS PUBLISHED BY ARKTOS

Ludwig Klages	*The Biocentric Worldview* *Cosmogonic Reflections*
Pierre Krebs	*Fighting for the Essence*
John Bruce Leonard	*The New Prometheans*
Stephen Pax Leonard	*The Ideology of Failure* *Travels in Cultural Nihilism*
William S. Lind	*Retroculture*
Pentti Linkola	*Can Life Prevail?*
H. P. Lovecraft	*The Conservative*
Norman Lowell	*Imperium Europa*
Charles Maurras	*The Future of the Intelligentsia & For a French Awakening*
Michael O'Meara	*Guillaume Faye and the Battle of Europe* *New Culture, New Right*
Brian Anse Patrick	*The NRA and the Media* *Rise of the Anti-Media* *The Ten Commandments of Propaganda* *Zombology*
Tito Perdue	*The Bent Pyramid* *Lee* *Morning Crafts* *Philip* *The Sweet-Scented Manuscript* *William's House* (vol. 1–4)
Raido	*A Handbook of Traditional Living*
Steven J. Rosen	*The Agni and the Ecstasy* *The Jedi in the Lotus*
Richard Rudgley	*Barbarians* *Essential Substances* *Wildest Dreams*
Ernst von Salomon	*It Cannot Be Stormed* *The Outlaws*

OTHER BOOKS PUBLISHED BY ARKTOS

Sri Sri Ravi Shankar	*Celebrating Silence* *Know Your Child* *Management Mantras* *Patanjali Yoga Sutras* *Secrets of Relationships*
George T. Shaw (ed.)	*A Fair Hearing*
Fenek Solère	*Kraal*
Oswald Spengler	*Man and Technics*
Richard Storey	*The Uniqueness of Western Law*
Tomislav Sunic	*Against Democracy and Equality* *Homo Americanus* *Postmortem Report* *Titans are in Town*
Hans-Jürgen Syberberg	*On the Fortunes and Misfortunes of Art in Post-War Germany*
Abir Taha	*Defining Terrorism* *The Epic of Arya* (2nd ed.) *Nietzsche's Coming God, or the Redemption of the Divine* *Verses of Light*
Bal Gangadhar Tilak	*The Arctic Home in the Vedas*
Dominique Venner	*For a Positive Critique* *The Shock of History*
Markus Willinger	*A Europe of Nations* *Generation Identity*
Alexander Wolfheze	*Alba Rosa*

Lightning Source UK Ltd.
Milton Keynes UK
UKHW010636190820
368484UK00002B/367